Ecological Studies, Vol. 194

Analysis and Synthesis

Ecological Studies

Volumes published since 2002 are listed at the end of this book.

U. Lüttge (Ed.)

Clusia

A Woody Neotropical Genus of Remarkable Plasticity and Diversity

With 111 Figures, 2 in Color, and 38 Tables

Springer

Prof. Dr. Ulrich Lüttge
Institute of Botany
Department of Biology
Technical University of Darmstadt
Schnittspahnstraße 3–5
64287 Darmstadt
Germany

Cover illustration: Male flower of *Clusia hilariana* Schlecht. (Photo Ulrich Lüttge)

ISSN 0070-8356
ISBN-10 3-540-37242-3 Springer Berlin Heidelberg New York
ISBN-13 978-3-540-37242-4 Springer Berlin Heidelberg New York

Library of Congress Control Number: 2006934618

Springer is a part of Springer Science+Business Media
springer.com

Editor: Dr. Dieter Czeschlik, Heidelberg, Germany
Desk editor: Dr. Andrea Schlitzberger, Heidelberg, Germany
Cover design: WMXDesign GmbH, Heidelberg, Germany
Typesetting and production: Friedmut Kröner, Heidelberg, Germany

31/3152 YK – 5 4 3 2 1 0 – Printed on acid free paper

Preface

From emotional attachment to the inherent beauty and diversity and the curiosity and thought provoking challenges of *Clusia* the plan to compose a book-monograph on *Clusia* emerged a number of years ago when it became evident that the research interests of several groups internationally focusing on studies of this remarkable genus of neotropical dicotyledonous shrubs and trees would develop a very broad picture comprising all aspects of tree life. *Clusia* displays unique features because it is the only dicotyledonous tree genus with crassulacean acid metabolism (CAM). *Clusia* species are extraordinarily flexible and plastic and in some cases all variants of CAM can even be expressed in one given species. This has raised special attention to this genus among the vast diversity of tropical tree genera and put *Clusia* in the limelight of international research interest. The scope of these studies embraces morphology, anatomy and plant architecture, phytogeographical distribution and community ecology, phylogeny and genetic diversity, physiology and metabolism, physiological ecology and functional diversity. Thus, *Clusia* can serve as a general example covering all facets of tree biology.

To present this was not possible without a close co-operation of a team of authors. I hope that the result bears up to the expectations of creating a comprehensive and integrated picture of *Clusia,* and thus produce a unique story of the biological history and topical impact of an outstanding tropical tree genus. I thank all co-authors of the book for the joint efforts and their patience with repeated editorial requests. I am grateful to the entire international scientific *Clusia* community including many members who are not co-authors of this book for much stimulating exchange. Particular thanks are due to Dr. Annie M. Borland, Newcastle upon Tyne, UK, for reading several chapters of the book during their preparation and for making important comments. I am most grateful to Professor Dr. Dr. h.c. mult. Otto Ludwig Lange as member of the board of editors of Ecological Studies for his encouragement, support and valuable suggestions. I thank Doris Schäfer, Darmstadt, Germany, for her care with many of the illustrations.

Continuous and long lasting internationally co-operative *Clusia* research in the field in South America and elsewhere over the years has been particularly supported by the following institutions: Instituto Venezolano de Investigaciones Scientíficas (IVIC), Caracas, Venezuela; Brazilian Research Council (CNPq); the PROBRAL programme of the Brazilian Postgraduate Education Council (CAPES) and the German Academic Exchange Service (DAAD); and the international partnership programme of Volkswagen-Foundation, Hannover, Germany. Without their support much of the work reported in this book would not have been realized although in addition we must also remember support given by many other institutions to research groups whose work is quoted in this book from the published literature.

I am grateful to Springer-Verlag for taking up the idea to publish the book and I particulary thank Dr. Andrea Schlitzberger and Dr. Dieter Czeschlik for the wonderful cooperation in producing it.

Darmstadt, in October 2006 ULRICH LÜTTGE

Contents

Section I Background

Section II Phylogeny, Diversity and Ecology

Section III Functions and Physiological Ecology

Contributors

DOROTHY S.D. ARAUJO
Universidade Federal do Rio de Janeiro, CCS, IB, Depto. de Ecologia, Caixa postal 68 020, cep. 21941-970, Rio de Janeiro, RJ Brazil, e-mail: dorothya@biologia.ufrj.br

VOLKER BITTRICH
Departamento de Botânica, Universidade Estadual de Campinas, SP Brazil, e-mail: volker@unicamp.br

ANDRÉ T.C. DIAS
Universidade Federal do Rio de Janeiro, CCS, IB, Depto. de Ecologia, Caixa postal 68 020, cep. 21941-970, Rio de Janeiro, RJ Brazil, e-mail: atcdias@biologia.ufrj.br

HEITOR M. DUARTE
Rua Botucato 460, Apto. 803, bl. 4, Grajaú, 20 541 – 340 Rio de Janeiro, RJ Brazil, e-mail: hmduarte@click21.com.br

MATS H.G. GUSTAFSSON
Department of Biological Sciences, University of Aarhus, Ny Munkegade, Building 1540, 8000 Århus C, Denmark, e-mail: mats.gustafsson@biology.au.dk

RÜDIGER HAMPP
Universität Tübingen, Botanisches Institut, Auf der Morgenstelle 1, 72 076 Tübingen, Germany, e-mail: ruediger.hampp@uni-tuebingen.de

MEIKE KREUZER
Universität Tübingen, Botanisches Institut, Auf der Morgenstelle 1, 72076 Tübingen, Germany

Ulrich Lüttge

Institute of Botany, Technical University of Darmstadt, Schnittspahnstr. 3–5, 34287 Darmstadt, Germany, e-mail: luettge@bio.tu-darmstadt.de

Rogerio Margis

Universidade Federal do Rio Grande do Sul, ICBS, Depto. de Bioquímica, cep. 90035-003, Porto Alegre, RS, Brazil, e-mail: rogerio.margis@ufrgs.br

Rodrigo L. Martins

Universidade Federal do Rio de Janeiro, CCS, IB, Depto. de Ecologia, Caixa postal 68 020, cep. 21941-970, Rio de Janeiro, RJ Brazil, e-mail: rodr.lemes@gmail.com

Fabio R. Scarano

Universidade Federal do Rio de Janeiro, CCS, IB, Depto. de Ecologia, Caixa postal 68 020, cep. 21941-970, Rio de Janeiro, RJ Brazil, e-mail: fscarano@biologia.ufrj.br

Anja Vaasen

Bundesamt für Verbraucherschutz und Lebensmittelsicherheit, Mauerstrasse 39–42, 10117 Berlin, Germany, e-mail: anja.vaasen@bvl.bund.de

Tânia Wendt

Universidade Federal do Rio de Janeiro, CCS, IB, Depto. de Botânica, cep. 21941-590, Rio de Janeiro, RJ Brazil, e-mail: twendt@biologia.ufrj.br

Klaus Winter

Smithsonian Tropical Research Institute, P.O. Box 2072, Balboa, Republic of Panamá, e-mail: winterk@tivoli.si.edu

Section I Background

1 Historical Recollections

ULRICH LÜTTGE

Entdecker sein ist nicht genug, man muss auch publizieren.
"It is not enough to discover, one also must publish."
CARL FRIEDRICH PHILIPP VON MARTIUS [1794–1868]

1.1 Namesakes: CAROLUS CLUSIUS and *Clusia*

CHARLES DE L'ECLUSE (Fig. 1.1), latinized CAROLUS CLUSIUS, son of a noble family, was the greatest botanist of his time (Hunger 1927; Veendorp and Baas Becking 1938; Mägdefrau 1992). He was born in Arras in 1526. He studied law at Lovain and Marburg and later medicine at Wittenberg, Montpellier and Paris and also lived in various other European cities including Frankfurt, Strasbourg, Montpellier, Antwerp and London. He travelled in Spain and Portugal, in the European Alps and in Hungary. In Vienna he founded a medical garden – *Hortus medicus* – in 1573. In 1592, at the age of 66, he was appointed director of the *Hortus medicus* in Leyden (Veendorp and Baas Becking 1938), one of the six oldest gardens in Europe founded in 1587, where he died in 1609. Due to his extensive travels in Europe he was intimately familiar with the European flora. He discovered many new species, which he described and depicted thoroughly. He introduced tulips to the Netherlands and was involved in the introduction of potato to European gardens providing its first detailed description in 1601 although at the time the great nutritional value of potato for large populations was not yet appreciated.

With CLUSIUS' original descriptions of the alpine vegetation it is evident why one of the most beautiful alpine plants, the large flowered gentian *Gentiana clusii* Perr. et Song. was named after him. However, we do not know the background of the motivation to name the genus *Clusia* after him. *Clusia* L. is a genus of 300–400 neotropical woody plants in the family Clusiaceae (alternatively Guttiferae or Hypericaceae) in the order Malpighiales of the Rosidae (Eudicotyledoneae). The name *Clusia* was coined by the Franciscan CHARLES PLUMIER who was born in 1646 in Marseille and died in 1704 in Cadiz.

Ecological Studies, Vol. 194
U. Lüttge (Ed.)
Clusia: A Woody Neotropical Genus
of Remarkable Plasticity and Diversity

Fig. 1.1. Sculpture of Carolus Clusius in the Botanical Garden of Leyden in the Netherlands, where also a historical reconstruction of the *Hortus medicus* of Clusius can be seen

Plumier was one of the most outstanding scientific explorers of his time (Mägdefrau 1992). He travelled widely in America and the West Indies and knew *Clusia*. Linnaeus adopted the name for the genus from Plumier who cannot be named as authority for some rules of nomenclature. Linnaeus distinguished two species of *Clusia*, namely *C. major* with three "varieties" and *C. minor* (Bittrich and Stevens 1998).

1.2 Alexander von Humboldt: The First Ecophysiological Studies of *Clusia*

In the February of 1800 Alexander von Humboldt performed ecophysiological gas-exchange measurements on *Clusia rosea* Jacq. at Lake Valencia in Venezuela (Alexander von Humboldt's diaries; Faak 2000). At the time photosynthesis was measured observing gas bubbles emerging from green tissues submerged in water. Humboldt wrote:

> *In keiner Pflanze zirkuliert vielleicht eine so ungeheure Menge an Luft als in der* Clusia rosea. *Wenn man die Blätter dem Sonnenlicht aussetzt, so geben sie … nicht eine einzige Luftblase.*
>
> < *In no other plant perhaps as much air is circulating than in the* Clusia rosea. *If one exposes the leaves to the sunlight, they produce … not a single air bubble.*>

This implies that there is no obvious gas exchange in the light via the leaf surface. We now know that *C. rosea* is a plant with Crassulacean acid metabolism (CAM) (Ball et al. 1991a, b) where stomata remain closed during the light period and CO_2 from internal sources, i.e. organic acids, accumulated after dark fixation of CO_2 in the night, is assimilated (Chap. 8) in the light. Humboldt goes on writing:

> *Aus dem verwundeten Teil des Stengels fährt aber mit ungeheurer Geschwindigkeit ein Strom von perlartigen Luftbläschen aus; … Zerschneidet man das Blatt selbst, so bemerkt man zahllose ähnliche Luftströme. Das Blatt hat Gefäßbündel, die transversal verlaufen. Es scheint als fahre die Luft aus jedem dieser Gefäßbündel aus – ein herrliches Schauspiel.*
>
> <*However, at the cut end of the petiole with an immense velocity a stream of pearl-like gas bubbles is released; … When one cuts the leaf itself one observes numerous similar air streams. The leaf has transversely arranged veins. It appears that the air is released from each of these vascular bundles – a magnificent spectacle.*>

We now know that in the CAM cycle in the light period a high internal gas pressure of CO_2 and O_2 is building up behind closed stomata explaining HUMBOLDT's observation of a vigorous gas flow via wounds of the leaves. HUMBOLDT also made the appropriate control to show that his observations were due to photosynthesis:

> *Setzte ich den Apparat in den Schatten, so hörte der Luftstrom auf. Der Reiz des Sonnenlichtes fehlt.*
>
> *<Did I place the apparatus in the shade, the gas stream ceased. The stimulation by the sun light is missing.>*

He then tried to determine the chemical composition of the gas stream. He could not determine CO_2, but by using chemical reactions he could measure oxygen and nitrogen:

> *Diese Luft aus dem Innern der* Clusia rosea *besteht aus 0,35 Oxygen und 0,65 Stickgas.*
>
> *<The air from the interior of* Clusia rosea *consists of 0.35 oxygen and 0.65 nitrogen gas.>*

This means that the gas in the leaf air spaces had 35 % O_2 and 65 % N_2. As measurements in the field are always difficult, HUMBOLDT was prudent to check his chemical O_2-determinations later in the laboratory of L.J. GAY-LUSSAC in Paris and found them to be too high by a systematic error of 5 % (HUMBOLDT-quotations from Faak 2000; see also Krätz 2001; Lüttge 2002). The measurements of HUMBOLDT were confirmed much later by Spalding et al. (1979) using gas chromatography, and we now know that it is a feature of CAM that high internal O_2 pressures are building up in the light period during CO_2 assimilation behind closed stomata.

Thus, HUMBOLDT correctly described many aspects of the CAM cycle and we might consider him the discoverer of CAM in the trees of *Clusia*. However, he did not interpret his observations far enough and had no idea of the functions of the CAM cycle.

1.3 The Discovery of Crassulacean Acid Metabolism (CAM) in *Clusia*

Another researcher who made the right observations but did not realize that they were features of CAM was Hartenburg (1937), a Ph.D. student of the famous German ecophysiologist OTTO STOCKER. He studied gas exchange of *Clusia mexicana* Vesque in a glasshouse in summer and found that after an early peak in the morning CO_2-uptake ceased during the day and there was even a small CO_2 release from the leaves around midday (Fig. 1.2). He discussed in detail whether the latter was due to respiration but discarded this possibility:

> *Dass es sich um Atmungskohlensäure handelt, ist wenig wahrscheinlich. Eher kann an die Entbindung von CO_2 gedacht werden, das im* Clusia-*Blatt spezifisch irgendwie physikalisch oder chemisch gebunden ist.*
>
> *<It is unlikely that this is respiratory CO_2. More likely it is due to CO_2 specifically bound either physically or chemically in the* Clusia *leaf and set free.>*

Current work shows that in CAM during the day indeed so much CO_2 is released from malic acid, in which it is chemically bound during the night, that very high internal CO_2-concentrations may arise in the leaves (Chap. 8) so that some CO_2 can diffuse out along a steep downhill concentration gradient to the atmosphere via the epidermis despite closed stomata. Hartenburg (1937) discovered even another facet of CAM in *Clusia* that currently is exciting physiological ecologists (Chap. 9). The strange behaviour described

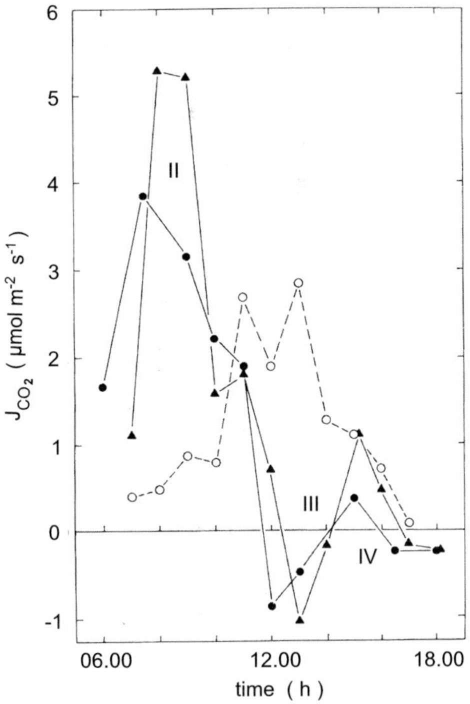

Fig. 1.2. Three gas exchange curves of *C. mexicana* after measurements of Hartenburg (1937) on a sunny day (*closed circles*, 29 June 1935), a partially cloudy day (*closed triangles*, 24 August 1935) and an overcast rainy day (*open circles*, 27 August 1935) in a glass house of the Botanical Garden of the Technical University Darmstadt. The *solid lines* (*closed symbols*) correspond to a CAM-gas exchange with phases II, III and IV, the *dashed line* (*open symbols*) represents C_3 gas exchange. (Net CO_2-uptake positive values, net CO_2-release negative values.) (See Lüttge 1995)

above, and which he could not explain, was only observed on a sunny day and on a partially cloudy day but not on a totally overcast rainy day, when he found quite conventional CO_2 uptake all over the day as in normal C_3-photosynthesis (Fig. 1.2). Thus, he saw the reactions of a plant that was intermediate between CAM and C_3-photosynthesis and could switch between the two modes in response to external conditions.

HARTENBURG, unlike ALEXANDER VON HUMBOLDT, could have much more advanced the interpretation of his observations. Since 1804 it has been known from experiments of DE SAUSSURE that some plants have a CO_2 gas exchange with nocturnal CO_2 uptake and largely suppressed gas exchange during the day. The associated acid rhythm with a nocturnal accumulation of acids and a remobilization during the day had been known since the early nineteenth century (see Wolf 1960). However, it remained left to the Mexicans Tinoco Ojanguren and Vazquez-Yanez (1983) to describe, explain and establish firmly the performance of CAM in trees of *Clusia*, and after a seminal publication of Ting et al. (1985) this then became general knowledge.

Why is it so important for the history of *Clusia* research that *Clusia*s perform CAM? There are several sizeable and tree-like plants which perform CAM. Among the monocotyledons these are the *Yucca*s, and among the dicotyledons giant stem succulent cacti and euphorbs and the Didieraceae of Madagascar exhibit CAM. One may range such species among the trees (Menninger1967) although they either have no secondary growth at all or the special type of secondary growth of monocotyledons. *Clusia*s are the only "real" CAM-trees with a typically dicotyledonous secondary growth based on the activity of a circumferential cambium (Chap. 2). Without the reported occurrence of CAM *Clusia* most likely would have remained just one of many other taxa in the large diversity of tropical trees. It is an open phylogenetic question why there are not more CAM plants among trees. However, it certainly was the unique performance of CAM in these trees that raised very pronounced interest in *Clusia* following the publications of Tinoco Ojanguren and Vazquez-Yanez (1983) and Ting et al. (1985), and it explains the burst of research activities in the last 20 years that now allows us to survey a large body of literature in this *Clusia* monograph.

References

Ball E, Hann J, Kluge M, Lee HSJ, Lüttge U, Orthen B, Popp M, Schmitt A, Ting IP (1991a) Ecophysiological comportment of the tropical CAM-tree *Clusia* in the field. I. Growth of *Clusia rosea* Jacq. on St. John, US Virgin Islands, Lesser Antilles. New Phytologist 117:473–481

Ball E, Hann J, Kluge M, Lee HSJ, Lüttge U, Orthen B, Popp M, Schmitt A, Ting IP (1991b) Ecophysiological comportment of the tropical CAM-tree *Clusia* in the field. II. Modes of photosynthesis in trees and seedlings. New Phytologist 117:483–491

Bittrich V, Stevens PF (1998) Lectotypification of the names *Clusia major* and *Clusia minor* (Clusiaceae). Taxon 47:117–122

Faak M (ed.) (2000) Alexander von Humboldt. Reise durch Venezuela. Auswahl aus den amerikanischen Reisetagebüchern. Beiträge zur Alexander von Humboldt-Forschung 12. Akademie-Verlag, Berlin, p 193

Hartenburg W (1937) Der Wasser- und Kohlensäurehaushalt tropischer Regenwaldpflanzen in sommerlicher Gewächshauskultur. Jahrbücher wissenschaftlicher Botanik 85:641–697

Hunger FWT (1927) Charles de l'Ecluse (Carolus Clusius). Nederlandsch Kruidkundige 1526–1609. Martinus NijHoff, s' Gravenhage

Krätz O (2001) Alexander von Humboldt (1769–1859) auf Pflanzenjagd. In: Palmengarten Heft 35. Palmengarten, Frankfurt M, pp 33–46

Lüttge U (1995) *Clusia*: Ein Modellfall der ökophysiologischen Plastizität in einer tropischen Gattung. Rundgespräche der Kommission für Ökologie, vol. 10, Bayerische Tropenforschung einst und jetzt. Dr. Friedrich Pfeil, München, pp 173–186

Lüttge U (2002) CO_2-concentrating: consequences in crassulacean acid metabolism J Exp Bot 53:2131–2142

Mägdefrau K (1992) Geschichte der Botanik. Leben und Leistung großer Forscher. 2nd edn. Gustav Fischer, Stuttgart

Menninger EA (1967) Fantastic trees. Viking Press, New York

Spalding MH, Stumpf DK, Ku MSB, Burris RH, Edwards GE (1979) Crassulacean acid metabolism and diurnal variations of internal CO_2 and O_2-concentrations in *Sedum praealtum* DC. Aust J Plant Physiol 6:557–567

Ting IP, Lord EM, Sternberg L da SL, DeNiro MJ (1985) Crassulacean acid metabolism in the strangler *Clusia rosea* Jacq. Science 229:969–971

Tinoco Ojanguren C, Vazquez-Yanez C (1983) Especies CAM in la selva humeda tropical de Los Tuxtlas, Veracruz. Bol Soc Bot Mex 45:150–153

Veendorp H, Baas Becking LGM (1938) Hortus Academicus Lugduno Batavus. The development of the gardens of Leyden University. Reprint 1990: Rijksherbarium/Hortus Botanicus, Leiden

Wolf J (1960) Der diurnale Säurerhythmus. In Ruhland W (ed) Handbuch der Pflanzenphysiologie, vol. XII/2. Springer, Berlin-Göttingen-Heidelberg, pp 809–889

Section II Phylogeny, Diversity and Ecology

Introduction

ULRICH LÜTTGE

On 26 March 1983 we climbed the highest elevation of the island of Trinidad, Cerro del Aripo, 941 m a.s.l. We were studying epiphytic C_3 and CAM bromeliads. It was the same year when Tinoco Ojanguren and Vazquez Yanes (1983) published their discovery of CAM in *Clusia*. However, at that time we did not know about it. Later we discovered that we even had a branch of *Clusia* in the cover photograph (Fig. II.1) depicting *Aechmaea aripensis* in the special issue number 5, vol. 9, Plant Cell and Environment (1986) with our bromeliad work. One other member of the Cerro del Aripo crew of 1983, HOWARD GRIFFITHS, Cambridge, UK, later also became a *Clusia* enthusiast, and therefore we now know that on Trinidad – in addition to the ubiquitous *Clusia minor* L. – there are three endemic *Clusia* species, namely *C. tocuchensis* Britt., *C. aripoensis* Britt. and *C. intertexta* Britt. (Borland et al. 1992, Sect. 9.4.2.9). In view of the endemism of the *Clusia* species in Trinidad it is interesting to note that in the phylogenetic tree of Fig. 6.1 (Chap. 6) the three branches which have *C. minor, C. tocuchensis* and *C. aripoensis* in them are separated at the very base of the tree. (*C. intertexta* is not contained in this tree.) Thus, the endemic species must have evolved separately from a basic original *Clusia* ancestor on the island.

In the six chapters of this Section we present a broad coverage of aspects of the plant life and organismic biology of *Clusia*. We begin by describing anatomy and morphology of *Clusia* life forms to introduce its physical constitution. However, this also leads us already into some considerations of function, especially where the hydraulic architecture is considered as a basis of some intrinsic aspects of water relations combining structure-function assessments (Chap. 2).

The following three chapters constitute a unit dealing with the occupation of space by *Clusia*. As an exclusively neotropical genus *Clusia* is spread widely in Central America and on the South American continent and Chap. 3 takes us on a journey around its phytogeographical locations. Seeds of *Clusia* may germinate terrestrially or epiphytically and in the latter case eventually reach the ground via adventitious aerial roots to become independent trees. Seedlings

Fig. II.1. *Aechmea aripensis*, an epiphytic CAM-bromeliad in the upper montane rain forest at about 850 m a.s.l. on Cerro del Aripo, Trinidad, with a branch of *Clusia* in the *lower right hand corner* of the photograph

Fig. II.2. Seedling of *Clusia amygdioi* Silva et Weinberg growing in the tank of a bromeliad on rock outcrops in the Atlantic rain forest of Estação Biologica de Santa Lucia, Santa Teresa, Espírito Santo State, Brazil

are also often found in the water and humus filled tanks of both terrestrial and epiphytic bromeliads (Fig. II.2) and are thus protected in their early stages of life by the bromeliads as their nurse plants before they grow out of the tanks to become independent trees. However, *Clusia*s themselves are often functioning as nurse plants where free standing shrubs and trees of *Clusia* allow the establishment of a diverse vegetation underneath their canopies. Thus, *Clusia*s are often pioneer species starting vegetation islands, e.g. on the bare sandy ground of coastal restingas in Brazil (Chap. 4). Most likely it is the enormous ecophysiological plasticity and flexibility of *Clusia*s as described below in Chaps. 8 and 9 of Sect. III, which make them particularly fit for such a function. Reproductive biology plays a central role in the conquest of space by plants which are firmly rooted in their substrate and need to combine distribution with reproduction. *Clusia*s have developed several unique features in their reproduction, such as dioecy, resin production as an award for pollination by bees, and a partially dominating asexual propagation. These are described and assessed in Chap. 5.

Chapters 6 and 7 form another unit, where the latest molecular approaches are applied to fathom species diversity, phylogeny and classification, and genetic variation in a defined geographical range. We realize that there is a very high speciation rate in the genus. The genotypic plasticity studied in these chapters is revisited again at the end of Chap. 9 in Sect. III, i.e. after ecophysiological plasticity has been described and when some more theoretical thoughts on the relations between genotypes and phenotypes in relation to diversity can be developed.

References

Borland AM, Griffiths H, Maxwell C, Broadmeadow MSJ, Griffiths MN, Barnes JD (1992) On the ecophysiology of the Clusiaceae in Trinidad: expression of CAM in *Clusia minor* L. during the transition from wet to dry season and characterization of three endemic species. New Phytol 122:349–357

Tinoco Ojanguren C, Vázquez-Yanes C (1983) Especies CAM en la selva húmeda tropical de los tuxtlas, Veracruz. Bol Soc Bot Mex 45:150–153

2 Morphology, Anatomy, Life Forms and Hydraulic Architecture

ULRICH LÜTTGE and HEITOR M. DUARTE

2.1 Hundreds of Species of One Morphotype

The 300 to 400 woody species of *Clusia* all display one typical morphotype (Lüttge 2000). *Clusia*s are branched shrubs and trees with dichasial cymes and opposite leaves (see also Sect. 6.1). Among the various species leaves vary in absolute size. However, the leaves of all species are morphologically and anatomically very similar, always entire, leathery and somewhat succulent (Fig. 2.1). In view of the important effects of leaf form and structure on photosynthesis and ecophysiological performance (Niinemets and Sack 2005), this is remarkable particularly with respect to the large photosynthetic flexibility of some species of *Clusia*. However, this has not been much explored for *Clusia*. On the other hand, floral morphology of *Clusia*s is rather variable (Sect. 6.5).

Leaf succulence of *Clusia* species is indicated by comparatively high fresh weight/dry weight and fresh weight or plant water/area ratios (Table 2.1). Although the data reveal some developmental influences (mature plants versus seedlings of *C. rosea*) and effects of nitrogen supply and irradiance (*C. minor*), the values obtained are rather homogenous. Two important exceptions are the conspicuously succulent *Clusia alata* Pl. et Tr. and *Clusia hilariana* Schlechtendal. These are obligate Crassulacean acid metabolism (CAM) plants. Succulence with highly vacuolated photosynthetically active cells is a general feature of CAM plants, which fix CO_2 in the dark period forming malate that is nocturnally accumulated in the vacuoles (see Chaps. 8 and 9). Another species, where comparatively high values of leaf succulence were obtained is *C. rosea*, which is also a CAM species.

Zotz et al. (1997) compared the leaf succulence attributes of *Clusia minor* L. with six other woody plants. *C. minor* is a C_3-photosynthesis/CAM-intermediate species. The ratios of fresh weight/dry weight of 4.7 and of plant water/area of 580 g m^{-2} measured by Zotz et al. (1997) in *C. minor* (Table 2.1)

Ecological Studies, Vol. 194
U. Lüttge (Ed.)
Clusia: A Woody Neotropical Genus
of Remarkable Plasticity and Diversity

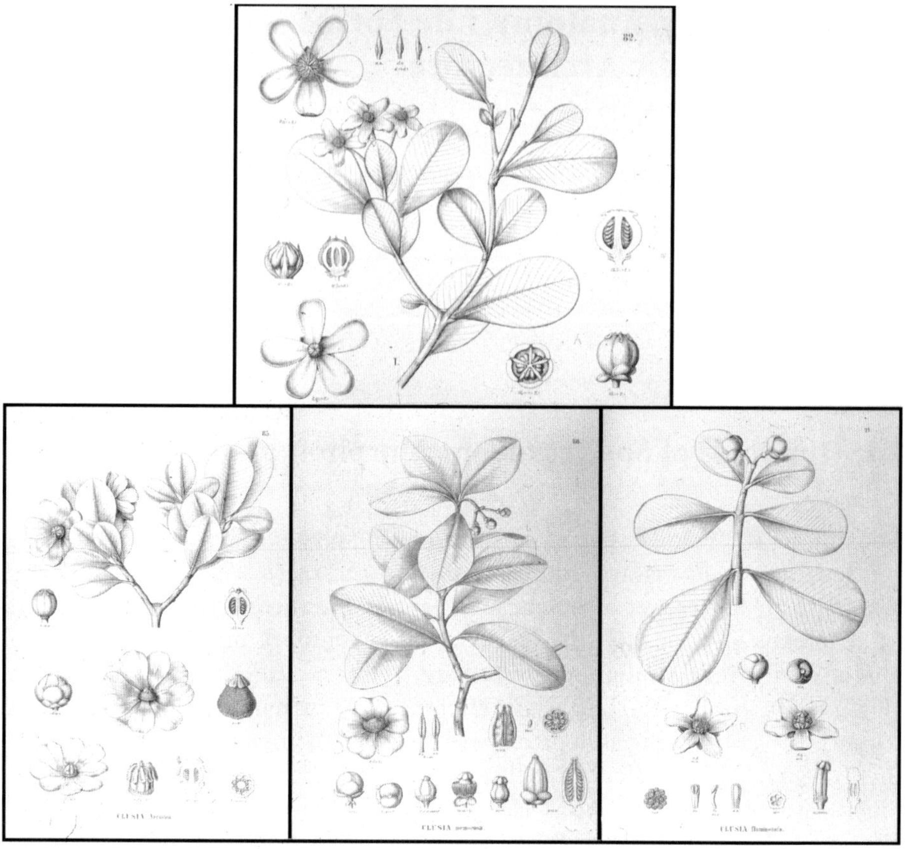

Fig. 2.1. Illustrations of *Clusias* in Flora Brasiliensis (Martii 1889). *Top*: *C. parviflora* Saldanha et Engl. *Bottom from left to right*: *C. arrudae* Planch. et Triana, *C. nemorosa* G. Mey., *C. fluminensis* Planch. et Triana

were the highest among the seven woody plants compared, where the lowest values were 1.6 and 150 g m^{-2}, respectively. All seven species were hemi-epiphytic life forms (see Sect. 2.3) in a tropical forest canopy. A comparison of evaporative loss of water and water storage showed that a net loss of 10 % of the leaf water in *C. minor* and *Clusia uvitana* Pittier, another C_3/CAM-intermediate plant, would sustain average transpiration rates for 7–9 h as compared to 0.2–0.8 h in the other species. In addition to leaf succulence this can be related to the performance of CAM as a water saving mode of photosynthesis (see Chaps. 8 and 9).

Table 2.1. Fresh weight (FW)/dry weight (DW) and plant water (PW, ml) or FW (g)/leaf area (m^2) ratios of four CAM species of *Clusia* (*C. alata, C. hilariana, C. major* and *C. rosea*) and five C_3/CAM-intermediate species (*C. aripoensis, C. criuva, C. minor, C. uvitana, C. venosa*) that can switch between both modes of photosynthesis. ± N, with and without supply of nitrogen; PAR=photosynthetically active radiation

Species	Comments	FW/DW	PW[a] or FW[b]/area	Reference
C. alata Pl. et Tr.	dawn	4.1	750[a]	Popp et al. 1987
	dusk	4.0	785[a]	
C. rosea Jacq.	dawn	3.7	720[a]	
	dusk	3.8	705[a]	
C. rosea Jacq.	mature	3.8	795[b]	Ball et al. 1991
	seedlings	5.2	465[b]	
C. venosa Jacq.		4.5	375[a]; 480[b]	Franco et al. 1990
C. minor L.		6.5	565[a]; 670[b]	
C. major L.		5.3	505[a]; 625[b]	
C. alata Pl. et Tr.		7.1	1 355[a]; 1 575[b]	
C. minor L.		5.6		Holbrook and Putz 1996
C. minor L.		4.7	580[a]	Zotz et al. 1997
C. minor L.			735[b]	Borland et al. 1998
C. rosea Jacq.			805[b]	
C. aripoensis Britt.			585[b]	
C. hilariana Schl.			1 390[b]	Berg et al. 2004
C. minor L.	–N, low PAR	6.8	640[a]	Franco et al. 1991
	–N, high PAR	4.9	710[a]	
	+N, low PAR	8.2	605[a]	
	+N, high PAR	6.1	665[a]	
C. uvitana Pittier	shade leaves	5.2	655[b]	Zotz and Winter 1994
	sun leaves	4.3	780[b]	
C. criuva Camb.			420[a]	Herzog et al. 1999

2.2 Leaf Anatomy

Leaf anatomy was studied in *C. hilariana* and *Clusia spiritu-sanctensis* G. Mariz et Weinberg (Schneider 1985; da Silva et al. 2005), and in *Clusia rosea* Jacq. and *C. alata* (Popp et al. 1987; Borland et al. 1998) all of which are CAM species, in *Clusia mexicana* Vesque (Hartenburg 1937) and *Clusia minor* L. (Borland et al. 1998; Duarte 2006) which are C_3/CAM-intermediate and in *Clusia aripoensis* Britt. (Borland et al. 1998) which is weak CAM inducible. The succulent leaves of CAM plants are often uniformly composed of isodiametric and nearly spherical cells, e.g. in the genus *Kalanchoë* with an internal air space of 3 % of the total leaf volume in *K. daigremontiana* (Duarte et al. 2005). The large vacuoles of this succulent leaf tissue provide the storage capacity for nocturnal accumulation of malate, which in turn is related to the capacity for CAM. However, the leaves of all of the *Clusia*s are clearly bifacial and differentiated into a palisade parenchyma of two to four cell layers and a spongy parenchyma with large intercellular air spaces (Fig. 2.2A–D) of, e.g., 9.3 % of the total leaf volume in *C. minor* (Duarte 2006). In the region of the major vein of leaves of *C. minor* there is no spongy parenchyma and there are one to four layers of palisade parenchyma on the adaxial side of the bundle which is thinner but continuous with the palisade parenchyma of the interveinal lamina tissue (Fig. 2.2F,G). The bundle is surrounded by isodiametric parenchyma cells which also have contact with the palisade parenchyma. In the tissue around the bundle in the major vein there are scarcely any intercellular spaces, and thus, lateral gas diffusion shall be highly limited in this part of the leaf (Fig. 2.2F,G). The size and the architecture of the internal air spaces of leaves are very important for lateral gas diffusion in leaves and the role and signalling functions of CO_2 and O_2 in synchronizing photosynthetic activities within leaves (Sects. 8.5 and 11.2.2), which have been studied in *C. minor* (Duarte and Lüttge 2007). *C. aripoensis* and *C. minor* have three to four adaxial hypodermal cell layers. *C. rosea* may have one to two adaxial and one abaxial hypodermal cell layers (Ting et al. 1985; Ball et al. 1991).

Dimensions were measured in leaves of *C. rosea* sampled in the field and in *C. rosea*, *C. minor* and *C. aripoensis* grown in a phytotron (Table 2.2). Leaves of seedlings of *C. rosea* were thinner than those of mature plants and in the latter sun exposed leaves were thicker than shaded leaves. These differences were due to differences in the thickness of the photosynthetically active mesophyll, i.e. the thickness of both palisade cell layers and spongy parenchyma (Table 2.2). Dimensions are also presented for *C. minor* by Holbrook and Putz (1996) but are not incorporated in Table 2.2 because these authors do not separate epidermis and hypodermis.

The high degree of vacuolization of the mesophyll cells with only a thin layer of cytoplasm along the walls seen in Fig. 2.2 is typical of CAM performing leaves.

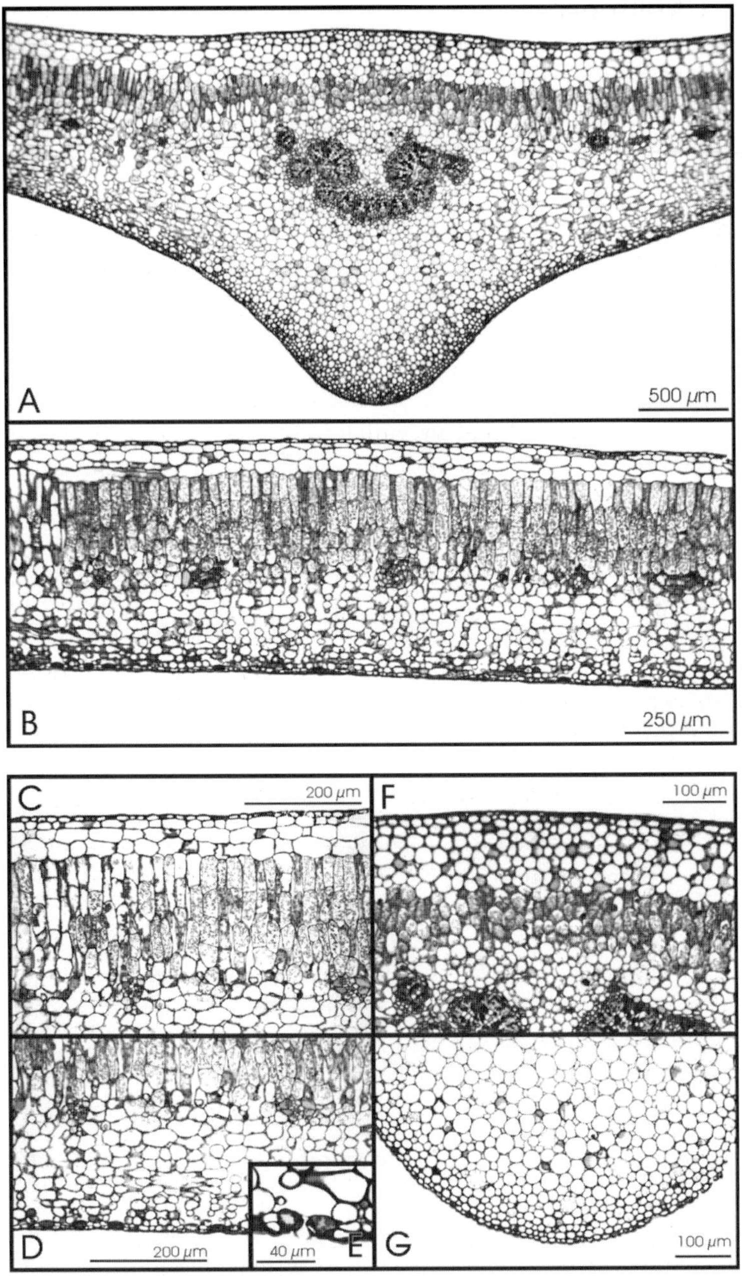

Fig. 2.2A–G. Leaf anatomy of *C. minor*: **A** cross section at the main vein; **B** cross section at an intercostal portion of the leaf; **C** detail of the adaxial epidermis, hypodermis and palisade parenchyma; **D** detail of the abaxial epidermis and spongy parenchyma; **E** stomate in the adaxial epidermis; **F** detail of the adaxial tissue of the main vein, there is collenchyma directly under the epidermis, the palisade parenchyma is not interrupted by the vein but there is no spongy parenchyma (see G) and the cells are tightly packed; **G** detail of the abaxial tissue of the main vein

Table 2.2. Thickness of cell layers measured in cross sections of leaves (A) of *C. rosea* sampled in the field on St. John Island, Lesser Antilles, and (B) of *C. rosea*, *C. minor* and *C. aripoensis* grown in a phytotron. Summarized and rounded from detailed analyses of (A) Ball et al. (1991) and (B) Borland et al. (1998)

(A) *C. rosea*	Shaded (μm)	Exposed (μm)	Seedlings (μm)
Adaxial epidermis	20	15	15
Hypodermis 1	15	15	25
Hypodermis 2	30	30	25
Palisade 1	125	130	75
Palisade 2	105	130	80
Palisade 3	70	110	–
Spongy parenchyma	440	500	330
Hypodermis	20	25	20
Abaxial epidermis	20	20	20
Sum	845	975	590
(B)	*C. rosea*	*C. minor*	*C. aripoensis*
	(% of total leaf thickness)		
Hypodermis	5	12	21
Palisade parenchyma	36	42	24
Spongy parenchyma	59	46	55

Table 2.3. Stomatal densities (number mm^{-2}) of *C. rosea* and *C. minor* in comparison to other woody hemi-epiphytes

Plants	Stomatal density	Reference
C. rosea	55	Ball et al. 1991
C. minor	186	Holbrook and Putz 1996
	180	Duarte 2006
C. minor	90	Zotz et al. 1997
Others	110 to 1035	

The leaves of *Clusia*s contain many secretory ducts filled with a latex-type secretion.

Clusia leaves are hypostomatic. The stomata of *C. mexicana* are at the bottom of an external cavity formed by a pair of peripheral horns (Hartenburg 1937). The guard cells of *C. minor* have an inner and an outer pair of horns (Fig. 2.2E). The stomatal apparatus of *C. rosea* forms several cavities by means of an outer pair of horns on the subsidiary cells and an outer and inner pair of horns on the guard cells (Popp et al. 1987). Stomatal densities are rather low. In comparison to a range of stomatal densities of other woody species in tropical forests values of *Clusia* are at the lowest end (Table 2.3).

2.3 Life Forms

As noted above, with respect to leaf morphology and anatomy all species of *Clusia* are very uniform and constitute one single morphotype. However, a large diversity between species as well as plasticity within species is apparent in other respects. These include the ecological amplitude of *Clusia* (Chap. 9), the different photosynthetic physiotypes (Chap. 8) and biochemical reactions of CAM (Chap. 8). Morphological diversity and plasticity are given by expression of different life forms. Seeds of given species of *Clusia* can facultatively germinate both terrestrially and epiphytically in small accumulations of humus in forks of tree branches, within the tanks of bromeliads and in epiphyte nests (Figs. 2.3 and 2.4). Terrestrial seedlings directly develop free standing shrubs and trees (Fig. 2.4). Epiphytic seedlings produce many adventitious aerial roots. Some of these adventitious roots serve as holdfasts and at the same time strangle the bark of their host trees. Some of the adven-

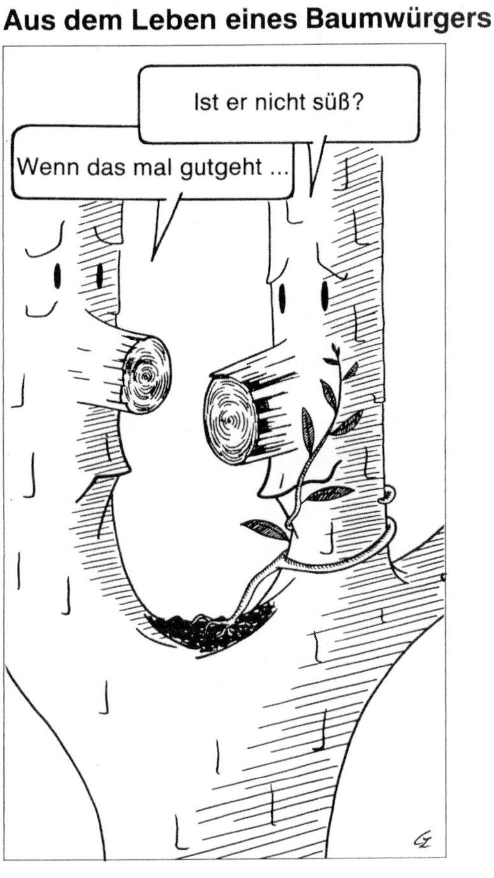

Fig. 2.3 Cartoon unexpectedly added to Lüttge (1991) by the publisher, headed *"From the life of a strangler"* where the right hand branch of the phorophyte says *"isn't he sweet?"* but the left hand branch replies *"if this may go well ... "*. Drawing by J. Czichos, www.joachim-czichos.de

Fig. 2.4A–L. Life forms of *Clusias*: **A** terrestrial seedlings of *C. multiflora* H.B.K. (area of Instituto Venezolano de Investigaciones Científicas, IVIC, Caracas, Venezuela); **B** terrestrial seedings of *C. spiritu sanctensis* Mariz et Weinberg (Estação Biologica de Santa Lúcia, Santa Teresa, Espírito Santo State, Brazil); **C** seedling of *C.* spec. rooting between the bark and the wood of a rotting tree (Serrania de San Luis, Falcon State, Venezuela); **D** seedling of *C. rosea* Jacq. in a tree fork (St. John, US-Virgin Islands); **E** seedling of *C. rosea* in the tank of the bromeliad *Aechmea lingulata* (St. John); **F** tank of *A. lingulata* cut open to show the roots of *C. rosea* (St. John);

G *C. rosea* in an epiphyte nest (St. John); **H** *C. rosea* strangling the phorophyte (St. John); **I** free standing shrubs of *C. minor* L. (*narrow leaves*) and *C. multiflora* (*broad leaves*) (area of IVIC); **J** free standing shrub of *C. spiritu sanctensis* (Santa Teresa); **K** *C. rosea* strangler getting a free standing tree (St. John); **L** *C. rosea* free standing tree (St. John)

titious roots grow positively gravitropically and eventually establish contact with the soil. The strangled host trees may die from interruption of the stream of assimilates in the phloem of their bark, and with their dead trunks rotting away the originally epiphytic *Clusias* become free standing trees. This life form is therefore called hemi-epiphytic.

Holbrook and Putz (1996) have compared several morphological/anatomical traits in rooted trees and epiphytes of *C. minor*. Specific leaf area (area per weight) was 30 % larger in the epiphytes compared to the trees. Leaf thickness, stomatal density and the diameter and length of stomata were similar in both life forms. The bulk elastic modulus was larger, i.e. leaves were more rigid in the rooted trees. The differences between the rooted trees and the epiphytes with respect to these traits were larger in species of *Ficus* than in *C. minor*, and the authors conclude that it is a physiological rather than a morphological/anatomical plasticity that is a more important attribute in *Clusia*. It is notable in this context that *Ficus* species are obligate C_3-plants (Ting et al. 1987) while *Clusias* have the CAM-option, and therefore, although hemi-epiphytic species of both genera are very similar life forms, in the absence of any CAM option *Ficus* must adapt differently in the epiphytic stage. In fact *Clusias* are the only known hemi-epiphytic stranglers having CAM (Ting et al. 1987).

2.4 Hydraulic Architecture

The physiological morphology and anatomy in relation to water transport in plants, i.e. the hydraulic architecture, among others is governed by parameters such as:

- Specific conductivity of stem segments, K_s, providing information about the hydraulic efficiency of xylem on a cross-sectional area basis
- Leaf specific conductivity of stem segments, K_l, providing information about the hydraulic sufficiency of a stem segment on a leaf area basis
- Ratio of wood cross section invested per unit leaf area, H_v, the so-called Huber-value

Hydraulic architecture relates to (i) life form as well as (ii) mode of photosynthesis. With respect to *Clusia* these are (i) free standing trees vs hemi-epiphytes and (ii) C_3-photosynthesis vs CAM. Hemi-epiphytes potentially need a smaller wood cross section than free standing trees to support their foliage and crown as they have the mechanical support of their host trees or phorophytes. CAM is a water saving variation of photosynthesis and CAM plants may operate with lower water conductivities than C_3-plants. Since species of *Clusia* are the only woody dicotyledonous plants performing CAM (Chap. 1) this may distinguish their hydraulic architecture from all other woody species in tropical forests.

Hence, variants of desirable comparison are:

1. Free standing trees vs hemi-epiphytes both with C_3-photosynthesis
2. Free standing trees vs hemi-epiphytes both with CAM
3. Free standing trees with C_3-photoynthesis vs CAM
4. Hemi-epiphytes with C_3-photosynthesis vs CAM

where 1 and 2 would indicate the role of hemi-epiphytism and 3 and 4 that of the mode of photosynthesis.

A comparison of various tropical trees and woody hemi-epiphytes including species of *Clusia* led to the general suggestion (Zotz et al. 1997) that:

- Hemi-epiphytes show significantly higher specific stem conductivity, K_s, compared to trees
- Hemi-epiphytes invest less wood cross-section per unit leaf area, H_v, compared to trees
- Hemi-epiphytes tend to have less conductive stems per unit leaf area, K_l, compared to trees

Table 2.4 shows a comparison of the hemi-epiphytic C_3/CAM-intermediate *C. uvitana* with a number of other tropical tree species (Zotz et al. 1994). In terms of hydraulic architecture the K_s values are rather similar, i.e. the xylem of *Clusia* is about as efficient on a wood area basis as that of the other species. However, K_l of *Clusia* was much lower than in the other species, i.e. its stems were less efficient on a unit leaf area basis and *Clusia* supports a large leaf area

Table 2.4. Comparison of hydraulic architecture parameters of hemi-epiphytic *C. uvitana* with other tropical woody species with respect to the parameters of hydraulic conductivity K_s (specific stem conductivity) and K_l (conductive stem per unit of leaf area) as well as xylem pressure at 50 % loss of hydraulic conductivity (Ψ_{xp50}) and evaporation (E). Rounded values from Zotz et al. (1994, 1997) and Patiño et al. (1995)

	C. uvitana	Others
K_s (kg s^{-1} m^{-1} MPa^{-1})	0.7 to 2.1	1.3 to 2.5 (4 to 14[a])
K_l (kg s^{-1} m^{-1} MPa^{-1}x10^4)	0.9 to 2.4	5 to 30
Ψ_{xp50} (MPa)	–1.3	–1.0 to –6.2
E:		
average maximum (mg s^{-1} m^{-2})	12	50 to 80
average (mg s^{-1} m^{-2})	2.3[b]	10 to 26
mean daily (g 24 h^{-1} m^{-2})	198[c]	902 to 2227

[a] Values for *Ficus*
[b] *Clusia minor* showed a value of 1.8
[c] *C. minor* showed a value of 153

per unit stem area. This may be explained by both the hemi-epiphytic life form and the performance of CAM of *C. uvitana* as compared to the other species, i.e. this work does not distinguish between variants 1 and 2 and variants 3 and 4 above. It was also shown, however, in the study of Zotz et al. (1994) that the vulnerability to cavitation, i.e. the relative loss of hydraulic conductivity under a pressure applied via a pressure chamber, was rather high in *C. uvitana*. In the other tree species the range of xylem pressure at which 50 % loss of the hydraulic conductivity occurred, Ψ_{xp50}, was large extending from –1.0 to –6.2 Pa (Table 2.4), depending on their ecological performance, where the less negative values were characteristic for drought evaders and the more negative values for drought tolerators. The high Ψ_{xp50}-value of *C. uvitana* of –1.3 Pa indicates that it is a drought evader. This performance is most likely based on the use of CAM as an adaptation to limited water supply as discussed in Chaps. 8 and 9. This is corroborated by the comparatively low evaporation rates observed in *C. uvitana* (Table 2.4). Although a systematic comparison with respect to points 1 to 4 above is not available in the literature, this certainly suggests a strong influence of the option to perform CAM on hydraulic architecture. Moreover, in another study, Patiño et al. (1995) compared the hemi-epiphytic C_3/CAM-intermediate *C. uvitana* with four C_3 hemi-epiphytic and three C_3 terrestrial species of *Ficus*. Ranges of values of K_s, K_l and H_v are given in Table 2.5. The ranges of K_s values for the hemi-epiphytic and the terrestrial *Ficus* species overlap, but as expected from the generalization given above, the range of values reaches higher levels in the former. However, for the hemi-epiphytic *C. uvitana* the K_s value is much lower and the fact that it can afford much lower specific stem conductivity than the C_3 hemi-epiphytes of *Ficus* must be due to its option to perform CAM. As also expected, K_l values are lower in hemi-epiphytic *Ficus* species than in the free standing ones, again meeting the expectations from the general pattern given above, but the values for *Clusia* are still much lower. This indicates that in addition to hemi-epiphytism hydraulic architecture is related to performance of CAM.

Table 2.5. Comparison of hydraulic architecture parameters of hemi-epiphytic *C. uvitana* with hemi-epiphytic and free standing trees of *Ficus*. Specific stem conductivity (K_s), cross section per unit leaf area (H_v), conductive stem per unit of leaf area (K_l). Rounded values from Patiño et al. (1995)

	K_s (kg s^{-1} m^{-1} MPa^{-1})	$H_v \times 10^4$	$K_l \times 10^4$ (kg s^{-1} m^{-1} MPa^{-1})
Ficus free standing	11 to 14	2.0 to 6.1	23 to 52
Ficus hemi-epiphytic	7 to 34	1.0 to 2.2	7 to 23
C. uvitana hemi-epiphytic	1.1	1.4	1.5

References

Ball E, Hann J, Kluge M, Lee HSJ, Lüttge U, Orthen B, Popp M, Schmitt A, Ting IP (1991) Ecophysiological comportment of the tropical CAM-tree *Clusia* in the field. I. Growth of *Clusia rosea* Jacq. on St.John, US Virgin Islands, Lesser Antilles. New Phytol 117:473–481

Berg A, Orthen B, de Mattos EA, Duarte HM, Lüttge U (2004) Expression of crassulacean acid metabolism in *Clusia hilariana* Schlechtendal in different stages of development in the field. Trees 18:553–558

Borland AM, Técsi LI, Leegood RC, Walker RP (1998) Inducibility of crassulacean acid metabolism (CAM) in *Clusia* species: physiological/biochemical characterization and intercellular localization of carboxylation and decarboxylation processes in three species which exhibit different degrees of CAM. Planta 205:342–351

Duarte HM (2006) Chronobiologie von *Clusia minor*: circadianer Rhythmus in einer Pflanze mit C_3/CAM-intermediärem photosynthetischen Verhalten. Dissertation, Darmstadt

Duarte HM, Jakovljevic I, Kaiser F, Lüttge U (2005) Lateral diffusion of CO_2 in leaves of the crassulacean acid metabolism plant *Kalanchoë daigremontiana* Hamet et Perrier. Planta 220:809–816

Duarte HM, Lüttge U (2007) Gas-exchange, photorespiration and spatiotemporal dynamics of relative quantum use efficiency in leaves of the C_3-photosynthesis/CAM-intermediate species *Clusia minor* L. (Clusiaceae) in both modes of photosynthesis. Trees (in press)

Franco AC, Ball E, Lüttge U (1990) Patterns of gas exchange and organic acid oscillations in tropical trees of the genus *Clusia*. Oecologia 85:108–114

Franco AC, Ball E, Lüttge U (1991) The influence of nitrogen, light and water stress on CO_2 exchange and organic acid accumulation in the tropical C_3-CAM tree, *Clusia minor*. J Exp Bot 42:597–603

Hartenburg W (1937) Der Wasser- und Kohlensäurehaushalt tropischer Regenwaldpflanzen in sommerlicher Gewächshauskultur. Jahrb wiss Bot 85:641–697

Herzog B, Grams TEE, Haag-Kerwer A, Ball E, Franco AC, Lüttge U (1999) Expression of modes of photosynthesis (C_3, CAM) in *Clusia criuva* Camb. in a cerrado/gallery forest transect. Plant Biol 1:357–364

Holbrook NM, Putz FE (1996) From epiphyte to tree: differences in leaf structure and leaf water relations associated with the transition in growth form in eight species of hemiepiphytes. Plant Cell Environ 19:631–642

Lüttge U (1991) *Clusia*. Morphogenetische, physiologische und biochemische Strategien von Baumwürgern im tropischen Wald. Naturw 78:49–58

Lüttge U (2000) Photosynthese-Physiotypen unter gleichen Morphotypen, Species und bei Klonen: Kann ökophysiologische Plastizität zur Entstehung von Diversität beitragen? Ber Reinh Tüxen-Ges 12:319–334

Martii CFP (1889) Flora Brasiliensis. Vol XII Pars I, CII Guttiferae. Lipsiae apud Frid. Fleischer in Comm, Oldenbourg

Niinemets Ü, Sack L (2005) Structural determinants of leaf light-harvesting capacity and photosynthetic potentials. Prog Bot 67:385–418

Patiño S, Tyree MT, Herre EA (1995) Comparison of hydraulic architecture of woody plants of differing phylogeny and growth form with special reference to free-standing and hemi-epiphytic *Ficus* species from Panama. New Phytol 129:125–134

Popp M, Kramer D, Lee H, Diaz M, Ziegler H, Lüttge U (1987) Crassulacean acid metabolism in tropical dicotyledonous trees of the genus *Clusia*. Trees 1:238–247

Schneider SZ (1985) Anatomia foliar de *Clusia hilariana* Schlechtendal e *Clusia spirito-sanctensis* G. Mariz et Weinberg (Guttiferae) ocorrentes no estrado do Espírito Santo. M.S.Thesis, 162 pp, Universidade Federal do Rio de Janeiro
Silva LC da, Oliva MA, Azevedo AA, Araújo JM, Aguiar RM (2005) Micromorphological and anatomical alterations caused by simulated acid rain in restinga plants: *Eugenia uniflora* and *Clusia hilariana*. Water Air Soil Pollut 168:129–143
Ting IP, Hann J, Holbrook NM, Putz FE, Sternberg L da SL, Price D, Goldstein G (1987) Photosynthesis in hemiepiphytic species of *Clusia* and *Ficus*. Oecologia 74:339–346
Ting IP, Lord EM, Sternberg L da SL, DeNiro MJ (1985) Crassulacean acid metabolism in the strangler *Clusia rosea* Jacq. Science 229:969–971
Zotz G, Tyree MT, Cochard H (1994) Hydraulic architecture, water relations and vulnerability to cavitation of *Clusia uvitana* Pittier: a C_3-CAM tropical hemiepiphyte. New Phytol 127:287–295
Zotz G, Patiño S, Tyree MT (1997) Water relations and hydraulic architecture of woody hemi epiphytes. J Exp Bot 48:1825–1833
Zotz G, Winter K (1994) A one-year study on carbon, water and nutrient relationships in a tropical C_3-CAM hemi-epiphyte, *Clusia uvitana* Pittier. New Phytol 127:45–60

3 Biogeographic Features of *Clusia*, with Emphasis on South American and Especially Brazilian Species

Dorothy Sue Dunn de Araujo and Fabio Rubio Scarano

3.1 Introduction

Cox and Moore (1993), in their classic "Biogeography" text book, broadly define this discipline as "the study of living things in a spatial and temporal context". They argue that biogeography studies will often provide answers to questions such as: "a) Why are there so many living things? b) Why are they distributed the way they are? c) Have they always occupied their current distribution patterns? d) Do man's activities today affect these patterns and if so, what are the prospects for the future?" Ideally, therefore, whenever a complete account of *Clusia* biogeography is available, all these questions should be answerable. In order to reach this stage of knowledge, one should have enough relevant information on the three main biogeographic processes: migration, evolution and extinction (Shrader-Frechette and McCoy 1993). Since very little is known about these processes for *Clusia*, this chapter is by no means an attempt to provide a complete study of the biogeography of the genus. However, we use the four questions above as guidelines for this text.

There are indeed at least two major obstacles that stand in the way of a complete biogeographic study of *Clusia* and, consequently, answers to the questions posed by Cox and Moore (1993). First, taxonomic and systematic difficulties within the genus are legendary, although much progress has been achieved in the past few years (e.g., Chap. 6). For instance, there has been no comprehensive treatment of the genus in Brazil. Second, we are still short of a more holistic approach to the genus, integrating ecophysiology, reproductive biology, phytosociology and other areas within a single framework. While this book is an obvious attempt to circumvent the latter, there is still much to do as regards the former.

Ecological Studies, Vol. 194
U. Lüttge (Ed.)
Clusia: A Woody Neotropical Genus
of Remarkable Plasticity and Diversity

Despite these obstacles, we attempt to provide here our biogeographic interpretation of the available data set on species geographic distribution, habitat occurrence and ecology, gathered from specialized literature and herbaria data banks. We focus mainly on South American, and in particular Brazilian species of *Clusia*, but we believe that many of the patterns and processes discussed here might also apply to *Clusia* species elsewhere.

3.2 The Survey

We compiled a list of 109 species that occur in South America for which we found information on habitat and geographic distribution. This list represents a considerable proportion of the ca. 250–400 species of this Neotropical genus (Bittrich and Amaral 1996; Pipoly et al. 1998). Moreover, it includes species from both species rich areas (i.e., the Andean region and the Guianan shield; Pipoly et al. 1998), and less rich areas, such as Brazil. Although the genus occurs from Mexico to southern Brazil, and the Brazilian territory covers much of this range, species richness in Brazil is estimated to be comparatively low (at least 70 species; see also a map of *Clusia* distribution in the Americas in Chap. 6). This is probably related to the fact that Gentry (1982) placed the Clusiaceae in the Gondwana Extra-Amazon group. According to this author, most Neotropical botanical families have affinities with continents in either the Northern (Laurasian) or the Southern (Gondwanan) Hemisphere, and the species of those in the Southern Hemisphere can be further classified as Amazonian or Extra-Amazonian. The distribution of the former group of species is centered in the Amazon region, with low diversity in the Andes, while the latter group has centers of diversity in the Andean foothills or lower elevations. Clusiaceae belongs to the latter group. However, this may also be the result of more published data on the taxonomy of Clusiaceae, and consequently of the genus *Clusia*, in this region as seen in the recent floras of the Guianas (Boggen et al. 1997) and Venezuela (Pipoly et al. 1998).

Our main sources for this survey were the Flora of the Venezuelan Guyana (Pipoly et al. 1998), internet sources such as the virtual herbaria of the New York Botanical Garden (http://sciweb.nybg.org), especially the Catalog of Vascular Plant Species of Eastern Brazil, and the Missouri Botanical Garden (www.mobot.org), Checklist of Plants from Northeast Brazil-CNIP (http://www.cnip.org.br/), Lista das Plantas Vasculares das Restingas do Litoral Norte da Bahia (http://www.ibge.gov.br/ home/geociencias/recursosnaturais) and species lists from various regions of Brazil (Lima and Guedes-Bruni 1997; Mendonça et al. 1998; Ribeiro et al. 1999; Araujo 2000), plus a survey of three herbaria in Rio de Janeiro (R, GUA, RFA). Valid species names were checked using the International Plant Names Index from Kew Gardens (http://

www.ipni.org) and published works on the genus (Bittrich 1996; Bittrich and Amaral 1996, 1997; Pipoly 1997; Pipoly and Graff 1995a, b; Pipoly et al. 1998; Nogueira et al. 2001). Only published species with data on habitat, elevation and geographic distribution were included in this study.

Species were grouped according to three distinct criteria. a) *Geopolitical*: species were classified as Extra-Brazilian (present in the South American countries of, Venezuela, the Guianas, Colombia, Ecuador, Peru, and Bolivia, but not present in Brazil) or Brazilian (all species found in Brazil alone or predominantly in Brazil with a few species ranging to other countries). b) *Biomes*: the Brazilian species were classified as belonging to the Amazon forest domain, the Atlantic forest domain or the Cerrado domain (including campos rupestres). There was a certain amount of species overlap in the three categories. These major biomes often comprise a range of distinct vegetation types. For instance, in the Amazon region there are forest formations (e.g., *terra-firme* rain forest, flooded *várzea* and *igapó* forests, etc. – Rizzini 1979) and open vegetation (e.g., *campina, campinarana*, savannas – Veloso et al. 1991). The Atlantic rain forest sensu lato also comprises forests (e.g., rain forest sensu stricto, semideciduous forest, swamp forest; Morellato and Haddad 2000) and open vegetation (e.g., restingas, rocky outcrops, high altitude grasslands; Scarano 2002). Thus, we used one further classification criterion: c) *Habitat type*: species were classified as occurring in forest, non-Forest or both.

3.3 Patterns

Appendices 1 and 2 show the list of Extra-Brazilian and Brazilian species surveyed plus information regarding respective geopolitical location, biome and habitat type. They show that 60 of the species surveyed do not occur within the Brazilian borders (Extra-Brazilian) while 49 occur in Brazil, some of which may extend beyond the borders of this country. These appendices comprise the entire data base from which we derived the tables that follow.

The first pattern to emerge from the data compiled is that Brazilian species of *Clusia* tend to have wider ecological amplitude than Extra-Brazilian species (see also Chap. 6). While 47 % of the Brazilian species surveyed are found in both forests and non-forests, extra-Brazilian species are predominantly strictly forest species, i.e., they are found in rain forests, cloud forests, dwarf or elfin forests (Table 3.1). Extra-Brazilian species classified as non-forest are found mostly on rocky outcrops and white sands and in savannas and tepui scrub.

These data suggest that either migration processes were more intense for Brazilian species of *Clusia*, or that they were more adaptable to distinct environments, or most likely both. Dioecious species, such as the majority of those

Table 3.1. Percentage of *Clusia* species occurring in forest, non-forest and both types of habitat in Brazil vs extra-Brazil

Habitat type	Extra-Brazil (n=60)	Brazil (n=49)
Forest	65 %	33 %
Non-forest	20 %	20 %
Non-forest+forest	15 %	47 %

belonging to the genus *Clusia* (see Chap. 5), are often associated with fleshy fruit formation (Weller and Sakai 1999; Vamosi et al. 2003; Vamosi and Vamosi 2004; Matallana et al. 2005) and consequent long distance dispersal by birds (for *Clusia* see Gonzaga et al. 2000). Thus, there is no reason to believe that there should be relevant differences in dispersal potential of *Clusia* species of Brazil as compared to extra-Brazil.

Another interesting aspect related to *Clusia*'s sexual system has to do with the theory that dioecy is favoured in stressful environments. This was first suggested by Darwin (1877) and later supported by models and empirical surveys (Bawa and Opler 1975; Freeman et al. 1997; Thompson and Edwards 2001). It has been argued that division of labour in unisexual plants may increase male and female fitness due to a compensation effect (e.g., Lloyd and Webb 1977; Sutherland and Delph 1984), unless physiological constraints are so severe as to generate low plant density or lack of pollinators or reduced fertility (see Chap. 5). Once again, since almost all *Clusia* species are dioecious, they would all potentially display this compensation effect which would not explain the Brazilian vs Extra-Brazilian differences in ecological amplitude.

When we examine the geographic distribution of these species in relation to altitude, there are some interesting data that help clarify possible reasons behind Brazilian vs extra-Brazil differences. Table 3.2 shows that extra-Brazilian forest and non-forest species are more often limited to only one altitudinal band. The pattern is similar for Brazilian species (Table 3.3), but a smaller proportion of the species are found in the two altitudinal-band range (31 % vs 42 %). More interestingly, up to a third of the extra-Brazil species (19) are strictly upper montane (>1500 m), while there are no Brazilian species restricted to this altitude band. This suggests that the Brazilian species would fit easier in the label "generalists", while the extra-Brazilian species are often more specialized.

In the case of the Brazilian species, their overall high ecological amplitude, as assessed by the high proportion (47 %) of species that grow both in forest and non-forest environments, can be better understood by examining Table 3.4. *Clusia* in Brazil is most species rich in the Amazon region (65 % of the 49 Brazilian species surveyed) and Atlantic forest (39 %). Some species overlap the two regions. These large biomes comprise predominantly rain for-

Table 3.2. Distribution bands of extra-Brazil species in forest, non-forest vegetation and both. The altitude bands are L=lowland <500 m; M=montane 500–1500 m; UM=upper montane >1500 m

Habitat type	Narrow (one altitude band)				Wide (two altitude bands)			Extra wide (three altitude bands)		Total
	L	M	UM	Total	L/M	M/UM	Total	L/M/UM	Total	
Forest (F)	5	5	13	23	3	12	15	1	1	39
Non-forest (NF)	2	0	6	8	1	3	4	0	0	12
NF+F	2	0	0	2	3	3	6	1	1	9
Total	9	5	19	33	7	18	25	2	2	60

Table 3.3. Distribution bands of Brazilian species in forest, non-forest vegetation and both. The altitude bands are L=lowland <500 m; M=montane 500–1500 m; UM=upper montane >1500 m

Habitat type	Narrow (one altitude band)				Wide (two altitude bands)			Extra wide (three altitude bands)		Total
	L	M	UM	Total	L/M	M/UM	Total	L/M/UM	Total	
Forest (F)	7	5	–	12	2	–	2	3	3	17
Non-forest (NF)	2	5	–	7	1	1	2	1	1	10
NF+F	7	2	–	9	9	2	11	2	2	22
Total	16	12	–	28	12	3	15	6	6	49

Table 3.4. Distribution of *Clusia* species in major Brazilian biomes as well as forest and non-forest habitat types. There is species overlap in these domains (i.e. sums are over 100 %)

	Total number of species	Forest/ non-forest	Forest only	Non-forest only
Amazon forest domain	32	11 (34 %)	15 (47 %)	6 (19 %)
Atlantic forest domain	21	10 (48 %)	6 (28 %)	5 (24 %)
Cerrado/campo rupestre	8	2 (25 %)	0	6 (75 %)

est vegetation, but they also include open, non-forest vegetation. Conversely, the genus is apparently not well represented in Central Brazil, south of the Amazon, where open vegetation predominates, such as the savanna-like *cerrados* (see Oliveira and Marquis 2002 for a monograph on this large Brazilian biome). Only three species are listed for the cerrados of central Brazil by Mendonça et al. (1998). There is, however, a general lack of information about *Clusia* species of Central Brazil, there being no reliable taxonomic treatments on *Clusia* in this region.

About 48 % of the rainforest species from both the Amazon and Atlantic forest domains occur also in neighbouring non-forest habitats. Only 15 % of the species in these two large biomes are strictly non-forest, which in the Amazon domain includes savannas/cerrados, Rio Negro caatinga, white sand scrub, campinarana, granite outcrops, and rocky outcrops, while in the Atlantic domain this includes restingas and rocky outcrops. The proportion of forest species that also occur in non-forest habitats is higher for the Atlantic forest domain (48 % vs 34 %) as is the proportion of non-forest species only (24 % vs 19 %); this may be due to available habitat in the restingas and rocky slopes on the coast, contiguous to the Atlantic forest, as well as the potential for species interchange between these two habitats during climate change and sea level oscillations of the Quaternary (Scarano 2002; see also Sect. 3.4.2).

3.4 Discussion

We considered that the most objective way to portray the current status of the knowledge regarding *Clusia* biogeography, at the light of the survey produced here, would be to discuss the data guided by the four questions posed in the Introduction, quoting Cox and Moore (1993), bearing in mind the three processes central to biogeographic studies: migration, evolution and extinction.

3.4.1 Why Are There So Many Species?

An answer to this question would demand understanding of the speciation process and speciation rates within the genus, as well as the extinction process and extinction rates, all of which are unknown. Theoretically, speciation can be either gradual or rapid, as summarized by Niklas (1997): gradual speciation, or "phyletic gradualism", takes place by gradual accumulation of selectively favourable phenotypic traits due to selection pressures acting on population variation at the genetic level; whereas rapid speciation, or "genetic revolution", results from random processes of mutation or genetic drift. Similarly, extinction might occur gradually due to natural selection or rapidly due to profound environmental change.

In terms of speciation, the concept of *centres of dispersal* is much used by biogeographers. It is based on the hypothesis that areas where a taxonomic group is represented by high species richness are likely to be the same areas from which the group dispersed. Conversely, they can also be seen as refugia where certain groups were confined during periods of glaciation, and later dispersed as climatic conditions improved.

In the case of *Clusia* at least two centres of dispersal are recognized: the Andean region (Gentry 1995) and the Guianan shield (Pipoly et al. 1998; see also Chap. 6). Although the Atlantic rain forest of southeast Brazil is less expressive in terms of *Clusia* species richness, the family Clusiaceae ranks among the 12 most species-rich families in these forests, especially at higher altitudes (Oliveira-Filho and Fontes 2000). Prance (1987) recognized three centres of endemism on the east coast of Brazil, from Pernambuco in the Northeast to São Paulo in the Southeast, and the richness of endemic species has been used as an indication of former refugia. Whether the relatively high number of *Clusia* species in the Atlantic forest indicates a minor centre of dispersal, or is the result of climate changes during the Quaternary – when areas of refugia remained relatively stable through the drier periods – can only be answered when we have a better understanding of this group of species, both in terms of taxonomy and ecology. The case of the Atlantic rain forest is discussed next, to cast light on the question of why *Clusia* is distributed in such a way.

However, before we move on, this topic requires again that we highlight the enormous taxonomic problems of the genus. If taxonomy is problematic, we are clearly very limited when it comes to answering the question posed at the beginning of this section. Take for instance the case of *Clusia criuva* vs *C. parviflora* Engl. nom. illeg. The name *Clusia criuva* has been attributed to a number of morphotypes throughout Brazil, later to be corrected. One such case was that of *Clusia parviflora*: the population of Sugar Loaf mountain, in the city of Rio de Janeiro, was originally described as *C. fluminensis* and then *C. criuva* (see Meirelles et al. 1999 for a study on that particular vegetation). This identification was later corrected to *C. parviflora*, and the latter is now also confirmed by molecular biology data, as seen in Chap. 7. Interestingly, the phylogenetic tree produced by Vaasen et al. (2002) does not even place the two species at proximal positions. *Clusia criuva* was also the name given to the C_3 species found in Serra de São José, Tiradentes Municipality (Minas Gerais, southeastern Brazil), but a recent leaf anatomy study (D.G. Ribeiro and A.C. Franco unpublished data) shows that these specimens have a completely different leaf anatomy from the specimens determined as *C. criuva* in gallery forests of Brasília, again raising doubts about the correctness of the identification of the former. The species from the restingas of Rio de Janeiro was first identified as *C. parviflora* (Araujo and Henriques 1984) later to be changed to *C. criuva*. (Araujo et al. 1998). Here we used only the epithet *C. criuva* in our analysis of *Clusia* distribution (see Appendix 2) so as not to increase the confusion.

Perhaps even more confounding was a wrong determination of *Clusia aemygdioi* as *Clusia hilariana* in the Herbarium of the Mello Leitão Museum (Santa Teresa Municipality, Espírito Santo). For several years, this led us to believe that the restinga species, *Clusia hilariana*, also occurred in the neighbouring rainforests of that State (see Scarano 2002). The confirmation of those specimens as *C. aemygdioi* now suggests that *C. hilariana* might be a species restricted to open restinga habitats. This case is further discussed in Sects. 3.4.3 and 3.4.4.

3.4.2 Why Are They Distributed the Way They Are?

This is again another very difficult question to answer, which lies at the heart of most ecological studies until today (see Scarano et al. 2005). The wide habitat range of the genus *Clusia* (see also Chaps. 6 and 9) is most likely associated with variation in habit (hemi-epiphytes, shrubs, trees and climbers; see also Chap. 6) and in physiotype expression (C_3, CAM, C_3-CAM intermediate; Chaps. 8 and 9), which indicate a high ecological plasticity for the genus.

However, although plasticity and ecological resistance to change and to stressful conditions guarantee local survival, they do not necessarily imply in a wide distribution range, which is the result of a combination of ecological resistance and high dispersal capacity. As discussed in Sect. 3.3 above, this also seems to be the case for the genus.

The case of *Clusia* distribution within the Atlantic rain forest complex (sensu lato) illustrates well how a combination of ecophysiological performance and dispersal capacity results in a broad distribution pattern for a given taxon. The Brazilian Atlantic forest is a typical tropical rain forest on mountain slopes, which has been designated as a biodiversity hotspot due to high species richness and high level of species endemism (Myers et al. 2000). Plant communities subjected to more adverse environmental conditions surround the mesic rain forest. They often face flooding, drought, oceanicity and/or cold winter temperatures. For instance, between the mountain slopes and the sea, the coastal plains have swamp forests (Scarano et al. 1997), dry semi-deciduous forests (Araujo 1997), and open thicket vegetation on sandy marine deposits called restinga (Araujo 1992). At the other extreme, on mountain tops (>2000 m a.s.l.), high altitude grasslands and open scrub vegetation on rocky outcrops substitute the rain forest (Medina BMO et al. 2006). Despite these more adverse conditions, the habitats marginal to the Atlantic rain forest often present strikingly high plant species diversity, although always lower than the rain forest itself (Scarano 2002).

Scarano (2002) reviewed palaeoecological, biogeographic and ecological data and forwarded the hypothesis that the origin and maintenance of high diversity in these marginal habitats is due to past migration of canopy species from the rain forest to these habitats. Establishment of these canopy plants on the ground of the marginal habitats created conditions for the subsequent entry of a broad range of species through the process of ecological facilitation. These would have happened similarly on two fronts. Lowland habitats such as the restingas and the swamps, are geologically young, dating from the Quaternary (3000 to 120,000 years BP – Martin et al. 1993), while rocky outcrops and the dry forest consist of vegetation possibly older than the rain forests bearing several relict species. Since epiphytic and hemi-epiphytic habit has low or no requirements for underground nutrients for establishment and growth, when such plants arrived in these nutrient-poor habitats they could occupy open niches and trigger land colonisation and succession.

Thus, several epiphytic rain forest plants are actually terrestrial nurse plants in the marginal habitats. Many such nurse plants possess the crassulacean acid metabolism (CAM) mode of photosynthesis and perhaps the main ones belong to the genus *Clusia* (see Chap. 4).

If this hypothesis is true, it would appear that ecophysiological and dispersal features of *Clusia* might explain not only the broad distribution of the genus throughout the range of habitats within the Atlantic forest domain, but also of entire vegetation types, particularly in the case of the restingas (see Chap. 4).

3.4.3 Have They Always Occupied Their Current Distribution Patterns?

The above example clearly indicates that this is not the case for all *Clusia* species. It is worthwhile, in this topic, to discuss the alternatives of whether epiphytic bromeliads in the canopies of moist tropical forests originated from understorey plants in a struggle for light as proposed by Schimper (1888) or if they developed from terrestrial ancestors of open, sun-exposed vegetation pre-adapted to the stress of irradiance and low water and nutrient availability of the epiphytic habitat as discussed by Medina (1974). This latter hypothesis was strongly supported by a census of the bromeliad species of Trinidad (Pittendrigh 1948; Griffiths and Smith 1983) and the phylogenetic relationships in the family (Smith 1989), where epiphytism clearly evolved independently several times in the three subfamilies (Pitcarnioideae, Bromelioideae, Tillandsioideae) as is now clearly revealed by molecular phylogenetic studies (Crayn et al. 2000, 2004; Horres et al. 2000). Evidently Schimper's hypothesis is not valid for the Bromeliaceae. It may still apply, however, to other taxa, where one finds not only shade-tolerant species but also species which are clearly shade-demanding, reminiscent of their forest-floor past, e.g. among the epiphytic ferns and also orchids (Lüttge 1985, 1997; Lüttge et al. 1986; Goh and Kluge 1989). Scarano (2002) argued that an origin of epiphytism from pre-adapted terrestrial plants of sun-exposed habitats more plausibly occurred in the geologically younger lowlands and restingas, whereas the former may have occurred on rocky outcrops and in high altitude areas. In the case of the Atlantic forest sensu lato, then, both trajectories seem likely.

Another interesting case regarding the question posed by Cox and Moore (1993) that we used as title for this topic is that of *Clusia hilariana*. This species, discussed thoroughly in Chaps. 4 and 5, constitutes a puzzling case as regards its distribution pattern. As discussed in Sect. 3.4.1, it was believed to be a rain forest species that later migrated to Quaternary restinga terrain. Since rainforest specimens originally designated as *C. hilariana* were revised and identified as *C. aemygdioi*, there was growing suspicion that the former might be strictly a restinga species. Interestingly, only a few plant species are known to be restricted to the restingas (Araujo 2000). *Clusia hilariana* might then be an example of a recently originated *Clusia* species, which became dominant in the restingas of northern Rio de Janeiro. The ongoing thesis of R.L. Martins (unpublished data) examines some evidences from reproductive biology that might reinforce this hypothesis.

3.4.4 How Are These Biogeographic Patterns Affected by Man?

The rapid destruction of tropical vegetation in the past decades is most likely a threat to biogeographic patterns of *Clusia*. However, since these are still very poorly known, we cannot estimate at this point how populations of *Clusia* species have been affected by man. One can expect, nevertheless, that in some cases, these effects might have been very large. For instance, there are estimates indicating that the Atlantic rain forest has now been reduced to only 7.5% of its original area (Myers et al. 2000). Considering other Brazilian examples, the Amazon forest suffered its highest ever deforestation rate in 2004, but there are still an estimated 16% unharmed (Fearnside 2005); the Cerrado of Central Brazil has also been reduced to 41% of its original area (Ribeiro et al. 2005). Whether such a dramatic impact on some of *Clusia*'s natural habitats implies reduction or even extinction of populations, we can only guess.

The impacts of forest fragmentation on plant populations have been extensively documented (e.g., Tabarelli et al. 1999); however, there are no such studies that we are aware of regarding *Clusia* populations. One interesting study by Gonzaga et al. (2000) indicates that, although fragmentation has often removed connections between the Atlantic rain forest sensu stricto and its marginal habitats, many bird species from the rain forest are reported to use the restingas as a buffer zone and as a geographic extension of their distribution, including species reported to disperse *Clusia* seeds. Once again, long distance dispersal in this genus seems to be a key survival feature.

Chapter 4 discusses the relative importance of one species of *Clusia*, *C. hilariana* for local biodiversity and ecosystem processes in restinga vegetation. Moreover, the fact that this locally abundant plant is performing CAM makes this a vegetation of great interest as regards patterns of carbon sequestration: while the CAM canopy cover of *Clusia* confers night assimilation of carbon, the C_3 cover provided by the other shrubs assimilates carbon during the day time. This then might be in contrast to pure CAM canopies, e.g. in a pure cactus forest sensu Vareschi (1980), as it was shown recently that in a cactus mesocosm established in a sizeable glasshouse growth chamber the mean carbon budget was negative with a release of 11.5 mmol CO_2 m^{-2} day^{-1} and the model community exclusively with CAM did not develop the capacity to recycle CO_2 from plant and soil respiration (Rascher et al. 2006). Thus, unlike most vegetation types of the world, this restinga might sequester significant amounts of carbon on a 24-h basis. Therefore, if such relevance of a single *Clusia* species for local and global ecological processes repeats itself for other species of this genus, it would appear that negative anthropic effects on *Clusia* biogeographic patterns might prove particularly grave.

Acknowledgements. We thank Ulrich Lüttge and Otto Lange for critical reading of the manuscript and suggestions, Renata Bacellar Mello and Thadeu Teruaki S. Shibue for their assistance in gathering data in the early stages of this work, former and current graduate students (H.L.T. Zaluar, A.T.C. Dias, R.L. Martins) for exciting discussions and ideas about *Clusia*'s biogeographic patterns, and the Brazilian Research Council (CNPq) for our productivity grants.

Appendix 1. List of *Clusia* species that occur in South America but not in Brazil

Species	Country/Region	Source	Habitats	Elevation (m)	
Clusia alata Planch. et Triana	Colombia, Ecuador, Venezuela	Gonzalez and Jarvis[a]	Dry forest, montane forest, cloud forest, Andes slope forest	900–2250	Forest
Clusia araracuarae Pipoly	Colombia	Pipoly and Graff 1995b; NYBG[c]	Igapó vegetation		Forest
Clusia asymmetrica Pipoly	Venezuela	Pipoly et al. 1998	Granitic outcrops, montane forests	200–1200	Forest, non-forest
Clusia aymardii Pipoly	Venezuela	Pipoly et al. 1998	Scrub forests on tepui summits and slopes	700–1200	Forest
Clusia bernardoi Pipoly et Cogollo	Colombia	Pipoly and Cogollo 1998	Andes slope forest; premontane pluvial forest – cloud forest	1600–2000	Forest
Clusia brachystyla Maguire	Venezuela, Guyana	Pipoly et al. 1998	Exposed ridges on tepuis, meadows, along streams, low montane forests	300–1700	Forest, non-forest
Clusia bracteosa Cuatrec.	Colombia	Gonzalez and Jarvis[a]	Andes slope forest	1600	Forest
Clusia cardonae Maguire	Venezuela, Guyana	Pipoly et al. 1998	Tepui summits, montane forests, tepui scrub, dwarf forests	1000–1900	Forest, non-forest
Clusia caudata (Planch. et Triana) Pipoly	Colombia, Choco Floristic Province	Pipoly 1997	Endemic to Choco Floristic Province	1100–1700	Forest
Clusia chiribiquetensis Maguire	Venezuela, Colombia, Guyana	Pipoly et al. 1998	Lowland riparian forests, dwarf montane forests	0–2000	Forest
Clusia cochlitheca Maguire	Venezuela	Pipoly et al. 1998	Montane forests, scrub forests	600–2000	Forest
Clusia colombiana Pipoly	Colombia, N.Equador	Pipoly 1997	Sub-paramo	1600–2200	Non-forest
Clusia columnaris Engl.	Venezuela, Guyana, Suriname, Colombia	Pipoly et al. 1998	Lowland tropical forest; semideciduous forest on granitic outcrops, forest edges, evergreen lowland forests, Rio Negro caatinga; Gran Sabana	100–800	Forest, non-forest

Clusia crassifolia Planch, et Triana	Venezuela, Guyana	Pipoly et al. 1998	Gallery forests on upland and highland savannas or meadows, tepui scrub	1300-2300	Forest, non-forest
Clusia crenata Cuatrec.	Colombia, Ecuador	UCDAVIS[b]	Montane rainforest and cloud forest	1200–2200	Forest
Clusia decussata Ruiz et Pavon	Colombia (Amazonia)	Duque et al. 2002	Amazon rain forest		Forest
Clusia duartei Maguire	Venezuela	Pipoly et al. 1998	Montane scrub, forests along rivers	1500–1600	Forest
Clusia duidae Gleason	Venezuela, Guyana	Pipoly et al. 1998	Montane to upper montane mossy elfin forests	1100–2200	Forest
Clusia engleriana Pipoly	Peru	Pipoly 1997	Andes slopes	2300–2400	Forest
Clusia fabiolae Pipoly	Venezuela	Pipoly et al. 1998	Tepui scrub, rocky savannas below	1600–2500	Non-forest
Clusia grammadenioides Pipoly	Venezuela, Colombia	Pipoly et al. 1998	Forest margins on deep white sands, on terraces beside streams	10–200	Non-forest
Clusia gratula Maguire	Venezuela	Pipoly et al. 1998	Riparian forests, Rio Negro caatinga, dense montane thickets	100–1200	Forest, non-forest
Clusia guayanae Pipoly	Venezuela, Guyana	Pipoly et al. 1998	Montane scrub and cloud forests along sandstone bluffs	1200–1900	Forest
Clusia hexacarpa Gleason	Venezuela	Pipoly et al. 1998	Tepui forests, rock outcrops and exposed slopes of tepuis	1400–1800	Forest, non-forest
Clusia hirsuta Hammel	Colombia	Gonzalez and Jarvis[a]	Andes slope forest	1200	Forest
Clusia huberi Pipoly	Venezuela, Guyana	Pipoly et al. 1998	Upland forests	1000–2200	Forest
Clusia hylaeae Pipoly	Peru, Ecuador	Pipoly 1997	Lowland wet and premontane forest interface in Amazonia	0–500	Forest
Clusia imbricata Steyerm.	Venezuela	Pipoly et al. 1998	Dwarf montane forests, tepui summit forests, talus forests	800–2100	Forest
Clusia lineata (Benth.) Planch. et Triana	Panama, Ecuador	UCDAVIS[b]	Montane rainforest and cloud forest	1200–1950	Forest

Appendix 1. (*Continued*)

Species	Country/Region	Source	Habitats	Elevation (m)	
Clusia longistyla Cuatrec.	Ecuador	UCDAVIS[b]	Montane rainforest and cloud forest	1900–2250	Forest
Clusia lopezii Maguire	Venezuela, Colombia	Pipoly et al. 1998	Seasonally flooded (black-water) riparian forests, savanna edges	50–200	Forest, non-forest
Clusia loranthacea Planch. et Triana	Equador, Colombia	Gonzalez and Jarvis[a]	Montane rainforest and cloud forest; Andes slope forest	1200–1800	Forest
Clusia macropoda Klotzsch ex Engl.	Venezuela, Guyanas	Pipoly et al. 1998	Riparian forests	0–50	Forest
Clusia magnifolia Cuatrec.	Colombia	Duque et al. 2002	Amazon rain forest	0–500	Forest
Clusia maguireana Pipoly	Venezuela, Guyanay	Pipoly et al. 1998	Rocky tepui outcrops, overhanging ledges	1100–2300	Non-forest
Clusia minor L.	Venezuela, Guyanas, Cental Am., Mexico, W. Indies	Pipoly et al. 1998	Exposed igneous outcrops, semidry savannas and semideciduous scrub	100–700	Non-forest
Clusia multiflora Kunth	Bolivia, Colombia, Ecuador, Venezuela, Am. Central	INBIO[d]	Elfin cloud forest; montane rainforest	1900–1950	Forest
Clusia multilineata Pipoly	Venezuela	Pipoly et al. 1998	Montane river canyons	700–800	Forest
Clusia myriandra (Benth.) Planch. et Triana	Venezuela, Guyana	Pipoly et al. 1998	Lowland to lower montane forests, granitic outcrops	50–400	Forest, non-forest
Clusia niambiensis Pipoly, Cogollo et Gonzalez	Colombia, Ecuador (Choco)	Pipoly et al. 1998	Premontane and montane pluvial forest, understory	650–1650	Forest
Clusia pachyphylla Gleason	Venezuela	Pipoly et al. 1998	Tepui meadows, on rocks	1700–2200	Non-forest
Clusia parvifolia Maguire	Venezuela, Guyana	Pipoly et al. 1998	*Clusia* scrub forests	1600–2000	Non-forest
Clusia parvula (Maguire) Pipoly	Venezuela	Pipoly et al. 1998	Montane and high tepui scrub	1300–2100	Non-forest
Clusia pentandra Cuatrec.	Colombia	Gonzalez and Jarvis	Andes slope forest	2300–2300	Forest

Clusia phelpsiae Lasser et Maguire	Venezuela	Pipoly et al. 1998	Open rocky areas and tepui summit thickets	1800–2200	Non-forest
Clusia phelpsiana Maguire	Venezuela	Pipoly et al. 1998	Along streams in montane to upper montane forests	1500–2300	Forest
Clusia polyandra (Vesque) Pipoly	Colombia	Pipoly 1997	Premontane wet forests of Choco Floristic Province	1600–2200	Forest
Clusia polystigma Little	Ecuador	UCDAVIS[b]	Montane rainforest and cloud forest	1750	Forest
Clusia ptaritepuiensis (Steyerm) Pipoly	Venezuela	Pipoly et al. 1998	South facing slopes	1700–1800	Forest
Clusia radiata Maguire et K. D. Phelps.	Venezuela	Pipoly et al. 1998	Rocky outcrops along steep tepui slopes	1800–2000	Non-forest
Clusia rosea Jacq.	Venezuela, Antilles, Mexico/Chiapas, Colombia, Peru, Costa Rica, Panama	INBIO[d]	Tall cloud forest; dry forest, evergreen lowland forests, semideciduous forests	0–900	Forest
Clusia rotundifolia Gleason	Venezuela	Pipoly et al. 1998	Scrub savanna margins, open rocky areas	1000–2000	Non-forest
Clusia schomburgkii Vesque	Venezuela, Guyana	Pipoly et al. 1998	Along streams in forests on steep slopes	1000–1300	Forest
Clusia sipapoana (Maguire) Pipoly	Venezuela, Guyana, Peru, Panama	Pipoly et al. 1998	Evergreen lowland to montane forests	100–800	Forest
Clusia sphaerocarpa Planch. et Triana	Bolivia, Colombia, Ecuador, Peru	UCDAVIS[b]	Montane rainforest and cloud forest	1550–3000	Forest
Clusia steyermarkii Maguire	Venezuela	Pipoly et al. 1998	Montane gallery forests	1200–1400	Forest
Clusia tabulamontana Maguire	Venezuela, Guyana, Suriname	Pipoly et al. 1998	Scrub forests	500–1100	Forest
Clusia tetragona Pipoly et Cogollo	Colombia	Pipoly and Cogollo 1998	Premontane pluvial and cloud forest (margins)	1325–1900	Forest
Clusia thurifera Engl.	Ecuador	INBIO[d]	Montane rainforest and cloudforest	1500–2000	Forest
Clusia troncosii Maguire	Venezuela	Pipoly et al. 1998	Lowland savannas and forest edges	100–200	Non-forest

[a] http://gisweb.ciat.giar.org/
[b] http://maqui.ucdavis.edu/newest_maqui_dicots.html
[c] http://sciweb.nybg.org/science2/VirtualHerbarium.asp
[d] http://www.inbio.ac.cr/es/default.html

Appendix 2. List of *Clusia* species that occur in Brazil and elsewhere, together with habitat

Species	Country/Region	Source	Habitats	Elevation	Region	
Amazon forest domain						
Clusia annularis Maguire	Brazil/Serra Neblina, Venezuela	Pipoly et al. 1998	Venezuela, dry open slopes, meadows, rocky outcrops, shrub islands, low montane forest	100–2100	AM	Forest/non-forest
Clusia candelabrum Planch. et Triana	Brazil/AM, Venezuela, Colombia, Ecuador, Peru	Pipoly et al. 1998	Riparian forests	50–1600	AM	Forest
Clusia comans (Meisn.) Pipoly	Brazil/AP/PA, Venezuela, French Guyana	Pipoly et al. 1998	Riparian forests	0–100	AM	Forest
Clusia fockeana Miq.	Brazil/PA, Venezuela, Guyanas	Pipoly et al. 1998	Lowland tropical forest, upland shrublands	1300	AM	Forest/non-forest
Clusia gaudichaudii Choisy ex. Planch. et Triana	Brazil/AM, Suriname	Boggen et al. 1997	Amazon forest	0–200	AM	Forest
Clusia hammeliana Pipoly	Brazil, Venezuela, Ecuador, Guyana, Peru, Colombia, Panama	Pipoly et al. 1998	Evergreen lowland to montane forests	100–1300	AM	Forest
Clusia insignis Mart.	Brazil/AM-Rio Negro, Venezuela	Pipoly et al. 1998	Evergreen lowland forests, riparian forests; Amazon forest	100–200	AM	Forest
Clusia leprantha Mart.	Brazil/AM/PA, Venezuela, French Guyana, Colombia	Pipoly et al. 2001	Riparian forests; Amazon forest	100	AM	Forest
Clusia martiana Engl.	Brazil (AM), Venezuela, Colombia (Amazon basin), Peru, Bolivia	Pipoly et al. 1998	Along riverbanks	50–1700	AM	Forest

Clusia microstemon Planch. et Triana	Brazil/AM, Venezuela, Guyana, Suriname, Colombia, Central Brazil/gallery forest	Mendonça et al. 1998; Pipoly et al. 1998	Seasonally flooded (black-water) riparian forests, Rio Negro caatinga, savanna edges	50–200	AM	Forest/non-forest
Clusia spathulaefolia Engl. in Mart.	Brazil/AM, Venezuela, Colombia	Pipoly et al. 1998	Gallery forests along blackwater rivers, Rio Negro caatinga, montane forests on tepui slopes	100–1700	AM	Forest/non-forest
Clusia viscida Engl.	Brazil, Venezuela, Guyana, Suriname, Colombia, Ecuador, Peru	Pipoly et al. 1998	Rio Negro caatinga, seasonally flooded forests	100–200	AM	Forest/non-forest
Clusia obovata (Spruce ex Planch. et Triana) Pipoly	Brazil, Venezuela, Guyana, Suriname, Colombia	Pipoly et al. 1998	Dwarf tepui forests, granitic and sandstone outcrops, savannas, shrublands, evergreen lowland forests	50–200	AM	Forest/non-forest
Clusia opaca Maguire	Brazil/AM, Venezuela, Colombia	Pipoly et al. 1998	White sand savannas, meadow edges, Rio Negro Caatinga, white sand scrub	50–200	AM	Non-forest only
Clusia renggerioides Planch. et Triana	Brazil/AM, Venezuela, Colombia, Guyana, Peru	Pipoly et al. 2001	White sand savannas, riparian forests; Amazon forest, campinarana	100–200	AM	Forest/non-forest
Clusia schomburgkiana (Planch. et Triana) Benth. ex Engl.	Brazil (N. Amazonia), Venezuela, Guyana, Colombia	Pipoly et al. 1998	Savanna-forest ecotones, granitic outcrops, rocky forested slopes, along streams in rocky montane forest, iron-rich outcrops; Gran Sabana; community with *Humiria*	100–2000	AM	Non-forest only
Clusia penduliflora Engl.	Brazil (Central and Western Amazonia), Bolivia, Colombia, Equador, Peru, Central Am.	Pipoly and Graff 1995b	Primary lowland forest	50–500	AM	Forest
Clusia pusilla Steyerm.	Brazil, Venezuela, Guyana, Suriname	Pipoly et al. 1998	Edges of montane forests, streamsides, rocky meadows; tepui forests, rocky outcrops; Gran Sabana	600–2000	AM	Non-forest only

Appendix 2. (*Continued*)

Species	Country/Region	Source	Habitats	Elevation	Region	
Clusia scrobiculata Benoist	Brazil/AM, Venezuela, Guyanas	Boggen et al. 1997; Ribeiro et al. 1999	Amazon forest, slopes	500–1500	AM	Forest
Clusia amabilis Maguire	Brazil/AC/RO, Venezuela	Pipoly et al. 1998	Upper montane shrublands, rocky outcrops	1200–2300	AM	Non-forest
Clusia amazonica Planch. et Triana	Brazil/AM, Venezuela, Guyana, Colombia, Ecuador, Peru, Bolivia, Central Am.	Ribeiro et al. 1999	Amazon forest, black water habitats	0–1600	AM	Forest
Clusia octandra (Poepp.) Pipoly	Brazil, Venezuela, French Guyana, Ecuador, Peru, Colombia	Pipoly et al. 1998	Seasonally flooded (black-water) riparian forests, river banks, montane forests	100–200	AM	Forest
Clusia wurdackiana Pipoly	Brazil, Venezuela, Guyana, Suriname	Pipoly 1995	Common on rocky outrcrops	330–2000	AM	Non-forest only
Clusia columnaris Engl.	Brazil/PA, Venezuela, Guyana, Suriname, Colombia	Costa-Neto et al. 1996; Pipoly et al. 1998	Lowland tropical forest; semidecid forest on granitic outcrops, forest edges, evergreen lowland forests, Rio Negro caatinga; Gran Sabana; restinga	100–800	AM	Forest/non-forest (restinga)
Clusia grandiflora Splitg.	Brazil/AM/PA, Venezuela, Guyanas, Costa Rica	Bastos 1988; Ribeiro et al. 1999	Gallery forests, montane forests; Amazon forest, restinga	0–1300	AM	Forest/non-forest (restinga)
Atlantic forest domain						
Brazilian Northeast and other domains						
Clusia burchellii Engl.	Brazil/PB/PA/GO/MA	Oliveira-Filho and Carvalho 1993; Mendonça et al. 1998	Terra firme forest, cerrado, restinga (dunes)	0–800	NE/AM	Forest in AM; non-forest only in NE

[illegible] Pipoly	[illegible] Venezuela, Bolivia, Colombia, Peru, Ecuador, French Guiana, Guyana, Am.Central	Pipoly et al. 1998	Lowland (terra firme) to montane evergreen forests, swamps, semideciduous and seasonally flooded forests, campinarana; restinga	50–1500	NE/AM	Forest in AM; non-forest only in NE
Clusia nemorosa G.Mey	Neotropical; in Brazil/Amazonia to Bahia	Pipoly et al. 1998	Lowland forest edges, rocky places; Amazon forest, cerrado, caatinga, Atlantic forest, upland forest, restinga	50–1600	NE/AM	Forest/non-forest
Clusia palmicida Rich. ex Planch. et Triana	Brazil/PA/PE/BA/MA, Venezuela, Guyanas, Colombia, Ecuador, Peru	Pipoly et al. 1998	Lowland forest; riparian forest, Atlantic forest; on white sands, restinga	50–1300	NE/AM	Forest/non-forest
Clusia panapanari (Aubl.) Choisy	Brazil/AM/RO/PI/BA, Venezuela, Guyanas	Pipoly et al. 1998, CNIP[a]	Lowland forests, riparian forests; Amazon forest; Atlantic forest	0–400	NE/AM	Forest
Clusia melchiori Gleason	Brazil/BA/AM, Venezuela, Guyana, Surinam, Colombia, Guyanan shield portion of these countries	Pipoly et al. 2001	Montane to upper montane elfin forests often dominated by *Bonnetia*, tepuis; campo rupestre (BA)	600–3000	NE/AM	Forest in Guyanas; non-forest only in NE
Clusia savannarum Maguire	Brazil/AM/BA, Venezuela, Guyana	Pipoly et al. 1998	Shrub islands in white-sand savannas; BA-cerrado, campo rupestre	1000	NE/AM	Non-forest only
Northeast only						
Clusia dardanoi G. Mariz et Maguire	Brazil/BA/PE	Bittrich 1996; CNIP[a]	Atlantic forest, wet forest, upland forest	>700	NE	Forest
Clusia paralicola G. Mariz	Brazil/PE	Nogueira et al. 2001	Caatinga, litoral, rocky places	0–470	NE	Forest/non-forest
Clusia sellowiana Schltdl.	Brazil/BA/GO	Mendonça et al. 1998; CNIP[a]	Atlantic forest, *Humiria* communities, restinga	0–1000	NE	Forest/non-forest
Southeast/South & Northeast						
Clusia intermedia G. Mariz	Brazil/CE/PE/BA	NYBG[c]	Campo rupestre, cloud forest, Atlantic forest	600–1000	SE/NE	Forest/non-forest
Clusia criuva Cambess	Brazil/MG/GO/BA-RS	Bittrich 2003	Atlantic forest, semideciduous forest, gallery forests, restinga	0–1800	SE/NE	Forest/non-forest

Appendix 2. (*Continued*)

Species	Country/Region	Source	Habitats	Elevation	Region	
Clusia hilariana Schltdl.	Brazil/RJ/PE/BA	Nogueira et al. 2001	Restinga	0–30	SE/NE	Non-forest only
Southeast/South						Forest/non-forest
Clusia lanceolata Cambess.	Brazil/RJ-SP	Bittrich 2003	Atlantic forest, restinga	0–1100	SE	
Clusia spiritu-sanctensis G. Mariz et Weinberg	Brazil/RJ-ES	Nogueira et al. 2001	Atlantic forest, restinga	0–100	SE	Forest/non-forest
Clusia fluminensis Planch. et Triana	Brazil/RJ	Nogueira et al. 2001	Atlantic forest, restinga	0–100	SE	Forest/non-forest
Clusia aemygdioi Gomes da Silva et Weinberg	Brazil/ES	SYSTAX[b]	Atlantic forest	900	SE	Forest
Clusia marizii Gomes da Silva et Weinberg	Brazil/ES/RJ	Vieira and Silva 1994	Atlantic forest	30–1100	SE	Forest
Clusia studartiana C.Vieira et Gomes da Silva	Brazil/RJ	Vieira and Silva 1994	Atlantic forest	1100	SE	Forest
Clusia fragrans Gardner	Brazil/BA/MG/RJ	Vieira and Silva 1994	Montane Atlantic forest, high altitude grasslands, occasionally in campos rupestres, often rupicolous or saxicolous	950–1600	SE	Forest/non-forest
Clusia organensis Planch. et Triana	Brazil/MG/ES/RJ/SP	Bittrich 2003	Atlantic forest	1100	SE	Forest
Cerrado/Campo rupestre						
Clusia burlemarxii Bittrich	Brazil/BA	Bittrich 1996	Campo rupestre	850–1400	NE	Non-forest only
Clusia diamantina Bittrich	Brazil/MG	Bittrich 1996	Campo rupestre	1200–1400	SE	Non-forest only
Clusia obdeltifolia Bittrich	Brazil/MG/BA	Bittrich 1996	Campo rupestre	1400	SE/NE	Non-forest only

[a] http://www.cnip.org.br/
[b] http://www.biologie.uni-ulm.de/systax/
[c] http://sciweb.nybg.org/science2/VirtualHerbarium.asp
Abbreviations of Brazilian states are AC-Acre; AM-Amazonia; AP-Amapá; BA-Bahia; CE-Ceará; ES-Espirito Santo; GO-Goias; MA-Maranhão; MG-Minas Gerais; PA-Pará; PB-Paraíba; PE-Pernambuco; PI-Piauí; RJ-Rio de Janeiro; RS-Rio Grande do Sul; RO-Rondonia; SP-São Paulo
[illegible]

References

Araujo DSD (1992) Vegetation types of sandy coastal plains of tropical Brazil: a first approximation. In: Seeliger U (ed) Coastal plant communities of Latin America. Academic Press, San Diego, pp 337–347

Araujo DSD (1997) Cabo Frio region, southeastern Brazil. In: Davis SD, Heywood VH, Herrera-Macbryde O, Villa-Lobos J, Hamilton AC (eds) Centres of plant diversity: a guide and strategy for their conservation 3. The Americas. WWF, IUCN, Washington DC, pp 373–375

Araujo DSD (2000) Análise florística e fitogeografica das restingas do Estado do Rio de Janeiro. PhD Thesis. Universidade Federal do Rio de Janeiro, Rio de Janeiro, BrazilAraujo DSD, Henriques RPB (1984) Análise florística das restingas do Estado do Rio de Janeiro. In: Lacerda LD, Araujo, DSD, Cerqueira R, Turcq B (eds) Restingas: origem, estrutura, processos. CEUFF, Niterói, pp 159–193

Araujo DSD, Scarano FR, Sá CFC, Kurtz BC, Zaluar HLT, Montezuma RCM, Oliveira RC (1998) Comunidades vegetais do Parque Nacional da Restinga de Jurubatiba. In: Esteves FA (ed) Ecologia das Lagoas Costeiras do Parque Nacional da Restinga de Jurubatiba e do Município de Macaé (RJ). Nupem-UFRJ, Rio de Janeiro, pp 39–62

Bastos MNC (1988) Levantamento floristico em restinga arenosa litorânea na Ilha de Maiandeua, PA. Bol Mus Para Emilio Goeldi 4:159–173

Bawa KS, Opler PA (1975) Dioecism in tropical forest trees. Evolution 29:167–179

Bittrich V (1996) Três novas espécies do gênero *Clusia* (Guttiferae) do Brasil. Bol Bot USP 15:733–782

Bittrich V (2003) Clusiaceae. In: Wanderley MGL, Shepherd GJ, Melhem TS, Giulietti AM, Kirizawa M (eds) Flora fanerogâmica do Estado de São Paulo, vol 3. Editora Rima, São Paulo, pp 45–62

Bittrich V, Amaral MCE (1996). Floral morphology and pollination biology of some *Clusia* species from the Gran Sabana (Venezuela). Kew Bull 51:681–694

Bittrich V, Amaral MCE (1997) Floral biology of some *Clusia* species from Central Amazonia. Kew Bull 52:617–635

Boggen J, Funk V, Kelloff C, Hoff M, Cremers G, Feuillet C (1997) Checklist of the plants of the Guianas (Guyana, Surinam, French Guiana), 2nd edn. Smithsonian Institution (online version), Washington, DC

Costa-Neto SV, Bastos MNC, Lobato LCB (1996) Composição florística e fitofisionomia da Restinga de Crispim, Município de Marapanim, Pará. Bol Mus Para Emilio Goeldi 12:237–249

Cox CB, Moore PD (1993) Biogeography: an ecological and evolutionary approach. Blackwell, Oxford

Crayn DM, Terry RG, Smith JAC, Winter K (2000) Molecular systematic investigations in Pitcairnioideae (Bromeliaceae) as a basis for understanding the evolution of crassulacean acid metabolism (CAM). In: Wilson KL, Morrison DA (eds) Monocots: systematics and evolution. CSIRO, Melbourne, pp 569–579

Crayn DM, Winter K, Smith JAC (2004) Multiple origins of crassulacean acid metabolism and the epiphytic habit in the Neotropical family Bromeliaceae. Proc Natl Acad Sci USA 101:3703–3708

Darwin C (1877) The different forms of flowers on plants of the same species. John Murray, London

Duque A, Sánchez M, Cavelier J, Duivenvoorden JF (2002) Different floristic patterns of woody understorey and canopy plants in Colombian Amazonia. J Trop Ecol 18:499–525

Fearnside PM (2005) Deforestation in Brazilian Amazonia: history, rates and consequences. Conserv Biol 19:680–688

Freeman DC, Doust JL, El-Keblawy A, Miglia KJ, McArthur ED (1997) Sexual specialization and inbreeding avoidance in the evolution of dioecy. Bot Rev 63:65–92

Gentry AH (1982) Neotropical floristic diversity: phytogeographical connections between Central and South America, Pleistocene climatic fluctuations, or an accident of the Andean orogeny? Ann Missouri Bot Gard 69:557–593

Gentry AH (1995) Patterns of diversity and floristic composition in Neotropical montane forests. In: Churchill SP, Balslev H, Forero E, Luteyn JL (eds) Biodiversity and conservation of Neotropical montane forests. New York Botanical Gardens, New York, pp 103–126

Goh CJ, Kluge M (1989) Gas exchange and water relations in epiphytic orchids. In: Lüttge U (ed) Vascular plants as epiphytes. Evolution and ecophysiology. Ecological studies, vol 76. Springer, Berlin Heidelberg New York, pp 137–166

Gonzaga LP, Castiglioni GDA, Reis HBR (2000) Avifauna das restingas do sudeste: estado do conhecimento e potencial para futuros estudos. In: Esteves FA, Lacerda LD (eds) Ecologia de restingas e lagoas costeiras. NUPEM/UFRJ, Macaé, pp 151–163

Griffiths H, Smith JAC (1983) Photosynthetic pathways in the Bromeliaceae of Trinidad: relations between life forms, habitat preference and the occurrence of CAM. Oecologia 60:176–184

Horres R, Zizka G, Kahl G, Weising K (2000) Molecular phylogenetics of Bromeliaceae: evidence from *trn*L(UAA) intron sequences of the chloroplast genome. Plant Biol 2:306–315

Lima HC, Guedes-Bruni RR (1997) Diversidade de plantas vasculares na Reserva Ecológica de Macaé de Cima. In: Lima HC, Guedes-Bruni RR (eds) Serra de Macaé de Cima: Diversidade Florística e Conservação em Mata Atlântica. Instituto de Pesquisas Jardim Botânico do Rio de Janeiro, Rio de Janeiro, pp 29–39

Lloyd DG, Webb CJ (1977) Secondary sex character in seed plants. Bot Rev 43:177–216

Lüttge U (1985) Epiphyten: Evolution und Ökophysiologie. Naturwissenschaften 72:557–566

Lüttge U (1997) Physiological ecology of tropical plants. Springer, Berlin Heidelberg New York

Lüttge U, Ball E, Kluge M, Ong BL (1986) Photosynthetic light requirements of various tropical vascular epiphytes. Physiol Vég 24:315–331

Martin L, Suguio K, Flexor JM (1993) As flutuações de nível do mar durante o quaternário superior e a evolução geológica de "deltas" brasileiros. Bol Inst Geoc USP, Publ Esp 15:1–186

Matallana G, Wendt T, Araujo DSD, Scarano FR (2005) High abundance of dioecious plants in a tropical coastal vegetation. Am J Bot 92:1513–1519

Medina BMO, Ribeiro KT, Scarano FR (2006) Plant-plant and plant-topography interactions on a rock outcrop at high altitude in southeastern Brazil. Biotropica 38:27–34

Medina E (1974) Dark CO_2 fixation, habitat preference and evolution within the Bromeliaceae. Evolution 28:677–686

Meirelles ST, Pivello VR, Joly CA (1999) The vegetation of granite rock outcrops in Rio de Janeiro, Brazil, and the need for its protection. Environ Conserv 26:10–20

Mendonça RC, Felfili JM, Walter BMT, Silva-Junior M, Rezende AV, Filgueiras TS, Nogueira PE (1998) Flora vascular do cerrado. In: Sano SM, Almeida SP (eds) Cerrado: ambiente e flora. EMBRAPA, Planaltina, pp 289–556

Morellato LPC, Haddad CFB (2000) Introduction: the Brazilian Atlantic forest. Biotropica 32:786–792

Myers N, Mittermeier RA, Mittermeier CG, Fonseca GAB, Kent J (2000) Biodiversity hotspots for conservation priorities. Nature 403:853–858

Niklas KJ (1997) The evolutionary biology of plants. University of Chicago Press, Chicago

Nogueira PCL, Bittrich V, Shepherd GJ, Lopes AV, Marsaioli AJ (2001) The ecological and taxonomic importance of flower volatiles of *Clusia* species (Guttiferae). Phytochemistry 56:443–452

Oliveira PS, Marquis RJ (eds) (2002) The Cerrados of Brazil. Columbia University Press, New York

Oliveira-Filho AT, Carvalho DA (1993) Florística e fisionomia da vegetação no extremo norte do litoral da Paraíba. Rev Bras Bot 16:115–130

Oliveira-Filho AT, Fontes MAL (2000) Patterns of floristic differentiation among Atlantic forests in southeastern Brazil, and the influence of climate. Biotropica 32:1–16

Pipoly JJ (1995) *Clusia fabiolae*, a new species, with a synopsis of *Clusia* section Anandrogyne (Clusiaceae) in Guayana. Sida 16:737–755

Pipoly JJ (1997) Nomenclatural notes on Neotropical Clusieae (Clusiaceae). Sida 17:765–767

Pipoly JJ, Cogallo AP (1998) New species and nomenclatural notes in *Clusia* (Clusiaceae) from Andean Colombia and Venezuela. Sida 18:401–409

Pipoly JJ, Graff A (1995a) The genus *Clusia* section *criuva* (Clusiaceae) in Guayana. Sida 16:649–678

Pipoly JJ, Graff A (1995b) A synopsis of the genus *Clusia* sections *Criuvopsis* and *Brachystemon* (Clusiaceae) in northern South America. Sida 16:505–528

Pipoly JJ, Kearns DM, Berry PE (1998) *Clusia* L. In: Berry PE, Holst BK, Steyermark JA, Yatkievych K (eds) Flora of the Venezuelan Guayana; vol 4 . Caesalpiniaceae – Ericaceae. Missouri Botanical Garden Press, St Louis, pp 248–295

Pittendrigh CS (1948) The bromeliad-*Anopheles*-malaria complex in Trinidad. I. The bromeliad flora. Evolution 2:58–89

Prance GT (1987) Biogeography of neotropical plants. In: Whitmore TC, Prance GT (eds) Biogeography and Quaternary history of tropical America. Clarendon Press, Oxford, pp 46–65

Rascher U, Bobich EG, Osmond CB (2006) The "Kluge-Lüttge-Kammer": a preliminary evaluation of an enclosed, crassulacean acid metabolism (CAM) mesocosm that allows separation of synchronized and desynchronized contributions of plants to whole system gas exchange. Plant Biol 8:167–174

Ribeiro JELS, Hopkins MJG, Vicentini A, Sothers CA, Costa MAS, Brito JM, Souza MAD, Martins LHP, Lohmann LG, Assunção PACL, Pereira EC, Silva CF, Mesquita MR, Procópio LC (1999) Flora da Reserva Ducke: guia de identificação das plantas vasculares de uma floresta de terra-firme na Amazônia Central. INPA, Manaus

Ribeiro JF, Bridgewater S, Ratter JA, Sousa-Silva JC (2005) Ocupação do bioma Cerrado e conservação da sua diversidade vegetal. In: Scariot A, Sousa-Silva JC, Felfili JM (eds) Cerrado: ecologia, biodiversidade e conservação. Brasília: Ministério do Meio Ambiente, pp 385–399

Rizzini CT (1979) Tratado de fitogeografia do Brazil. V.2. São Paulo: HUCITEC-EDUSP

Scarano FR (2002) Structure, function and floristic relationships of plant communities in stressful habitats marginal to the Brazilian Atlantic rain forest. Ann Bot 90:517–524

Scarano FR, Ribeiro KT, Moraes LFD, Lima HC (1997) Plant establishment on flooded and unflooded patches of a freshwater swamp forest of southeastern Brazil. J Trop Ecol 14:793–803

Scarano FR, Duarte HM, Franco AC, Geßler A, de Mattos EA, Rennenberg H, Lüttge U (2005) Physiological synecology of tree species in relation to geographic distribution and ecophysiological parameters at the Atlantic forest periphery in Brazil: an overview. Trees 19:493–496

Schimper AFW (1888) Botanische Mitteilungen aus den Tropen. II. Epiphytische Vegetation Amerikas. G. Fischer, Jena

Shrader-Frechette KS, McCoy ED (1993) Method in ecology. Cambridge University Press, Cambridge
Smith JAC (1989) Epiphytic bromeliads. In: Lüttge U (ed) Vascular plants as epiphytes. Evolution and ecophysiology. Ecological studies, vol 76. Springer, Berlin Heidelberg New York, pp 109–138
Sutherland S, Delph LF (1984) On the importance of male fitness in plants: patterns of fruit-set. Ecology 65:1093–1104
Tabarelli M, Mantovani W, Peres CA (1999) Effects of habitat fragmentation on plant guild structure in the montane Atlantic forest of southeastern Brazil. Biol Conserv 91:119–127
Thompson DI, Edwards TJ (2001) Breeding biology, resource partitioning and reproductive effort of a dioecious shrub, *Clutia pulchella* L. (Euphorbiaceae). Plant Syst Evol 226:13–22
Vaasen A, Begerow D, Lüttge U, Hampp R (2002) The genus *Clusia* L.: molecular evidence for independent evolution of photosynthetic flexibility. Plant Biol 4:86–93
Vamosi JC, Vamosi SM (2004) The role of diversification in causing the correlates of dioecy. Evolution 58:723–731
Vamosi JC, Otto SP, Barrett SCH (2003) Phylogenetic analysis of ecological correlates of dioecy in angiosperms. J Evol Biol 16:1006–1018
Vareschi V (1980) Vegetationsökologie der Tropen. Ulmer, Stuttgart
Veloso HP, Rangel-Filho ALRR, Lima JCA (1991) Classificação da vegetação brasileira, adaptada a um sistema universal. IBGE, Rio de Janeiro
Vieira CM, Silva AG (1994) Clusiaceae. In: Lima MPM,Guedes-Bruni RR (eds) Reserva Ecológica de Macaé de Cima - RJ: aspectos florísticos das espécies vasculares, vol 1. Jardim Botânico do Rio de Janeiro, Rio de Janeiro, pp 113–127
Weller SG, Sakai AK (1999) Using phylogenetic approaches for the analysis of plant breeding systems evolution. Ann Rev Ecol Syst 30:167–199
http://www.mobot.org/mobot/Research/ven-guayana/clusiaceae/ Flora of the Venezuelan Guayana – Clusiaceae
http://sciweb.nybg.org/science2/VirtualHerbarium.asp NYBG Virtual Herbarium
http://sciweb.nybg.org/science2/hcol/sebc/sebcindex2.asp Catalog of Vascular Plant Species of Eastern Brazil
http://www.cnip.org.br/ Checklist das Plantas do Nordeste (Versão 1.3 - 2006)
http://florabrasiliensis.cria.org.br/search?search_string=*Clusia* Flora brasiliensis 31 May 2006
http://www.ipni.org/index.html International Plant Names Index, Royal Botanic Gardens, Kew
http://www.mnh.si.edu/biodiversity/bdg/checklst.html Checklist of the Plants of the Guianas
http://www.ibge.gov.br/home/geociencias/recursosnaturais/levantamento/florarestinga
Lista das Plantas Vasculares das Restingas do Litoral Norte da Bahia

4 *Clusia* as Nurse Plant

André Tavares Corrêa Dias and Fabio Rubio Scarano

4.1 Introduction

The nurse-plant syndrome (see Franco and Nobel 1989) takes place when plant species shelter seedlings, young and/or adult individuals of other species through their ontogeny. The nurse-plant might then enhance fitness, survival and/or growth of associated species (Callaway et al. 2002; Bruno et al. 2003). However, positive and negative interactions are unlikely to occur separately in nature (Holmgren et al. 1997; Brooker and Callaghan 1998; Dickie et al. 2005). This balance is affected by spatial and temporal shifts (Morris and Wood 1989; Tielbörger and Kadmon 1997; Callaway 1998) related to plant ontogenetic development and/or changes in resource availability (Callaway and Walker 1997). For instance, the overall importance of positive interactions on community structure, such as the nurse-plant syndrome, is claimed to be higher in resource-poor environments (Callaway et al. 2002; Lortie and Callaway 2006).

This syndrome often results in the formation of vegetation clumps or islands (Pugnaire et al. 1996; Kikvidze and Nakhutsrishvili 1998; Weltzin and McPherson 1999) and is well known for arid and alpine zones. However, there are still a few examples from tropical environments. One such example of a nurse plant is *Clusia hilariana* Schltdl., a species from the Brazilian restingas. Despite some field observations suggesting that a nurse plant effect might occur in the case of other CAM *Clusia* species in the Brazilian restingas (e.g., *C. fluminensis* Pl. and Tr. and *C. spiritu-sanctensis* Mariz and Weinberg), *C. hilariana* has been more thoroughly studied in this respect. This chapter revises such studies, aiming to discuss the extent and consequences of such effect on restinga vegetation. We also discuss which features of *C. hilariana* are likely to be responsible for such positive effects, which may serve as indication for other possible nurse species of *Clusia*.

Ecological Studies, Vol. 194
U. Lüttge (Ed.)
Clusia: A Woody Neotropical Genus
of Remarkable Plasticity and Diversity

4.2 Restinga de Jurubatiba: Phytosociology

Clusia hilariana occurs, often with high abundance, along the coast of southeastern Brazil, particularly in northern Rio de Janeiro state and southern Espírito Santo state. The studies reviewed here were conducted in the Restinga de Jurubatiba National Park (22°00'–22°23'S; 41°15'–41°45'N), more specifically at the so-called *Clusia* scrub, a vegetation physiognomy that covers some 40 % of the ca. 14,000 ha of this Park and consists of hemispheric vegetation islands of various sizes surrounded by white sand (Chap. 3 and Sect. 9.4.2.1 provide detailed accounts of the restinga vegetation as a whole and of the particular vegetation of this Park).

Clusia hilariana is the *dominant* plant of this vegetation (Araujo et al. 1998, 2004). In phytosociological terminology, this means that this species has the highest *importance value* (IV), i.e. the sum of relative frequency, relative density and relative basal area, among plant species within a given sampled area (Müller-Dombois and Ellenberg 1974). Pimentel (2002) did a thorough survey of 12 sample areas using the line intercept method and sampled all woody plants ≥50 cm tall, on three parallel beach ridges that varied in respect to their distances from the sea. She found that the vegetation has an oligarchic structure (i.e., when the IV of only a few species add up to a high percentage of the total IV, in this case, the top 6 out of a total 62 species accounted for 49 % of the total IV of the vegetation), which is a pattern commonly found also in tropical forest formations, especially in disturbed or early successional communities (Pascal and Pélissier 1996), and in other open vegetation types marginal to the Atlantic rain forest (Scarano 2002).

Additionally, Pimentel (2002) found a Shannon diversity index (H'; Magurran 1988) of 3.07. This index is given by

$$H' = -\Sigma p_i \times \ln p_i$$

where p_i is the ratio of the number of individuals (n_i) of a species i to the total number of individual plants (N) on the selected area ($p_i = n_i / N$).

This diversity value is lower than in typical Atlantic (e.g., Sanchez et al. 1999) or Amazon rain forests (e.g., Campbell et al. 1992), where it often ranges between 4.00 and 5.00. However, it is surprisingly high considering the extreme environmental conditions these plants are often subjected to (e.g., Scarano et al. 2005). This value was also higher than those obtained for other open woody restinga communities in southeastern Brazil, i.e. from 2.63 to 2.93 (Pereira et al. 2001), sampled by the same method. Pimentel (2002) argued that this high diversity resulted from the high number of species with low abundances and restricted distributions.

The fact that Liebig et al. (2001) found H' values of 2.70 underneath the canopy of male and female *C. hilariana* trees is an important indirect evidence of the nursing role of this species. It has to be mentioned, however, that this study included all plants whereas Pimentel (2002) sampled only woody plants.

4.3 Evidences for Nurse Plant Effects

In order to irrefutably confirm the key role of *C. hilariana* as a nurse plant species, we would necessarily have to create experimental set ups in the field to simulate a situation where this species is not present, i.e. we would have to perform removal experiments. The removal and the eventual 'disappearance' of the species from a given point in space simulate a disturbance of such extent as to locally extinguish the species. Such experiments demand a huge logistic effort for set up, monitoring and analysis (Diaz et al. 2003 and Kareiva and Levin 2003 reviewed removal experiments). They also require special permits from environmental officials. However, before we designed such experiments, we decided first to gather additional observational data on plant-plant association and coexistence, on spatial variation regarding gender and ontogenetic stage of *C. hilariana* plants, and, more recently, to apply structural equation modelling (SEM) to assess which are the main causal factors related to the nursing effect. Thus, in the next two sections (Sects. 4.3.1 and 4.3.2) we review previous work, while in the last one (Sect. 4.3.3) we present original data analyses and discuss the applications of SEM as a non-destructive method to analyse patterns of species interactions.

4.3.1 Association, Coexistence and Facilitation

The M.Sc. dissertation of Correia (1998) was the first study to propose a role for *C. hilariana* in facilitating germination and growth of other restinga plants. She examined population structure and interspecific pairwise associations between four woody species within a 0.5-ha plot in an open restinga site: *C. hilariana*, *Protium icicariba* (DC.) Marchand (Burseraceae), *Andira legalis* (Vell.) Toledo (Leg. Faboideae) and *Vernonia crotonoides* Sch.Bip. ex Baker (Asteraceae). While *Clusia* and *Protium* were the two most abundant species, *Vernonia* was intermediary in abundance and *Andira* was rare. She showed that adult *Clusia* had a positive association with their own juveniles and those of *Protium*, and that this pattern was clearer as the vegetation islands dominated by *Clusia* increased in size. This indirect evidence of facilitation was corroborated by a parallel dissertation conducted by Zaluar (1997) who demonstrated that underneath the canopy of *Clusia* there was higher species richness than under any other woody species.

Another interesting finding of Correia (1998) was that a significant proportion of *Clusia* seedlings were found inside bromeliad tanks (see also Chap. 5

and Sect. 9.3). Thus, it would appear that the 'suspended soil' found within the tanks formed by bromeliad rosettes favours seed germination and seedling growth. This has also been found in the restinga of Barra de Maricá for *Clusia fluminensis* by Macêdo and Monteiro (1987) and Zaluar (2002). In short, our nurse plant is itself nursed by bromeliads. Scarano et al. (2004) presented a model describing the consequences of such plant-plant interactions to succession in the open restingas of the Restinga de Jurubatiba National Park.

Dias et al. (2005) produced the most conclusive evidence regarding the nursing effect of *Clusia*. This study compared the nurse role of patches with *Clusia* dominance vs patches without *Clusia* dominance (*Clusia* patches vs non-*Clusia* patches, henceforth), while examining differences in patch architecture (above-ground vegetation density and stratification) and woody species composition and size structure. Additionally, we performed an experiment of seed introduction for three woody species underneath these two types of patches and monitored germination, mortality and seedling growth over a one-year period. There was a positive association between the presence of adult *Clusia* and juvenile density of other woody species. This was attributed to architectural differences between the two patch types, where *Clusia* patches were more stratified, while non-*Clusia* patches were flatter and with a higher vegetation density (Fig. 4.1). These differences in patches could lead to distinct environmental conditions underneath the canopy and also distinct attractiveness to potential seed dispersers. However, in spite of the higher juvenile density in *Clusia* patches, there were no differences in germination rate and seedling mortality between the two patch types. This suggested that the understorey of *Clusia* patches is not necessarily a better environment for germination and seedling survival than non-*Clusia* patches. The higher juvenile density and higher species richness of seedlings and juveniles on *Clusia* patches could result from a higher visitation of potential seed dispersers to this patch type. *Clusia* patches might provide preferential shelter and nesting sites for animals such as birds and bats in the restinga, because *Clusia*'s height and architecture makes it the most conspicuous plant in this vegetation type.

To investigate possible effects of *C. hilariana* on dispersal, we used the species similarity between the canopy (i.e., adult shrubs and trees that composed the canopy of the vegetation patches) and the understorey (i.e., seedlings and juveniles of shrubs and trees) as an indirect assessment of between-patch seed dispersal, assuming that lower similarity between canopy and understorey is a consequence of higher dispersal activity. The lower similarity between canopy and understorey species in *Clusia* patches as compared to non-*Clusia* patches led us to conclude that there is a higher species invasion into the former patch type, which probably resulted from a greater activity of dispersers.

These results, however, required further examination on the mechanisms that drive plant interactions within patches, since some ambiguity was still

Fig. 4.1A–D. Different types of vegetation patches in the open vegetation of the Restinga de Jurubatiba National Park: **A** mature *Clusia*; **B** senescent *Clusia*; **C** dead *Clusia* with well developed understorey; **D** plant patch without *Clusia* dominance

unsolved. For instance, the higher growth rates of introduced plants in non-*Clusia* patches seemed to contrast with the general observation that *Clusia* acts as a nurse plant. It thus appeared that despite possibly favouring seed arrival by improving the activity of dispersers, *Clusia* may later suppress plant growth in its understorey. Indeed, it has often been shown that preferred sites for germination are not always the best sites for plant growth (Morris and Wood 1989; Walker and Vitousek 1991) as competitive interactions may increase with time (Callaway and Walker 1997). A parallel can be traced to adaptive behaviour of individual species at different phenophases, where a change occurring during a critical period may be adaptive for a given phenophase while negative for later developmental stages (Amzallag 2005; see also Chap. 9, Sect. 5).

4.3.2 Gender and Ontogenetic Variation

As the evidences suggesting a nursing role of *Clusia* accumulated, we began to ask how repeatable this could be in space. This question seemed particularly relevant for two reasons.

First, our study species is dioecious, i.e. it has separate male and female individuals, while most Angiosperms are hermaphrodites (see Chap. 5). Several examples in the literature suggest that for dioecious plants, females tend to establish on sites with better growing conditions (Urbanska 1992; Crawley 1997), which can affect establishment of other species in the understorey. More obviously, female plants bear fruits that are also likely targets of potential dispersers of seeds of other species. Thus, Liebig et al. (2001) tested the hypothesis that female *Clusia* plants would nurse a higher diversity of species than males. The hypothesis was rejected, since the understory of both male and female plants had Shannon diversity values (H') around 2.7 indistinctly. At that time, we interpreted that potential seed dispersers were not visiting *Clusia* plants in search for their fruits (that would mean a higher diversity associated to females), but in search for a resting place. In this case, there was no difference between genders, since their morphology was proven equal. In summary, the dioecious nature of *Clusia* did not imply spatial variation regarding its nurse role, i.e. the nurse plant effect occurs irrespective of gender.

Second, two other 'forms' of *Clusia* are distinguishable in the landscape: mature adults with a dense canopy vs senescent adults with canopy gaps (Fig. 4.1). Phenological studies showed that *Clusia* is an evergreen plant, with no marked timing for leaf production or leaf mortality (Rosado 2006). This suggests that senescent individuals are also older, or at least they are closer to the end of their life cycle. The M.Sc. dissertation of Ramos (2003) tested the hypothesis that senescent *Clusia*s would nurse smaller species diversity than mature ones, due to plant death in the understorey caused by the changes in environmental conditions. She sampled all plants >20 cm in the understorey of 12 mature *Clusia* plants and 12 senescent ones. Again, our hypothesis was rejected: there was no significant difference in plant species richness or diversity between patch types. This suggests that the nursing role of *Clusia* for most species probably takes place in the initial phases of seed germination and seedling establishment or, instead, that the effect of *Clusia*'s senescence might act only on a longer temporal scale not covered by our sampling.

However, the understorey of the senescent *Clusia*s was on average taller and with a higher basal area than the understorey of the mature plants. These results conformed to the higher growth of seedlings on non-*Clusia* patches found by Dias et al. (2005) and fostered two alternative hypotheses: 1) When understorey plants reach a certain size and age, their growth is inhibited by *Clusia*, and is later resumed when *Clusia* senesces or dies. This had already been proposed by Dias et al. (2005) who suggested that the senescence and death of *Clusia* in a given patch could promote the growth of understorey juveniles and a change in architecture to a short and dense vegetation type. 2) When understorey plants reach a certain size and number they exert a strong competition over *Clusia*, which senesces and dies. Removal experiments would be essential to test which of the two hypotheses applies. Hypothesis 1 would require an experiment of *Clusia* removal, while Hypothesis 2 would require an experiment of understorey removal.

4.3.3 Structural Equation Modelling (SEM): Mechanisms Behind the Nursing Effect

Although the empirical data above helped to elucidate a number of processes related to the nurse plant effect of *Clusia* and to formulate new hypotheses, we still had a major gap as regards understanding of which mechanisms would be driving this effect. Thus, we needed an analytical method to allow hypothesis testing involving multiple interacting variables. The multivariate tool called "structural equation modelling" (SEM) seemed to be the appropriate alternative. We can trace the origins of SEM back to "path analysis", which was developed by Sewall Wright (1889–1988). However, its development until present is less than straightforward and results from various scholarly contributions (Bollen 1989). Although Wright was one of the most influential evolutionary biologists of the twentieth century, path analysis was largely ignored by his fellow colleagues. Further developments on this philosophical and methodological underpinning, leading to SEM, were mostly promoted under the scope of social sciences (Bollen 1989; Hoyle 1995; Pearl 2000; Shipley 2000).

To some extent, path analysis resembles multiple regressions in that it consists of a system of linear equations representing interactions between variables. However, unlike multiple regressions, path analysis deals with more complex causal schemes that have more than one dependent variable and with interactions between these dependent variables. This method allows for the estimation of direct effects and indirect effects (i.e., the effect that is completely transmitted by some other variable, also called intervening variable) between variables, which can be pictorially represented by a path diagram, providing a synthetic scheme of the hypothesised relation between variables. The resulting partial correlations of path analysis are assumed to represent causal relations between variables, but they do not test causal links. Only with the development of new estimation techniques it was possible to test the whole causal model, with its implied statistical constraints, solving the main weakness of the path analysis as proposed by Wright. This allowed to test whether the statistical constraints implied by the hypothesised causal model agreed with observations, allowing falsification of the hypothesised causal structure (see Shipley 2000 for a brief SEM historical review).

The estimated coefficients of SEM represent the relation between two variables when all other variables of the model are kept constant at their means (i.e., partial covariance). At the same time, possible effects between variables not specified by the model are restricted to zero (i.e., covariance equals zero), thus providing a method for control of confoundings (Pearl 2000; Shipley 2000). The system of structural equations can be expressed on the matrix form as follows:

$$y = \mathrm{B}y + \Gamma x + \zeta$$

where y is the column vector of dependent variables; x is the column vector of independent variables, B is the matrix of the effect coefficients between dependent variables, Γ is the matrix of the effect coefficients of independent variables on dependent variables and ζ is the column vector of errors in the equations (Bollen 1989).

Although the use of SEM is still restricted in ecology, one can already find publications in different sub-disciplines such as ecophysiology (Shipley et al. 2005), interaction ecology (Cariveau et al. 2004), community and ecosystem ecology (Grace and Pugesek 1997; Kubota et al. 2004; Taylor and Irwin 2004; Weiher et al. 2004). In this section we reanalysed data of Dias et al. (2005) using SEM, in order to identify possible mechanisms by which *C. hilariana* facilitates the establishment of other species in the restinga. We also discuss the potential of this analytical tool for further studies on plant-plant interactions and ecosystem ecology.

As already discussed above (Sect. 4.3.1), *Clusia* patches provide a more important site for plant establishment as compared to non-*Clusia* patches, and this nurse effect is possibly related to dispersal activity (Dias et al. 2005). Here we exchanged the approach of comparing *Clusia* vs non-*Clusia* patches by one of measuring relative abundance of *Clusia*, i.e. a continuous variable. This allowed us to use SEM and test the possible mechanisms by which *Clusia* dominance affects seedling density and richness. The specified models test whether these effects are direct, or indirect through changes on dispersal activity. We also evaluated if the effect of *Clusia* on dispersal activity was due to patch height. *Clusia* patches are more conspicuous, which could promote higher visitation of potential dispersers (e.g., birds and bats). Our models also investigated the possible effect of *Clusia* on seedling density and richness through changes in litter accumulation. Since litter may physically impair seedling establishment (Berendse 1999) and *Clusia* promotes litter accumulation in plant patches (Dias et al. 2006), we expected an indirect negative effect of *Clusia* dominance on recruitment by increasing litter stock.

We tested the effect of *Clusia* separately for seedling density and seedling species richness. For both seedling density and richness we tested two alternative nested models; with (models B and D) and without (models A and C) a direct effect of *Clusia* (Fig. 4.2).

Maximum likelihood was used to estimate structural equation parameters with the R 2.0.1 statistical program (Bates et al. 2006) and SEM package (Fox 2004). We tested which of the nested competing models for seedling density (models A and B) and for seedling richness (models C and D) provided the most appropriated fit with data using chi-square test (χ^2), goodness-of-fit index (GFI), and adjusted goodness-of-fit index (AGFI). Models with a higher number of parameters to be estimated also have a higher probability to fit with data due to a lower number of degrees of freedom. The use of AGFI might correct this due to its penalties to model complexity (Hu and Bentler 1995).

As described by Dias et al. (2005), 30 vegetation patches, of at least 5 m diameter, were randomly chosen within the site. A 1x2 m plot was set in the middle of each vegetation patch. The central position of plots was chosen as to avoid edge effects. For assessment of recruitment process, all rooted plants within the plots were counted, identified and classified in two stage classes. Seedlings were plants <50 cm in height. For measurements of litter mass, one sub-plot (20 cm side) was randomly placed in each plot. All litter was collected within sub-plots and oven-dried at 50 °C until constant weight. A pin-frame approach (adapted from Kent and Coker 1992) was used to determine *Clusia*'s relative abundance. A thin stick (0.8 cm diameter), sub-divided into four 90-cm sections, was positioned vertically in the plots to record: (1) the number of times the stick was touched by vegetation, (2) at which height interval each plant recorded touched the stick and (3) which species was touching the stick. This procedure was repeated every 0.5 m in each plot, adding up to 15 samples per plot. These 15 samples were pooled to comprise the total number of touches of the canopy, where only measurements above the first section (90 cm in height) were considered. The relative abundance of *Clusia* in the canopy was determined as the number of touches of this species divided by the total number of touches of the canopy.

We assumed that species similarity between canopy and understorey of a given vegetation patch negatively correlates to dispersal activity (Dias et al.

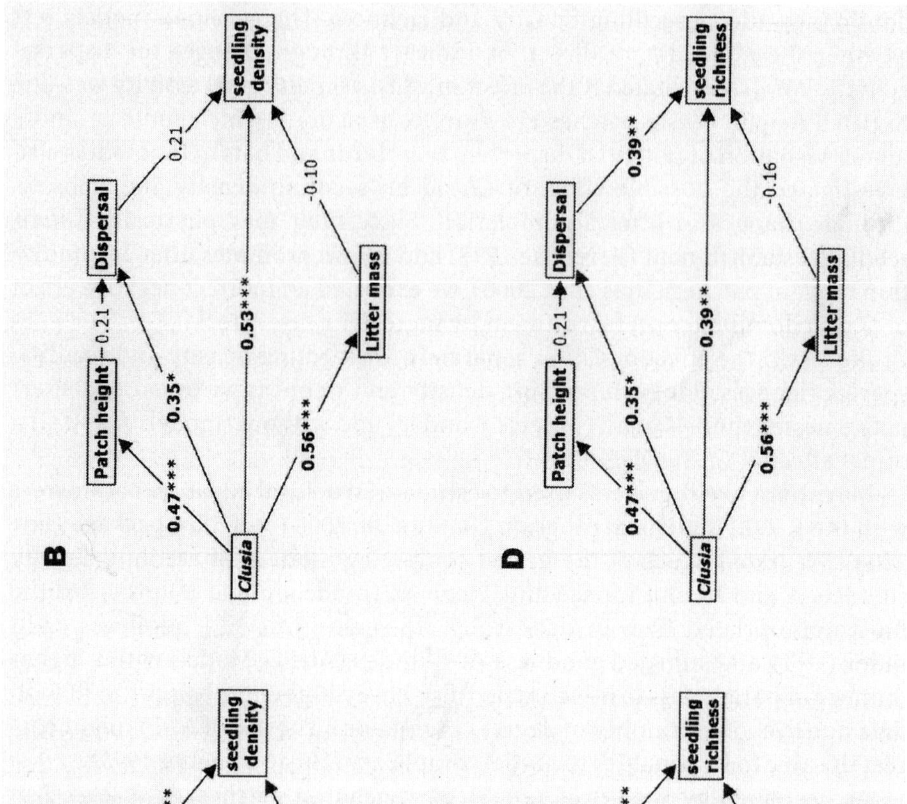

Fig. 4.2A–D. Path diagrams representing hypotheses about the effect of dominance of *Clusia hilariana* on seedling density and richness. For both seedling density and richness two alternative nested models were tested: **A,C** without a direct effect of *Clusia*; **B,D** with a direct effect of *Clusia*. *Values in the arrows* are the standardized parameters estimated by the models (*$p<0.06$; **$p<0.05$; ***$p<0.01$)

2005). Thus, we used the inverse of the Jaccard similarity index (Magurran 1988) as an index of dispersal ($I_{dispersal}$). This index varies from zero (i.e., no species occur in both locations compared) to one (i.e., all species occur in both locations). For each vegetation patch we calculated:

$$I_{dispersal} = 1 - C_j$$

while $C_j = j/(a+b-j)$

where C_j is the Jaccard index, j is the number of species common to both canopy and understorey, a is the number of species in the canopy and b is the number of species in the understorey.

The models for seedling density and richness, which predicted both direct and indirect effects of *Clusia*, provided a better fit with data. For seedling density, model A, which did not specify a direct effect of *Clusia*, did not fit the data. Its covariance matrix showed a marginally significant deviance from the observed covariance matrix (p=0.07; Table 4.1), while model B, which specified direct and indirect effects of *Clusia*, showed good fit to data (Table 4.1). For seedling richness we observed a similar pattern. Although the covariance matrix of model C did not show a significant difference from the observed covariance matrix (p=0.15; Table 4.1), model D, which predicted both direct and indirect effect of *Clusia*, showed a better fit to observations. Even AGFI, which penalises for model complexity, showed higher values for models that predict both direct and indirect effects of *Clusia* (Table 4.1). This suggests that there is an effect of *Clusia* via mechanisms that are not predicted by our models, which might play an important role in recruitment process.

The direct effect of *C. hilariana* on seedling richness was higher than the indirect effect due to the increase of dispersal (Table 4.2, Fig. 4.3). Perhaps more interesting is the fact that dispersal only affected seedling richness,

Table 4.1. Chi-square parameters (χ^2), goodness-of-fit-index (GFI) and adjusted goodness-of-fit-index (AGFI) for the models testing the effect of *Clusia hilariana* on the recruitment process. Models B and D include direct effect on seedling density and richness respectively, while models A and C do not include direct effects

	χ^2	Degrees of freedom	Probability	GFI	AGFI
Seedling density					
Model A	8.65	4	0.07	0.90	0.62
Model B	1.68	3	0.64	0.97	0.89
Seedling richness					
Model C	6.72	4	0.15	0.92	0.69
Model D	2.59	3	0.46	0.97	0.83

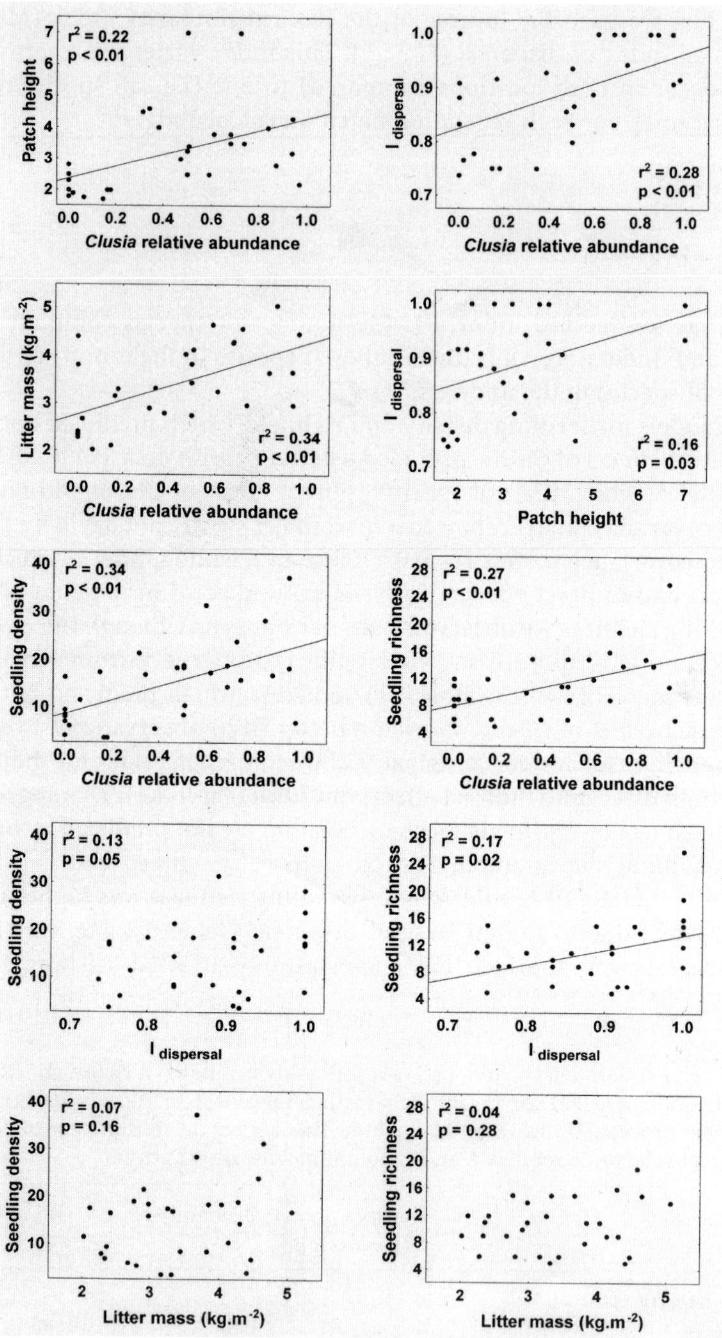

Figure 4.3. Bivariate plots of the variables (non-transformed data) used in the structural equation models. *Linear fittings* are shown for the relationships that were found significant

Table 4.2. Direct effect (DE), indirect effect (IE) and total effect (TE) of *Clusia hilariana* on seedling density and richness based on the accepted models

	DE	IE	TE
Seedling density	0.53	–	0.53
Seedling richness	0.39	0.14	0.53

since the effect of dispersal on seedling density disappeared when the direct effect of *Clusia* was included in the model. This shows that distinct processes regulate seedling density and seedling species richness. Between-patch dispersal might be important to increase within-patch seedling diversity but not to increase seedling density. Thus, we confirmed previous speculations (Liebig et al. 2001; Dias et al. 2005) and concluded that the effect of *C. hilariana* on community structure is, at least partly, due to its effect on seed dispersal.

In the model, dominance of *Clusia* also increased patch height, which is in harmony with our field observations that *C. hilariana* is the most conspicuous species locally. However, height alone is not responsible for enhancing activity of dispersers, indicating that there are probably other factors that might attract dispersers to *Clusia* patches. Although there was no effect of patch height on $I_{dispersal}$ in our models, there was a significant relationship between these variables in the simple regression analysis (Fig. 4.3). This is a good example of how SEM can control for confoundings. The spurious relation between patch height and $I_{dispersal}$ is probably due to the common cause shared by these two variables; i.e. *Clusia* dominance (Fig. 4.2). When *Clusia* dominance is statistically controlled, the relation between patch height and $I_{dispersal}$ was no longer found.

Several works addressed the identity of dominant species as an important factor determining ecosystem properties, but there is still a lack of information of how these changes in ecosystem functioning can affect recruitment in natural communities (Suding et al. 2004; Weiher et al. 2004). We showed that dominance of *C. hilariana* positively affected litter stock; however, this ecosystem's property did not show any effect on seedling density or richness. The litter stock beneath *C. hilariana* in the restinga is high when compared to other open woody vegetation (Dias et al. 2006) but it does not act as an important physical barrier to seedling establishment. Considering that 70 % of the litter underneath *Clusia* canopy is composed by its own leaves (Silva 2003), it is to be expected that this species markedly affects nutrient cycling in the restinga. The effect of low litter quality of *Clusia* on decomposition can possibly influence nutrient availability to seedlings. The effect of *Clusia* on these and other ecosystems properties, and consequent effects on recruitment is the issue of future works. The use of SEM can help to disentangle such effects and provide a promising approach to study such indirect interactions.

4.4 Conclusions

Clusia hilariana is an important nurse plant in the restingas of northern Rio de Janeiro state. Previous field evidences were corroborated by the use of SEM and showed that this species has a positive effect on both understorey seedling density and richness. This process is partly related to the activity of seed dispersers that use male and female plants indistinctly. It remains unclear what happens later in the successional process: as understorey plants grow, they either get suppressed by *Clusia* or outcompete it. Removal experiments in the future shall prove useful to uncover further mechanisms and causal factors for such nursing effect.

In addition to the positive role played by *Clusia* on local biodiversity, recent evidence also indicates that *Clusia*, despite its conservative strategy of carbon acquisition via CAM, gives a high contribution to biomass stock in this nutrient-poor restinga (Dias et al. 2006). Thus, it might strongly affect ecosystems processes such as productivity and nutrient cycling that are also likely to affect recruitment process and, consequently, species composition. The use of SEM on future works will help to provide a synthetic framework of community and ecosystem dynamics.

Acknowledgements. The data reviewed here is partly the product of the effort of several excellent ex-students - C.M.B. Correia, M. Pimentel, J.V.R. Ramos and H.L.T. Zaluar – to whom we are thankful. The studies on *Clusia hilariana* have been receiving long standing financial support of the Brazilian Long Term Ecological Research Project (LTER-PELD, site 5, CNPq) and of the Research Center of the Brazilian Oil Company (Cenpes-Petrobras). These projects have the kind permission and logistic support of the Brazilian Forestry Service (IBAMA) and the Restinga de Jurubatiba National Park. CNPq (Brazilian Scientific Council) and CAPES (Brazilian Council for Graduate Education) provide research grants for the authors.

References

Amzallag GN (2005) Perturbed reproductive development in salt-treated *Sorghum bicolor*: a consequence of modifications in regulation networks. J Exp Bot 56:2821–2829

Araujo DSD, Scarano FR, Sá CFC, Kurtz BC, Zaluar HLT, Montezuma RCM, Oliveira RC (1998) Comunidades vegetais do Parque Nacional da Restinga de Jurubatiba. In: Esteves FA (ed) Ecologia das Lagoas Costeiras do Parque Nacional da Restinga de Jurubatiba e do Município de Macaé (RJ). Nupem-UFRJ, Rio de Janeiro, pp 39–62

Araujo DSD, Pereira MCA, Pimentel MCP (2004) Flora e estrutura de comunidades na Restinga de Jurubatiba – síntese dos conhecimentos com enfoque especial para a formação aberta de *Clusia*. In: Rocha CFD, Esteves FA, Scarano FR (eds) Pesquisas de longa duração na Restinga de Jurubatiba: ecologia, história natural e conservação. Editora Rima, São Carlos, pp 59–76

Bates D, Chambers J, Dalgaard P, Gentleman R, Hornik K, Iacus S, Leisch F, Lumley T, Maechler M, Murdoch D, Murrell P, Plummer M, Ripley B, Lang DT, Tierney L, Urbanek S (2006) R: a language and environment for statistical computing. Royal Foundation for Statistical Computing, Vienna

Berendse F (1999) Implications of increased litter production for plant diversity. Trends Ecol Evol 14:4–5

Bollen KA (1989) Structural equations with latent variables. Wiley, New York

Brooker W, Callaghan TV (1998) The balance between positive and negative plant interactions and its relationship to environmental gradients: a model. Oikos 81:196–207

Bruno JF, Stachowicz JJ, Bertness MD (2003) Inclusion of facilitation into ecological theory. Trends Ecol Evol 18:119–125

Callaway RM (1998) Are positive interactions species-specific? Oikos 82:202–209

Callaway RM, Walker LR (1997) Competition and facilitation: a synthetic approach to interactions in plant community. Ecology 78:1958–1965

Callaway RM, Brooker RW, Choler P, Kikvidze Z, Lortie CJ, Michalet R, Paolini L, Pugnaire FI, Newingham B, Aschehoug ET, Armas C, Kikodze D, Cook BJ (2002) Positive interactions among alpine plants increase with stress. Nature 417:844–848

Campbell DG, Stone JL, Rosas A (1992) A comparison of the phytosociology and dynamics of 3 floodplain (*várzea*) forests of known ages, Rio-Juruá, western Brazilian Amazon. Bot J Linn Soc 108:213–237

Cariveau D, Irwin RE, Brody AK, Garcia-Mayeya LS, von der Ohe A (2004) Direct and indirect effects of pollinators and seed predators to selection on plant and floral traits. Oikos 104:15–26

Correia CMB (1998) Estrutura de populações, associação entre quatro espécies-chaves e formação de moitas em restinga de *Clusia*, no norte fluminense. MSc Dissertation, Universidade Federal do Rio de Janeiro, Rio de Janeiro, Brasil

Crawley MJ (1997) Sex. In: Crawley MJ (ed) Plant ecology, 2nd edn. Blackwell, Oxford, pp 156–213

Dias ATC, Zaluar HTL, Ganade G, Scarano FR (2005) Canopy composition influencing plant patch dynamics in a Brazilian sandy coastal plain. J Trop Ecol 21:343–347

Dias ATC, de Mattos EA, Vieira SA, Azeredo JV, Scarano FR (2006). Aboveground biomass stock of native woodland on a Brazilian sandy coastal plain: estimates based on the dominant tree species. For Ecol Manage 226:364–367

Díaz S, Symstad AJ, Chapin FS III, Wardle DA, Huenneke LF (2003) Functional diversity revealed by removal experiments. Trends Ecol Evol 18:140–146

Dickie IA, Schnitzer SA, Reich PB, Hobbie SE (2005) Spatially disjunct effects of co-occurring competition and facilitation. Ecol Lett 8:1191–1200

Fox J (2004) SEM: Structural equation models. R package version 0.8-0. http://www.r-project.org, http://www.socsci.mcmaster.ca/jfox/

Franco AC, Nobel PS (1989) Effect of nurse plants on the microhabitat and growth of cacti. J Ecol 77:870–886

Grace JB, Pugesek BH (1997) A structural equation model of plant species richness and its application to a coastal wetland. Am Nat 149:436–460

Holmgren M, Scheffer M, Huston MA (1997) The interplay of facilitation and competition in plant communities. Ecology 78:1966–1975

Hoyle RH (ed) (1995) Structural equation modeling: concepts, issues, and applications. Sage Publications, Thousand Oaks

Hu L, Bentler PM (1995) Evaluating model fit. In: Hoyle RH (ed) Structural equation modeling: concepts, issues and applications. Sage Publications, Thousand Oaks, pp 76–99

Kareiva P, Levin SA (eds.) (2003) The importance of species: perspectives on expendability and triage. Princeton University Press, Princeton

Kent M, Coker P (1992) Vegetation description and analysis. Wiley, New York
Kikvidze Z, Nakhutsrishvili G (1998) Facilitation in subnival vegetation patches. J Veg Sci 9:261–264
Kubota Y, Murata H, Kikuzawa K (2004) Effects of topographic heterogeneity on tree species richness and stand dynamics in a subtropical forest in Okinawa Island, southern Japan. J Ecol 92:230–240
Liebig M, Scarano FR, de Mattos EA, Zaluar HLT, Lüttge U (2001) Ecophysiological and floristic implications of sex expression in the dioecious neotropical CAM tree *Clusia hilariana* Schltdl. Trees 15:278–288
Lortie CJ, Callaway RM (2006) Re-analysis of meta-analysis: support for the stress-gradient hypothesis. J Ecol 94:7–16
Macedo MV, Monteiro RF (1987) Germinação e desenvolvimento de plântulas em tanque de *Neoregelia cruenta* (Bromeliaceae) na Restinga de Barra de Maricá, RJ. Simpósio sobre Ecossistemas da Costa Sul e Sudeste Brasileira 2:188–190
Magurran AE (1988) Ecological diversity and its measurement. Croom Helm, London
Morris WF, Wood DM (1989) The role of lupine in succession on Mount St Helens: facilitation or inhibition? Ecology 70:697–703
Müller-Dombois D, Ellenberg H (1974) Aims and methods of vegetation ecology. Wiley, New York
Pascal JP, Pélissier R (1996) Structure and floristic composition of a tropical evergreen forest in south-west India. J Trop Ecol 12:191–210
Pearl J (2000) Causality: models, reasoning and inference. Cambridge University Press, Cambridge
Pereira MCA, Araujo DSD, Pereira OJ (2001) Estrutura de uma comunidade arbustiva da restinga de Barra de Maricá–RJ. Revta Bras Bot 24:273–281
Pimentel M (2002) Variação espacial na estrutura de comunidades vegetais da formação aberta de *Clusia* no Parque Nacional da Restinga de Jurubatiba, RJ. DSc Thesis. Universidade Federal do Rio de Janeiro, Rio de Janeiro, Brazil
Pugnaire FI, Haase P, Puigdefabregas J, Cueto M, Clark SC, Incoll LD (1996) Facilitation and succession under the canopy of a leguminous shrub, *Rentama sphaerocarpa*, in a semi-arid environment in south-east Spain. Oikos 76:455–464
Ramos JVR (2003) Facilitação ou competição? O efeito da senescência de *Clusia hilariana* sobre o seu sub-bosque em ilhas de vegetação de restinga. MSc Dissertation. Universidade Federal do Rio de Janeiro, Rio de Janeiro, Brazil
Rosado BP (2006) A importância da inclusão de diferentes dimensões de variação de características morfo-fisiológicas e de crescimento para o entendimento dos padrões de abundância de espécies de restinga. MSc Dissertation. Universidade Federal do Rio de Janeiro, Rio de Janeiro, Brazil
Sanchez M, Pedroni F, Leitão-Filho HF,César O (1999) Composição florística de um trecho de floresta ripária na Mata Atlântica em Picinguaba, Ubatuba, SP. Revta Bras Bot 22:31–42
Scarano FR (2002) Structure, function and floristic relationships of plant communities in stressful habitats marginal to the Brazilian Atlantic Rainforest. Ann Bot 90:517–524
Scarano FR, Cirne P, Nascimento MT, Sampaio MC, Villela D, Wendt T, Zaluar HLT (2004) Ecologia Vegetal: integrando ecossistema, comunidades, populações e organismo. In: Rocha CFD, Esteves FA, Scarano FR (eds) Pesquisas de Longa Duração na Restinga de Jurubatiba: Ecologia, História Natural e Conservação. RiMa Editora, São Carlos, pp 77–97
Scarano FR, Duarte HM, Franco AC, Geßler A, de Mattos EA, Nahm M, Rennenberg H, Zaluar HLT, Lüttge U (2005) Ecophysiology of selected tree species in different plant communities at the periphery of the Atlantic forest of SE Brazil I. Performance of three different species of *Clusia* in an array of plant communities. Trees 19:497–509

Shipley B (2000) Cause and correlation in biology. Cambridge University Press, Cambridge

Shipley B, Vile D, Garnier E, Wright IJ, Poorter H (2005) Functional linkages between leaf traits and net photosynthetic rate: reconciling empirical and mechanistic models. Funct Ecol 19:602–615

Silva AP (2003) Solos e produção de serrapilheira em moitas no Parque Nacional da Restinga de Jurubatiba, RJ. MSc Dissertation. Universidade Estadual do Norte Fluminense, Campos dos Goytacazes, Brazil

Suding KN, Larson JR, Thorsos E, Steltzer H, Bowman WD (2004) Species effects on resource supply rates: do they influence competitive interactions? Plant Ecol 175:47-58

Taylor BW, Irwin RE (2004) Linking economic activities to the distribution of exotic plants. Proc Natl Acad Sci 101:17725–17730

Tielbörger K, Kadmon R (1997) Relationships between shrubs and annual communities in a sandy desert ecosystem: a three-year study. Plant Ecol 130:191–201

Urbanska KM (1992) Populationsbiologie der Pflanzen: Grundlagen, Probleme, Perspektiven. Fischer, Stuttgart, Jena

Walker LR, Vitousek PM (1991) An invader alters germination and growth of a native dominant tree in Hawaii. Ecology 72:1449–1455

Weiher E, Forbes S, Schauwecker T, Grace JB (2004) Multivariate control of plant species richness and community biomass in blackland prairie. Oikos 106:151–157

Weltzin J, McPherson GR (1999) Facilitation of conspecific seedling recruitment and shifts in temperate savanna ecotones. Ecol Monogr 69:513–534

Zaluar HLT (1997) Espécies focais e a formação de moitas na restinga aberta de *Clusia*, Carapebus, RJ. MSc Dissertation, Universidade Federal do Rio de Janeiro, Rio de Janeiro, Brazil

Zaluar HLT (2002) Dinâmica da vegetação em restingas abertas fluminenses: uma aproximação através da interação entre plantas. DSc Thesis, Universidade Federal do Rio de Janeiro, Brazil

5 Reproductive Biology

Rodrigo Lemes Martins, Tânia Wendt, Rogério Margis, and Fabio Rubio Scarano

5.1 Introduction

Most studies on the reproductive biology of species of the genus *Clusia* are limited in time, space and phenological scales that they cover. Floral biology descriptions are more abundant than studies on the possible ecological and evolutionary causes and consequences of sexual behaviour. For instance, Table 5.1, which is largely based on a previous list by Lopes and Machado (1998), shows information on 28 species more thoroughly studied, which adds up to ca. 10 % or less of the 250–400 species belonging to the genus (Bittrich and Amaral 1996a; Pipoly et al. 1998).

Despite these limitations, it is now well established that the reproductive biology of the genus *Clusia* often comprises three features which, individually, are relatively rare among angiosperms: dioecy, resin as reward for pollinators and floral mimetism. A fourth feature, equally rare, is agamospermy, the occurrence of which in the genus is still highly controversial. However, it is not always that all four features appear jointly for a given species. The various possible cases of single or combined occurrences of these features in a given species turns the reproductive biology of *Clusia* a very challenging subject, which might provide new insights into the ecology and evolution of sexual behaviour in plants.

These features began to emerge already in the earliest papers on *Clusia* reproductive behaviour. Janzen (1971) described the interaction of *Clusia* with resin-collecting bees, such as the Euglossini, but did not report on resin collection. Skutch (1971) first reported *Clusia* pollination by bees that collected secretion of the stamens, which prompted a series of studies examining the role of resin. The first report on agamospermy for the genus has been provided by Maguire (1976), who studied monoecious populations of *C. rosea*. However, agamospermy in *Clusia* is much less covered in the literature than the role of resin in pollination, perhaps because it now appears to be less com-

Ecological Studies, Vol. 194
U. Lüttge (Ed.)
Clusia: A Woody Neotropical Genus
of Remarkable Plasticity and Diversity

Table 5.1. Some of the main findings regarding aspects of the reproductive biology of *Clusia* species and respective sources

Species	Reward			Vector[a]	AM/MP	Agamospermy	RS	Comments	References
	Pollen	Nectar	Resin						
C. alata Pl. *et* Tr.	?	No	√	Mainly *Trigona corvina* (Cockerell 1913)	?	?	D	–	Ramirez and Gómez (1978)
C. aemygdioi Gomes da Silva et Weinberg	√	No	No	Mainly Trigonini and Halictidae	√	No	D	Ongoing research	G. Matallana – personal communication (Fig. 5.1E,F)
C. arrudae Pl. *et* Tr.	√	No	√	*Eufriesea* spp. *Melipona quadrifasciata* (Lepeletier 1836), *Trigona spinipes* (Fabricius 1793)	?	No	D	–	Carmo and Franceschinelli (2002)
C. burle-marxii Bittrich	√	No	No	Beetles	?	?	?	–	V. Bittrich apud Lopes and Machado (1998)
C. columnaris Engl.	√	No	√	Mainly Trigonini and Meliponini	?	?	D	–	Bittrich and Amaral (1996a)
C. criuva Camb.	√	No	No	Beetles	√	No	D	–	Correia et al. (1993)
C. fluminensis Pl. et Tr.	√	No	√	*Euglossa cordata* (Linnaeus 1758), *Plebeia mosquito* (Smith 1863), *Tetragonisca angustula* (Latreille 1811), and Trigonini	?	No	D	Pollen and resin mixture forms fruits, while pollen alone does not	V. Bittrich apud Carmo and Franceschinelli (2002), Correia (1983), Correia et al. (1989)
C. grandiflora Splitg.	√	No	√	Mainly Euglossini, but Trigonini also visit to collect pollen	?	?	D	Pollen in droplets of an oily liquid; differences in antibacterial resin effect by the genders	Bittrich and Amaral (1997), Lokvam and Braddock (1999)

C. gundlachii Stahl.	?	No	√	–	√	?	D	–	Gustafsson (2000)
C. hilariana Schlecht.	√	No	√	Mainly Halictinae	√	?	D	High pollen viability	Cruz et al. (1990), Faria et al. (2006), our own data
C. insignis Sem.	√	No	√	Mainly Euglossini	√	?	D	–	Bittrich and Amaral (1997)
C. intermedia Mariz	?	√	No	Flies	?	No	D	Ongoing research	G. Matallana – personal communication (Fig. 5.1A,B)
C. lanceolata Camb.	√	No	√	Mainly *T. spinipes*, and *E. cordata*	?	No	D	–	Correia et al. (1999)
C. leprantha Mart.	√	No	√	Trigonini	?	?	D	Pollen and resin mixture	Bittrich and Amaral (1997)
C. marizii Gomes da Silva	√	No	√	–	?	No	D	Small bees as pollinators	G. Matallana – personal communication (Fig. 5.1C,D)
C. minor L.	?	?	?	–	?	No	H	–	Hammel (1986), Bittrich (1994)
C. nemorosa G. Mey	√	No	√	Mainly Meliponini, and Trigonini	?	No	D/H	Maguire (1976) found indirect evidence of agamospermy	Maguire (1976); Bittrich and Amaral (1997); Lopes and Machado (1998)
C. odorata Sem.	?	No	?	Trigonini	?	?	D	First description of resin as reward	Croat (1978); Ramirez and Gómez (1978); Armbruster (1984)
C. organensis Pl. *et* Tr.	?	?	?	–	?	?	D	High pollen viability	Cruz et al. (1990)
C. pana-panari (Obley) Choisy	?	No	√	Mainly Trigonini	?	?	D	Pollen and resin mixture; thick staminodes; male flowers may present pistillode	Armbruster (1984); Bittrich and Amaral (1997)

Table 5.1. (*Continued*)

Species	Reward			Vector[a]	AM/MP	Agamo-spermy	RS	Comments	References
	Pollen	Nectar	Resin						
C. paralicola G. Mariz	?	?	√	*T. spinipes*	?	?	?	–	V. Bittrich apud Lopes and Machado (1998)
C. parviflora Engl. nom. illeg.	√	?	–	Mainly beetles and flies	?	?	?	Pollinated by beetles and flies	V. Bittrich apud Lopes and Machado (1998)
C. pusilla Steyerm.	√	No	–	*Euglossa* spp., Halictidae, and *Xilocopa* spp.	√	?	D	Mainly pollinated by small bees	Bittrich and Amaral (1996a)
C. renggerioides Pl. et Tr.	?	No	√	*Trigona* spp.	?	?	D	Pollen and resin mixture; apolar fluid on the stigmas	Bittrich and Amaral (1997)
C. rosea Jacq.	?	No	√	*Euglossa*	?	√	M	Indirect evidence of agamospermy	Maguire (1976)
C. schomburqkiana (Pl. *et* Tr.) Benth. ex Engl.	?	No	√	*Duckeola ghilianii* (Spinola 1853) and *Trigona* spp.	?	?	D/H	Hermaphroditism was found by Maguire (1959) and not found by Bittrich and Amaral (1996a)	Maguire (1959) apud Bittrich and Amaral (1996a)
C. aff. *spathulaefolia* Pl. *et* Tr..	?	?	?	Beetles	?	?	?	–	V. Bittrich apud Lopes and Machado (1998)
C. scrobiculata Benoist	?	?	?	–	?	?	H	Briefly mentioned in this reference	Bittrich and Amaral (1996a)

[a] Please note that the years following the name of authorities are in agreement with rules of zoological nomenclature, which differs in that sense from botanical rules. Therefore, these cannot be found in the list of references

AM/MP=automimicry/mistake pollination; RS=reproductive system; H=hermaphrodite; D=dioecious; M=monoecious; ?=information unavailable for the given species mon among *Clusia* species than initially expected (Lopes and Machado 1998; Correia et al. 1999; Carmo and Franceschinelli 2002).

This chapter reviews studies on the reproductive biology of *Clusia* and focuses particularly on the four main reproductive features of the genus: dioecy, resin as reward for pollinators, floral mimetism and the debate around agamospermy and vegetative reproduction. Finally, we use *C. hilariana* as a case study to discuss how such background might benefit, in the future, from a more integrated approach linking molecular biology and ecophysiology to population and community ecology.

5.2 Main Reproductive Features

5.2.1 Dioecy

Hermaphroditism is the dominant condition in flowering plants and dioecy is estimated to occur in only ca. 6 % of the entire angiosperm flora (Renner and Ricklefs 1995). Dioecious plants demand vectors for cross-pollination, and out-crossing avoids the consequences of inbreeding depression (Charlesworth and Charlesworth 1978; Thomson and Barrett 1981; Lloyd 1982; Sakai et al. 1995a; Freeman et al. 1997). Dioecy has been correlated with plant spatial distribution, tropical floras, oceanic islands and oligotrophic environments (Freeman et al. 1976; Bawa 1980a; Flores and Schemske 1984; Bawa et al. 1985; Sakai et al. 1995a, b; Thompson and Edwards 2001; Matallana et al. 2005), and since Darwin (1877) it has been suggested that resource allocation is a causal factor in the evolution and maintenance of dioecy.

Dioecy is the most common condition in the tribe Clusieae Choisy to which *Clusia* belongs. Thus, most *Clusia* species are dioecious and hermaphroditism (as in *C. scrobiculata*) is rare (Bittrich and Amaral 1996a). However, hermaphroditic flowers have been reported in occasional populations of predominantly dioecious species. Maguire (1966) was the first to report the occasional occurrence of hermaphroditic flowers in many scandent or epiphytic species. For instance, populations with dioecious and gynodioecious plants were found for *C. minor* (Maguire 1976), *C. nemorosa* (Mesquita and Franciscon 1995; Bittrich and Amaral 1997; Lopes and Machado 1998) and *C. schomburgkiana* (Bittrich and Amaral 1996a, 1997). In addition, Lopes and Machado (1998) found in *C. nemorosa* flowers of hermaphroditic individuals with a number of stamens smaller than average. These authors considered this as a type of 'female inconstancy'. Lloyd (1976) coined this term, which he defined as the presence of stamens in pistillate flowers. This peculiarity has been described for some species of *Clusia* by Maguire and Wurdack (1961), Maguire (1966, 1978) and Mariz (1974).

Dioecy requires cross-pollination. There are three classes of pollinator reward in *Clusia*: resin (which is probably the most common, if species number is considered), pollen and nectar, which are discussed next.

5.2.2 Resin

Resin is a rare floral reward among angiosperms. It occurs as a resource in *Dalechampia* (Euphorbiaceae; Armbruster 1984; Armbruster and Webster 1979), *Chrysochlamys, Tovomitopsis, Clusia* (tribe Clusieae at Clusiaceae; Gustafsson et al. 2002), *Clusiella* (Clusiaceae; Bittrich and Amaral 1996a, b), *Moronobea* (Clusiaceae; Vicentini and Fischer 1999) and also *Monstera* (Araceae), which produces a viscous stigmatic exudate (Ramírez and Gómez 1978). Early investigations focused on the roles of resin, primarily the *Dalechampia*-bees relationship (Armbruster and Webster 1979, 1981, 1982; Armbruster and Herzig 1984; Armbruster 1986; Armbruster and Steiner 1992; Armbruster et al. 1992). It has been claimed that resin provides adaptive advantages regarding attractiveness to bees, given that many such insects use resin for building nests (Armbruster 1984; Armbruster and Webster 1979; Bittrich and Amaral 1996a, b) and as a sticky defense against ant attack (Sakagami et al. 1989). Thus, resin attracts specific visitors, unlike pollen and nectar (Armbruster 1984). The use of resin as a food resource is suggested by Ramirez and Gomes (1978), but Armbruster (1984) argues that this hypothesis is improbable due to the toxicity, low nutritive constitution and resistance to decay and digestion. Furthermore, it is a predictable resource in time and space, since it takes longer than pollen and nectar to be entirely removed by visitors, and maintains viscosity for a long time (Armbruster 1984), conserving fungicidal and bactericidal properties (Lokvam and Braddock 1999).

Euglossini and some smaller bees are probably the main pollinators for *Clusia*. These typical forest bees cover large distances (up to 24 km) to collect resin, often in a traplining behaviour (Armbruster 1984; Janzen 1971; Roubik 1992; Lopes and Machado 1998). Thus, Janzen (1971) highlighted that

Fig. 5.1A–H. Flowers and pollinators of *Clusia*. A–F are sympatric plants at Santa Lúcia Biological Reserve, Santa Teresa municipality, state of Espírito Santo, SE-Brazil. **A** Diptera visiting a male flower of *C. intermedia* on a nectar based pollination system (see Table 5.1). **B** Female flower of *C. intermedia* visited by *Trigona spinipes* (Hymenoptera, Apidae). **C** Male flower of *C. marizii* visited by *T. spinipes*. **D** Female flower of *C. marizii*. **E** Male flower of *C. aemygdioi* visited by *T. spinipes*. **F** Female flower of *C. aemygdioi* visited by Halictinae in searching behaviour. **G** Male flower of *C. hilariana* **H** Female flower of *C. hilariana*. Photographs C–F were kindly provided by Glória Matallana, and G,H were kindly provided by Izar Aximoff

A
B
C
D
E
F
G
H

Euglossini is an important pollinator for plants that occur at low densities. These are the main pollinators of *C. arrudae* (Carmo and Franceschinelli 2002), *C. grandiflora* (Armbruster 1984, Bittrich and Amaral 1997), *C. rosea* (Armbruster 1984), *C. nemorosa* (Mesquita and Franciscon 1995; Bittrich and Amaral 1997; Lopes and Machado 1998), *C. insignis, C. leprantha, C. renggerioides, C. pana-panari* (Bittrich and Amaral 1997), *C. pusilla* (Lopes and Machado 1998) and *C. lanceolata* (Correia et al. 1989).

According to Bittrich and Amaral (1996a), flowers from different species have different traits to minimize the contact between resin and pollen when bees collect resin (Fig. 5.1). Thus, the infrageneric groups of *Clusia* with species offering resin, show a high diversity of forms. For further discussion on the evolution of resin-secretion see Armbruster (1984), Bittrich and Amaral (1996a, 1997), Gustafsson et al. (2002), Gustafsson and Bittrich (2003) and Chap. 6.

5.2.3 Automimetism and Mistake Pollination

Reproductive mimicry in plants is the utilization of false sensory cues to attract pollinators (Wiens 1978). One type of floral mimetism is the so-called automimicry syndrome, henceforth automimetism, where female flowers mimic male flowers that are the ones carrying food reward (Little 1983). Automimetism was first examined in more details by Baker (1976), while studying *Carica papaya* L.

In *Clusia*, automimetism is unmistakably recognized only in the case of pollen flowers. Female flowers do not produce pollen and in order to be pollinated they must look, or at least smell like the pollen-producing male ones (e.g., *C. criuva* and *C. gundlachii*; Correia et al. 1993; Gustafsson 2000). However, for some *Clusia* species with pollen flowers, resin and/or nectar are also produced by both sexes. In such cases, when pollen is the main reward tracked by pollinators, the visits in female flowers are described as mistake pollination (e.g., *C. insignis* and *C. pusilla*; Bittrich and Amaral 1996a, 1997). This poses a doubt as to whether mistake pollination is induced by automimetism or by a lack of specificity for pollen reward by the pollinator. For instance, Bawa (1980b) proposed that automimetic female flowers must offer rewards to the pollinators regardless of the degree of resemblance with male flowers, particularly when grouped in patches and spatially separated from males. This might apply to the *Clusia* cases discussed here.

Automimicry was first registered for the genus in *C. criuva* (Correia et al. 1993), which has pollen flowers pollinated by beetles. However, unlike *C. criuva*, all *Clusia* species reported to have mistake pollination are pollinated by small bees. While Euglossini are clearly important pollinating bees when resin is the floral reward (see Sect. 5.2.2), small bees are described as effective pollinators of some *Clusia* species (Lopes and Machado 1998). These pollinators regularly collect floral resin and pollen (Ramírez and Gómez 1978; Correia 1983; Correia et al. 1989, 1999) but in some *Clusia*, small bees collect only pollen as in *C. pusilla* pollinated by Halictinae (Bittrich and Amaral 1996a), and *C. fluminensis* pollinated mainly by *Plebeia mosquito* (Lopes and Machado 1998).

Clusia pusilla provided the first study in *Clusia* to consider differences on floral output between male and female plants (Bittrich and Amaral 1996a). This species conformed to the model proposed by Bawa (1980b), where automimetic plant populations often have abundant male flowers, in excess over mimetic female flowers, and a long floral display for both genders. This difference in floral output between genders might be promoted by competition between male plants; however the different genders may, on the other hand, bear the same total amounts of reward. The smaller number of female, mimetic flowers may result in a higher number of visits in these flowers without reward.

Clusia nemorosa, pollinated by the small bee *Trigona spinipes*, also has mimetic female flowers (Lopes and Machado 1998). Bawa and Opler (1975) argue that smaller bees are particularly effective pollinators when distances between male and female are small, since small bees will not fly long distances. Considering the ability of some *Clusia* species to colonize open areas and form large populations, pollination by small bees can be more common than described to date. One would then expect that automimicry in dioecious plants would be more common for large, dense populations at open areas. As expected, *C. nemorosa* forms large populations in open areas in the restingas of northeast Brazil. Automimicry in plants pollinated by small bees, such as most *Clusia* species, would then probably have evolved when bee deception by floral mimicry is facilitated by the small distances between male and female plants that are more likely to occur in dense populations, as will be discussed in Sect. 5.3 dedicated to *C. hilariana*.

5.2.4 Asexual Reproduction: Agamospermy and Vegetative Propagation

The term agamospermy refers to the production of fruits and seeds by non-fertilised flowers. It is a type of asexual reproduction that combines the possible advantages of perpetuating a well-succeeded genotype through time, with the advantages of the higher mobility of seeds as propagules, unlike other forms of asexual reproduction such as vegetative propagation (Crawley 1997; Martins and Oliveira 2003). In *Clusia* literature and elsewhere, agamospermy is often referred to as 'apomixis'. However, apomixis is more precisely defined as any type of asexual reproduction, either through agasmospermy or vegetative reproduction (Kearns and Inouye 1993; Harris and Harris 2001). Therefore, here we will use agamospermy whenever we refer to seed production without fertilisation.

There has been some controversy around the occurrence and the relevance of agamospermy for *Clusia*'s reproduction (Mesquita and Franciscon 1995). For instance, Correia (1983) reported agamospermy for *C. fluminensis*, in flowers that received resin in the stigmas. Since there were doubts about the purity (i.e., absence of pollen) of the resin, new tests were conducted by Correia et al. (1989) comparing *C. fluminensis*, *C. lanceolata* and *C. criuva*. They confirmed the occurrence of agamospermy triggered by resin for *C. fluminensis* and also for *C. lanceolata*. However, the same research group found no fruit formation by agamospermy for *C. lanceolata* ten years later (Correia et al. 1999). Instead, experimental pollination by resin and pollen mixed, showed the highest fruit production among all treatments undertaken. This phenomenon had already been previously described by Bittrich and Amaral (1996a) to occur in *C. renggerioides* and *C. pana-panari* who found that more-or-less apolar fluids secreted by stigmas probably serves to dilute resin drops mixed with pollen. In the case of *C. fluminensis*, Carmo and Franceschinelli (2002) refer to unpublished data of V. Bittrich that indicated that the application of filtered resin in stigmas resulted in no fruit production. Therefore, it would appear that so far undisputed agamospermy in *Clusia* only exists in *C. rosea* (Maguire 1976) and *C. minor* (Maguire 1976; Hammel 1986).

Vegetative propagation and clonal growth, however, have been less investigated. Some field evidences have been found for some of the restinga species of *Clusia*, (e.g., *C. hilariana*; Correia 1998; Scarano et al. 2004). This type of reproduction is a common feature to many other restinga shrubs (Cirne and Scarano 2001). In the next section we discuss the possible relevance of this type of reproduction to population maintenance and growth of *C. hilariana*.

5.3 The Case of *Clusia hilariana*

We have pinpointed *C. hilariana* for a case study, not only for the obvious fact that it has been our focal species for a number of years (Scarano 2002; Scarano et al. 2005), but mostly because our data set allows for the examination of the species' reproductive biology from a molecular perspective all the way to a plant community perspective. Reproductive biology studies for *Clusia* species have often focused on floral biology, as reviewed above, and time and spatial scales covered hardly ever permit a deeper ecological understanding of causes and consequences of *Clusia*'s sexual behaviour.

Our interest in the reproductive biology and population ecology of *C. hilariana* started shortly after we found evidence for the important ecological role played by this species in the restingas of northern Rio de Janeiro state (particularly at the Restinga de Jurubatiba National Park - RJNP), Brazil: it is the most abundant tree species in the open restingas and acts as a nurse plant to various other species (see Chap. 4).

One of the earliest studies was a M.Sc. Dissertation carried out by Correia (1998), who found 312 *Clusia* plants in 0.5 ha of the RJNP. They were divided in three age classes, based on diameter size: young (32 plants, i.e. ca. 10% of the total; those less than 20 cm in height), juveniles (211 plants) and adults (69 plants; those with a basal diameter =3.5 cm). This population structure suggests that regeneration and recruitment is not continuous for this species, which had more adults than seedlings. This was unexpected for a plant that is so abundant locally. Perhaps more interestingly, Correia could determine of the plants, which are less than 50 cm in height (n=59), those which had originated from seeds and those which originated from vegetative propagation. She found that 72 % of all plants less than 50 cm tall originated from seeds, where 2 % had germinated on the ground and 70 % within the tanks of terrestrial bromeliads (mainly *Aechmea nudicaulis* (L.) Griseb. and *Neoregelia cruenta* (Graham) L.B.Sm.; see also Scarano 2002). Forming adventitious roots the seedlings grow out from the tanks as they age and establish themselves as independent trees. The remaining 28 % of the young plants originated from vegetative propagation sprouting out from roots of mature trees.

These data, collected during one growing season only, indicated that abundant regeneration could be a periodic phenomenon (i.e., dependent on "good years") and that success apparently depended on tank bromeliads, which nurse seed-originated young plants (this pattern has also been found for *C. fluminensis* in the restinga of Barra de Maricá, further south; Macêdo and Monteiro 1987; Zaluar 2002; see also Chap. 4), and on the potential to reproduce vegetatively (see also Scarano et al. 2004).

We then decided to investigate this issue from a pollination biology perspective. *Clusia hilariana* is highly abundant locally. Antonovics and Levin (1980) proposed that plants at high densities are visited by pollinators more frequently than those at low densities (see also Larson and Barrett 2000). Therefore, one could expect high fruit and seed set locally. In contrast to this prediction, but in harmony with the findings of Correia (1998), we found a low production of fruit and viable seeds in natural open-pollinated as compared to hand-pollinated flowers (Faria et al. 2006). Plants commonly mature fewer fruits and seeds than could be produced given flower and ovule numbers (for reviews see Burd 1994; Larson and Barrett 2000). However, fruit set in our control plants was considered low in comparison to Sutherland and Delph (1984). They found a mean value for fruit set for 31 different dioecious species surveyed of 52.6%. Conversely, the values we found for open-pollinated *C. hilariana* for two reproductive seasons were ca. half that average and ranged from 22% to 26%. Other studies on dioecious *Clusia* species further confirmed that the fruit set values obtained for open-pollinated *C. hilariana* were indeed low: *C. criuva*, a forest tree, showed 90% fruit set in open- and hand-pollinated plants (Correia et al. 1993), and *C. nemorosa* in restingas showed 33.9% fruit set in open-pollinated plants vs 86.1% in hand-pollinated treatment (Lopes and Machado 1998).

We then attributed this low success to pollinator scarcity. The fact that there is no seed production via agamospermy in this species and that seed germination in situ is apparently almost restricted to the interior of bromeliad rosettes (Scarano 2002, Scarano et al. 2004) makes the high abundance of *C. hilariana* all the more puzzling.

Thus, the results of Correia (1998) and Faria et al. (2006) combined led us to two alternative explanations for the local persistence and dominance of *C. hilariana* despite low levels of fruit and seed set. 1) Successful years may occur irregularly and at long time intervals, such that longer-term studies are necessary to ensure sampling of such years. Low levels of reproduction sustained over time might be enough to ensure population stability. 2) Persistence and abundance may have been achieved by asexual reproduction.

At this point, it was clear that we needed to enhance both spatial and temporal scales of observations to verify whether any or both of the two alternatives above could explain *C. hilariana*'s high abundance in the RJNP. The study of Correia (1998) covered only one reproductive season and was conducted in a plot of 0.5 ha, while Faria et al. (2006) covered two seasons in a plot of 2 ha. Understanding the mechanisms of reproductive biology which ensure the high abundance of *C. hilariana* in this site seemed essential to foster conservation and management initiatives (Barbosa et al. 2004), since this very abundance seems to be responsible for much of the biodiversity and the functioning of the restinga ecosystem (Dias et al. 2005, 2006). Therefore, in order to deal with this issue, we decided to monitor two additional reproductive seasons and combine three approaches. 1) We increased the detail of pollinator observation and also of floral features to assess automimicry. 2) We increased the number of areas to observe fruit and seed formation. Since previous studies have described the existence of variation in terms of percent ground covered by this open restinga vegetation (low, intermediate and high; Pimentel 2002; Sampaio et al. 2005), we monitored aspects of reproductive biology in nine locations, each with a different vegetation cover. 3) In each of these locations we took leaf samples for DNA analysis and assessment of genetic diversity. These studies are the subject of an ongoing doctoral thesis (R.L. Martins, in preparation), but we point out below some of the main results. A large scale study of population structure and dynamics is also being carried out, but we still have no data available to discuss here.

5.3.1 Automimicry in *Clusia hilariana*

Clusia hilariana partly fits the pattern of automimicry proposed by Little (1983): (a) mimetic female flowers have features resembling male ones; (b) there is a higher floral output of male flowers than mimetic female flowers per area per day; (c) pollinators make shorter visits to the mimetic female flowers than to male flowers; and (d) male flowers produce pollinator reward (a high pollen production) and females do not (Fig. 5.1G). Additionally, the mimetic female flowers have a high stigmatic surface (Fig. 5.1H), which increases the chance of pollen arrival and fecundation. The only aspect that does not clearly fit Little's model is the fact that mimetic female flowers produce resin. This may imply that probably two alternative pollinating systems might co-occur. First, and mainly, a system driven by automimicry, where small Halictinae bees are the main pollinators of *C. hilariana*, aiming for pollen as reward. Second, and probably less effective, a system where Euglossini are the pollinators aiming for resin as a reward that is more abundantly available in male than in female flowers.

5.3.2 Effects of Population Spatial Distribution in Fruit Set

In order to test as to whether spatial distribution of males and females was somehow related to individual fruit set, we established circular plots of 35 m radius around each of 53 flowering females, within 9 sampling locations. For each plot we measured: (a) distance from each female to the nearest flowering male; (b) number of flowers of the nearest male; and (c) number of male and female flowers within the whole plot. We found that fruit set in *C. hilariana* was strongly related to the proximity of flowering female plants to flowering males ($t=-2.09$; $df=49$; $P<0.041$); however only 14 % of the fruit production might be explained by multiple correlation ($F_{3,49}=2.85$; $P<0.05$; $r^2=0.14$; $Y=16.9-0.31\ X_{a)}+0.02\ X_{b)}-1.92\ X_{c)}$). This is in harmony with the prediction of Bawa and Opler (1975) who proposed, based on body size and visiting behaviour, that smaller bees (such as the Halictinae) increase the chances of pollen flow when plant population is dense.

5.3.3 The Effects of Local Vegetation Cover on Fruit Set

The data discussed in Sect. 5.3.2 suggest an effect of the spatial arrangement of *C. hilariana* on its reproduction success. The findings of Pimentel (2002), indicating variation in terms of percent ground covered by the open restinga vegetation of the RJNP (low, intermediate and high; see also Sampaio et al. 2005), allowed us to investigate further the interference of spatial variation on reproductive success by analysing the correlations between vegetation cover and fruit set for plants of *C. hilariana*. We found considerable variation in fruit set of *C. hilariana* among nine locations, each with a different vegetation cover (Table 5.2). Further, we compared the patterns for two consecutive years. Fruit set was correlated to vegetation cover, however, while correlation was positive in 2004 ($F=7.83$; $P<0.05$; $R^2=0.52$), it turned out to be negative in 2005 ($F=5.76$; $P<0.05$; $R^2=0.45$). This seems to be related to the fact that while 7 out of 9 locations had a reduction in fruit set (more drastic in some than in others), 2 locations (10 and 11), conversely, had an increase in fruit set in 2005

Table 5.2. Percent vegetation cover (VC) of each of nine locations (numeration follows Pimentel 2002 and Sampaio et al. 2005) at the open restinga formation of RJNP, and respective fruit set (%) per location. Values in parenthesis are number of flowers assessed

Location	VC	2004	2005
1	34.6	27.4 (58)	16.2 (69)
2	33.0	30.4 (49)	17.8 (73)
3	38.2	45.7 (49)	19.4 (62)
4	56.4	46.0 (58)	10.0 (60)
5	28.5	13.3 (48)	11.6 (69)
6	29.5	24.4 (56)	17.7 (62)
8	37.0	28.7 (50)	15.9 (63)
10	20.0	27.0 (50)	33.3 (55)
11	28.0	15.2 (49)	32.2 (60)

despite lower rainfall. This dramatic difference in behaviour between years is puzzling, very difficult to interpret at this point, and obviously calls for further data collection. It is beyond the scope of this review to explore this matter any deeper, but it suffices to our goals in this chapter to highlight two points that emerge from this preliminary data: 1) spatial variation in fruit set is very high; and 2) vegetation cover alone does not explain this variation.

5.3.4 Population Genetics

Despite the scarcity of reports on the development, characterization and use of microsatellite loci in tropical plant species, Hale et al. (2002) had already characterized 13 polymorphic loci developed from *C. minor* (5) and *C. nemorosa* (8) and tested the transferability to 17 different *Clusia* species and assessed the degree of polymorphism only to *C. minor* and *C. ducu* Benth. Here, we tested the transferability of these same microsatellite markers to *C. hilariana* and assessed the degree of polymorphism for the amplified loci.

We collected leaves of 38 *C. hilariana* plants at the nine different locations described above at RJNP in order to perform microsatellite (SSR) analysis aiming to evaluate genetic diversity at the population level. DNA was extracted from leaves dried with silica gel by a modified CTAB method (Ferreira and Grattapaglia 1998). For all experiments, loci amplifications were performed as described by Margis et al. (2002) and sized by comparison to a 30–300-bp AFLP DNA ladder (Invitrogen).

Eight out of 13 primers pairs tested amplified loci of *C. hilariana*. This was possibly due to genome sequence homology, more specifically by the conservation of microsatellite flanking regions among closely related species (Dayanadan et al. 1997; White and Powell 1997). *Clusia hilariana* amplified

Table 5.3. Polymorphism information for *Clusia hilariana* and two other *Clusia* species studied by Hale et al. (2002) using microsatellite loci isolated from *C. minor* (Clm) and *C. nemorosa* (Cln) – = no amplification. Hale et al. (2002) data can be assessed in GenBank by numbers AY095353 to AY095365. Note that all products of amplification using primers from Clm for *C. hilariana* have a size range smaller than that found for the other 17 species, while those using primers from Cln fit on the same size range

Locus	Repeat in clone	Primer (5′ to 3′)	Hale et al. (2002)			This study	
			Size range (bp) for 17 *Clusia* species	No. of alleles		Size range (bp) for *C. hilariana*	No. of alleles
				C. minor (n=5)	*C. ducu* (n=5)		*C. hilariana* (n=38)
Clm1	$[TC]_{28}$	F: TCCAAAGGTATGCATCAGAGG R: GCAGAAACATGCACTCACAAA	166–262	6	6	–	–
Clm2	$[AG]_5AA[AG]_{13}$	F: GATCAACGTGCAGGAGCTAA R: CACTGTTTGCACAATCTTGCT	172–200	4	2	82–85	2
Clm3	$[CT]_{11}GTC[AC]_9$	F: CCTCTCTCTCCCACCTCCTT R: TTGGCTTCTTTCTTTTGCAG	99–151	4	–	–	–
Clm4	$[TC]_4TAC[TC]_{10}$	F: AAGTTTTGGACCCGATTCGT R: GGCGATGTCATCACATTTG	169–241	1 (5)	2	98	1
Clm5	$[GT]_{10}$	F: ACAATGGTCCTGCTCTGAGG R: CAAGAGCTGCTATCCCTGCTA	185–195	3	2	70	1
Cln1	$[GA]_{10}[GA]_4$	F: AGGCTGAGAGTTCCAACACG R: TCTTACCTTGAAGAATCTTTTCCTT	211–279	–	–	–	–
Cln2	$[TC]_{18}$	F: TGGTTGACCCCTCTAACAGTG R: GCAAACAACTCAATCAGAATGC	211–249	4	3	260–270	3

Table 5.3. (*Continued*)

Locus	Repeat in clone	Primer (5′ to 3′)	Hale et al. (2002)			This study	
			Size range (bp) for 17 *Clusia* species	No. of alleles		Size range (bp) for *C. hilariana*	No. of alleles
				C. minor (n=5)	*C. ducu* (n=5)		*C. hilariana* (n=38)
Cln3	$[TC]_6TT[TC]_{11}$	F: AGCAATGGTGAACCCGATAA R: GATCTTTATGTCAAACCAAATATACTG	136–168	3	4	137–145	2
Cln5	$[AG]_{29}A_4[AG]_5$	F: ATTGAACAAGCTGGGCACTC R: TCACCTTGGGCCTTTCTCTA	128–255	2	2	–	–
Cln6	$[TA]_4T_8[AC]_9[AT]_7$	F: GATCTCGGGTCAACCAACA R: TGATGCTACAATATAAACGAATGG	165–239	2	2	180	1
Cln7	$[TC]_{13}$	F: CTTACGCCGAGGTTGAAGTC R: GCTCACCAGCCTGGAAAATA	189–231	2	3	224	1
Cln8	$[AG]_{11}$	F: ATTTCCCGGAAGTTACATGA R: CACCACCGTGTAAGGGTTTT	175–233	5	6	180	1
Cln9	$[TC]_{14}T_9$	F: TTGGAAGTGAAAGGAATCCAA R: TCATAAATGAGGGGCAGGAC	137–247	3	5	–	–
Total	number of alleles			39	37		12

the least number of loci when compared with the 17 *Clusia* species tested by Hale et al. (2002), which might be an indication that our study species bears some phylogenetic distance from the 17 others. However, the data in Table 5.3 suggests that *C. hilariana* is indeed closer related to *C. nemorosa* than to *C. minor*, as shown by the phylogenetic trees of Vaasen et al. (2002) and Gustafsson et al. (2002) (Chap. 6) based on ITS sequence analysis. More importantly, from a population ecology viewpoint, we counted the number of alleles per loci amplified by the heterologous primers, and found a reduced polymorphism. Only *cln2, cln3*, and *clm3* were polymorphic displaying at maximum three alleles (*cln2*). These results suggest that *C. hilariana* might have a low genetic diversity in our site, at least when compared to *C. minor* and *C. ducu* studied by Hale et al. (2002). This is in agreement with the results described in Chap. 7, which also investigated genetic variation at the population level for *C. hilariana* with molecular markers. However, since the primers used in both studies were not primarily developed for *C. hilariana* specifically, the low diversity found could result from amplification of low diverse SSR regions of the DNA of our study species. Therefore, in order to reach a final conclusion about the genetic diversity of *C. hilariana* at our study site, we are currently performing new analysis using AFLP dominant markers that are independent of previous sequence knowledge and produce a larger number of information able to differentiate individuals even within a population (Kremer et al. 2005). To test this pattern further, we will compare the genetic diversity of the whole population of plants sampled for our nine sampling sites (n=90) with that of the local population of one sampling site (n=45) where the microsatellite analysis indicated the highest genetic diversity.

5.3.5 *Clusia hilariana*: A Synthesis of Ongoing Studies

In short, it appears that the high abundance of *C. hilariana* might be achieved by a combination of successful sexual reproduction in odd years with effective asexual reproduction. Eventual success in sexual reproduction is most likely related to automimicry properties, which, in the case of this plant highlights the importance of pollen as reward instead of resin. Once fruits are set and seeds dispersed, tank bromeliads will nurse seedlings and are apparently responsible for much of the success of regeneration and recruitment of seed-originated young plants. Although further confirmation is needed, there are indications that genetic diversity might be rather low, further confirming a possible important role for this type of reproduction in sustaining the high local abundance of *C. hilariana* at the RJNP.

This broad picture, however, hides processes at smaller spatial and temporal scale that account for a very large spatio-temporal variation in reproductive success of this species. For instance, spatial distribution of *C. hilariana*, particularly the distance from males to females, is inversely related to fruit set, i.e. the shorter the distance between plants of different genders, the higher the fruit set. At the community level, distinct overall vegetation cover is related to local fruit set, but whether these variables correlate positively or negatively depends on the sampling year.

5.4 Final Remarks

The understanding about the reproductive biology of the genus *Clusia* is still in its infancy, when we consider that less than 10 % of the species were studied in this respect, and the scales of space and time covered by such studies were always reduced. However, it is quite clear at this point that variation of possible behaviours is very high, and comprises different combinations of features per species, all of which are relatively rare in nature, such as dioecy, resin production, floral mimicry, and to a lesser extent agamospermy. This high diversity and peculiarity of reproductive strategies are matched by the high diversity on floral morphology, discussed by Gustafsson et al. in Chap. 6.5. It contrasts with uniformity of leaf morphology ("one morphotype") discussed by Lüttge in Chap. 2.1.

The case of *C. hilariana* calls for the relevance of enhancing spatio-temporal sampling scales of reproductive biology studies, and this is probably true not only to the *Clusia* genus. Moreover, it also shows how reproductive biology may provide essential information to deepen our understanding of processes operating at community and ecosystem levels and vice versa. Ecology definitely benefits from integration of hierarchies and scales (Pickett et al. 1994). This plant is an exceptional curiosity: dioecious, floral mimetic, resin producer, CAM and originally a possible migrant from rain forest canopies, later to become the dominant terrestrial plant in the restinga landscape of northern Rio de Janeiro (see Chap. 3), apparently by combining regular asexual reproduction with successful sexual reproduction only in odd years, assisted by tank-bromeliads which nurse their seeds and seedlings. That such an exception might sustain biodiversity and ecosystem processes in these restingas is a marvel of nature, fortunately well-preserved within the boundaries of the Restinga de Jurubatiba National Park.

Acknowledgments. We thank Mats Gustafsson and Ulrich Lüttge for important suggestions and additions to our manuscript; the Brazilian Research Council (CNPq) for productivity grants to TW, RM and FRS and a doctoral grant to RLM. The studies on *Clusia hilariana* have been receiving long standing financial support of the Brazilian Long Term Ecological Research Project (LTER-PELD, site 5, CNPq) and of the Research Center of the Brazilian Oil Company (Cenpes-Petrobras), and benefit from the kind permission and logistic support of the Brazilian Forestry Service (IBAMA) and the Restinga de Jurubatiba National Park.

References

Antonovics J, Levin DA (1980) The ecological and genetic consequences of density-dependence regulation in plants. Ecology 77:1779–1790

Armbruster WS (1984) The role of resin in Angiosperm pollination: ecological and chemical considerations. Am J Bot 71:1149–1160

Armbruster WS (1986) Reproductive interactions between sympatric *Dalechampia* species: are natural assemblages "random" or organized? Ecology 67:522–533

Armbruster WS, Herzig AL (1984) Partitioning and sharing of pollinators by four sympatric species of *Dalechampia* (*Euphorbiaceae*) in Panama. Ann Missouri Bot Gardens 71:1–16

Armbruster WS, Steiner KE (1992) Pollination ecology of four *Dalechampia* species (*Euphorbiaceae*) in Northern Natal, South Africa. Am J Bot 79:306–313

Armbruster WS, Webster GL (1979) Pollination of two species of *Dalechampia* (Euphorbiaceae) in Mexico by Euglossine Bees. Biotropica 11:278–283

Armbruster WS, Webster GL (1981) Sistemas de polinização de duas espécies simpátricas de *Dalechampia* (Euphorbiaceae) no Amazonas, Brasil. Acta Amazonica 11:13–17

Armbruster WS, Webster GL (1982) Divergent pollination systems in sympatric species of South American *Dalechampia* (Euphorbiaceae). Am Mid Nat 108:325–337

Armbruster WS, Webster GL, Thomas PC (1992) Pollination of two sympatric species of *Dalechampia* (Euphorbiaceae) in Suriname by male Euglossine Bees. Am J Bot 79: 1374–1381

Baker HG (1976) "Mistake" pollination as a reproductive system with special reference to the *Caricaceae*. In: Burley J, Styles BT (eds) Tropical trees: variation, breeding and conservation. Academic Press, London, pp161–169

Barbosa FAR, Scarano FR, Sabará MG, Esteves FA (2004) Brazilian LTER: Ecosystem and biodiversity information in support of decision-making. Environ Monit Assess 90:121–133

Bawa KS (1980a) Evolution of dioecy in flowering plants. Ann Rev Ecol Syst 11:15–39

Bawa KS (1980b) Mimicry of male by female flowers and intrassexual competition for pollinators in *Jacaratia dolichaula* (D. Smith) Woodson (Caricaceae) Evolution 34:467–474

Bawa KS, Opler PA (1975) Dioecism in tropical forest trees. Evolution 29:167–179

Bawa KS, Bullock SH, Perry DR, Coville RE, Grayum MH (1985) Reproductive biology of tropical lowland rain forest trees. II. Pollination systems. Am J Bot 72:346–356

Bittrich V (1994) Biologia da polinização de algumas espécies de *Clusia* da América do Sul. In: Resumos do XLV Congresso Nacional de Botânica. Sociedade Brasileira de Botânica, São Leopoldo, pp 180–181

Bittrich V, Amaral MCE (1996a) Flower morphology and pollination biology of some *Clusia* species from the Gran Sabana (Venezuela). Kew Bull 51:681–694

Bittrich V, Amaral MCE (1996b) Pollination biology of *Symphonia globulifera* (Clusiaceae). Plant Syst Evol 200:101–110

Bittrich V, Amaral MCE (1997) Flower biology of some *Clusia* species form Central Amazonia. Kew Bull 52:617–635

Burd M (1994) Bateman's Principle and plant reproduction: the role of pollen limitation in fruit and seed set. Bot Rev 60:83–139

Carmo RM, Franceschinelli EV (2002) Polinização e biologia floral de *Clusia arrudae* Planchon et Triana (Clusiaceae) na Serra da Calçada, Município de Brumadinho, MG. Revta Bras Bot 25:351–360

Charlesworth B, Charlesworth D (1978) A model for the evolution of dioecy and gynodioecy. Am Nat 112:975–997

Cirne P, Scarano FR (2001) Resprouting and growth dynamics after fire of the clonal shrub *Andira legalis* (Leguminosae) in a sandy coastal plain in southeastern Brazil. J Ecol 89:351–357

Correia CMB (1998) Estrutura de populações, associação entre quatro espécies-chaves e formação de moitas em restinga de *Clusia*, no norte fluminense. MSc Dissertation, Universidade Federal do Rio de Janeiro, Rio de Janeiro, Brasil

Correia MCR (1983) Contribuição ao estudo da biologia floral e do sistema de reprodução de *Clusia fluminensis* Pl.&Tr. (Guttiferae). MSc Dissertation. Universidade Federal do Rio de Janeiro, Rio de Janeiro, Brasil

Correia MCR, Ormond WT, Pimenta ML (1989) Biologia da reprodução de espécies de *Clusia* da Restinga de Marica – RJ, Brasil. Resumos do XL Congresso Nacional de Botânica, Cuiabá, MT, Brasil, p 100

Correia MCR, Ormond WT, Pinheiro MCB, Lima HA (1993) Estudos da biologia floral de *Clusia criuva* Camb. um caso de mimetismo. Bradea 24:209–219

Correia MCR, Ormond WT, Pinheiro MCB, Lima HA (1999) Biologia da reprodução de *Clusia lanceolata* Camb.. Hoehnea 26:61–73

Crawley MJ (1997) Sex. In: Crawley MJ (ed) Plant ecology. Blackwell, Oxford, pp 156–213

Croat T (1978) Flora of Barro Colorado Island. Stanford University Press, Stanford

Cruz ND, Boaventura YMS, Sellito YM (1990) Cytological studies of some species of the genus *Clusia* L. (Guttiferae). Rev Bras Genét 13:335–345

Darwin C (1877) The different forms of flowers on plants of the same species. John Murray, London

Dayanandan S, Bawa KS, Kesseli R (1997) Conservation of microsatellites among tropical trees (Leguminosae). Am J Bot 84:1658–1663

Dias ATC, Zaluar HLT, Ganade G, Scarano FR (2005) Canopy composition influencing plant patch dynamics in a Brazilian sandy coastal plain. J Trop Ecol 21:343–347

Dias ATC, de Mattos EA, Vieira SA, Azeredo JV, Scarano FR (2006) Aboveground biomass stock of native woodland on a Brazilian sandy coastal plain: estimates based on the dominant tree species. For Ecol Manage (in press)

Faria APG de, Matallana G, Wendt T, Scarano FR (2006) Low fruit set in the abundant dioecious tree *Clusia hilariana* (Clusiaceae) in a Brazilian restinga. Flora 201:606–611

Ferreira ME, Grattapaglia D (1998) Introdução ao uso de Marcadores Moleculares, 3rd edn. EMBRAPA-CENARGEN, Brasília

Flores S, Schemske DW (1984) Dioecy and monoecy in the flora of Puerto Rico and the Virgin Islands: ecological correlates. Biotropica 16:132–139

Freeman DC, Klikoff LG, Harper KT (1976) Differential resource utilization by the sexes of dioecious plants. Science 193:597–599

Freeman DC, Doust JL, El-Keblawy A, Miglia KJ, McArthur ED (1997) Sexual specialization and inbreeding avoidance in the evolution of dioecy. Bot Rev 63:65–92

Gustafsson MHG (2000) Floral morphology and relationships of *Clusia gundlachii* with a discussion of floral organ identity and diversity in the genus *Clusia*. Int J Plant Sci 161:43–53

Gustafsson MHG, Bittrich V (2003) Evolution of morphological diversity and resin secretion in flowers of *Clusia* L. (Clusiaceae): insights from ITS sequence variation. Nord J Bot 22:183–203

Gustafsson MHG, Bittrich V, Stevens PF (2002) Phylogeny of *Clusiaceae* based on *rbcl* sequences. Int J Plant Sci 163:1045–1054

Hale ML, Squirrell J, Borland AM, Wolff K (2002) Isolation of polymorphic microsatellite loci in the genus *Clusia* (Clusiaceae). Mol Ecol Notes 2:506–508

Hammel B (1986) New species of Clusiaceae form Central America with notes on *Clusia* and synonymy in the tribe Clusiae. Selbyana 9:112–120

Harris JG, Harris MW (2001) Plant identification terminology: an illustrated glossary, 2nd edn. Spring Lake Publishing, Spring Lake

Janzen DH (1971) Euglossine bees as long-distance pollinators of tropical plants. Science 171:203–205
Kearns CA, Inouye DW (1993) Techniques for pollination biologists. University Press of Colorado, Niwot
Kremer A, Caron H, Cavers S, Colpaert N, Gheysen G, Griebel R, Lemes M, Lowe AJ, Margis R, Salgueiro F (2005) Monitoring genetic diversity in tropical trees with multilocus dominant markers. Heredity 95:274–280
Larson BMH, Barrett SCH (2000) A comparative analysis of pollen limitation in flowering plants. Biol J Linn Soc 69:503–520
Little RJ (1983) A review of floral food deception mimicries with comments on floral mutualism. In: Jones CE, Little RJ (eds) Handbook of experimental pollination biology. Scientific and Academic Editions, New York, pp 294–309
Lloyd DG (1976) The transmission of genes via pollen and ovules in gynodioecious angiosperms. Theor Popul Biol 9:229–319
Lloyd DG (1982) Selection of combined versus separate sexes in seed plants. Am Nat 120:571–585
Lokvam J, Bradock JF (1999) Anti-bacterial function in the sexually dimorphic pollinator rewards of *Clusia grandiflora* (Clusiaceae). Oecologia 119:534–540
Lopes AV, Machado IC (1998) Floral biology and reproductive ecology of *Clusia nemorosa* (Clusiaceae) in northeastern Brazil. Plant Syst Evol 213:71–90
Macedo MV, Monteiro RF (1987) Germinação e desenvolvimento de plântulas em tanque de *Neoregelia cruenta* (Bromeliaceae) na Restinga de Barra de Maricá, RJ. Simpósio sobre Ecossistemas da Costa Sul e Sudeste Brasileira 2:188–190
Maguire B (1966) The genus *Clusia* (Guttiferae) in Suriname. Acta Bot Neerl 15:63–75
Maguire B (1976) Apomixis in the genus *Clusia* (Clusiaceae). A preliminary report. Taxon 25:241–244
Maguire B (1978) Notes on the *Clusiaceae*. Chiefly of Panama. II. Phytologia 38:203–214
Maguire B, Wurdack JJ (1961) The botany of the Guayana highland. Part III: Anatomy of Guayana Mutisieae. Part II. New York Botanical Garden Press, New York
Margis R, Felix D, Caldas JF, Salgueiro F, Araujo DSD, Breyne P, van Montagu M, de Oliveira DE, Margis-Pinheiro M (2002) Genetic differentiation among three neighboring Brazil-cherry (*Eugenia uniflora* L.) populations within the Brazilian Atlantic rain forest. Biodiversity Conserv 11:149–163
Mariz G (1974) Chaves para as espécies de *Clusia* nativas no Brasil. Memórias do Instituto de Biociências da Universidade Federal de Pernambuco, Brasil 1:249–314
Martins RL, Oliveira PEAM (2003) RAPD evidence for apomixis and clonal population in *Eriotheca* (Bombacaceae). Plant Biol 5:338–340
Matallana G, Wendt T, Araujo DSD, Scarano FR (2005) High abundance of dioecious plants in a tropical coastal vegetation. Am J Bot 92:1513–1519
Mesquita RCG, Franciscon CH (1995) Flower visitors of *Clusia nemorosa* G.F.W. Meyer (Clusiaceae) in an Amazonian White-Sand Campina. Biotropica 27:245–257
Pickett STA, Kolasa J, Jones CG (1994) Ecological understanding. Academic Press, New York
Pimentel MCP (2002) Variação espacial na estrutura de comunidades vegetais da Formação Aberta de *Clusia* no Parque Nacional da Restinga de Jurubatiba. DSc Thesis, Universidade Federal do Rio de Janeiro, Rio de Janeiro, Brasil
Pipoly JJ, Kearns DM, Berry PE (1998) *Clusia*. In: Berry PE, Holst BK, Steyermark JA, Yatkievych K (eds) Flora of the Venezuelan Guayana; vol 4. Caesalpiniaceae – Ericaceae. St Louis: Missouri Botanical Garden Press. Missouri, pp 248–295
Ramírez W, Gómez LD (1978) Production of nectar and gums by flowers of *Monstera deliciosa* (Araceae) and some species of *Clusia* (Guttiferae) collected by New Word *Trigona* bees. Brenesia 14/15:407–412

Renner SS, Ricklefs RE (1995) Dioecy and its correlates in the flowering plants. Am J Bot 82:596–606

Roubik DW (1992) Ecology and natural history of tropical bees. Cambridge University Press, Cambridge

Sakagami SF, Inoue J, Yamane S, Salmah S (1989) Nests of the myrmecophilous stingless bee *Trigona moorei*: how do bees initiate their nest within and arboreal ant nest? Biotropica 21:256–274

Sakai AK, Wagner LW, Ferguson DM, Herbst DR (1995a) Origins of dioecy in the Hawaiian flora. Ecology 76:2517–2529

Sakai AK, Wagner LW, Ferguson DM, Herbst DR (1995b) Biogeographical and ecological correlates of dioecy in the Hawaiian flora. Ecology 76:2530–2543

Sampaio MC, Picó FX, Scarano FR (2005) Ramet demography of a nurse bromeliad in Brazilian restingas. Am J Bot 92:674–681

Scarano FR (2002) Structure, function and floristic relationships of plant communities in stressful habitats marginal to the Brazilian Atlantic rain forest. Ann Bot 90:517–524

Scarano FR, Cirne P, Nascimento MT, Sampaio MC, Villela D, Wendt T, Zaluar HLT (2004) Ecologia Vegetal: integrando ecossistema, comunidades, populações e organismo. In: Rocha CFD, Esteves FA, Scarano FR (eds) Pesquisas de Longa Duração na Restinga de Jurubatiba: Ecologia, História Natural e Conservação. RiMa Editora, São Carlos, pp 77–97

Scarano FR, Duarte HM, Franco AC, Geßler A, De Mattos EA, Nahm M, Rennenberg H, Zaluar HLT, Lüttge U (2005) Ecophysiology of selected tree species in different plant communities at the periphery of the Atlantic Forest of SE-Brazil I. Performance of three different species of *Clusia* in an array of plant communities. Trees 19:497–509

Skutch AF (1971) A naturalist in Costa Rica. University of Florida Press, Gainesville

Sutherland S, Delph LF (1984) On the importance of male fitness in plants: patterns of fruit-set. Ecology 65:1093–1104

Thompson DI, Barrett SCH (1981) Selection of outcrossing, sexual selection, and evolution of dioecy in plants. Am Nat 118:443–449

Thompson DI, Edwards TJ (2001) Breeding biology, resource partitioning and reproductive effort of a dioecious shrub, *Clutia pulchella* L. (*Euphorbiaceae*). Plant Syst Evol 226:13–22

Vaasen A, Begerow D, Lüttge U, Hampp R (2002) The genus *Clusia* L.: molecular evidence for independent evolution of photosynthetic flexibility. Plant Biol 4:86–93

Vicentini A, Fischer EA (1999) Pollination of *Moronobea coccinea* (Clusiaceae) by the golden-winged parakeet in the Central Amazon. Biotropica 31:692–696

White GM, Powell W (1997) Cross-species amplification of SSR loci in the Meliaceae family. Mol Ecol 6:1195–1197

Wiens D (1978) Mimicry in plants. In: Hecht MK, Steere WC, Wallace B (eds) Evolutionary biology, vol.11. Plenum Publishing, New York

Zaluar HLT (2002) Dinâmica da vegetação em restingas abertas fluminenses: uma aproximação através da interação entre plantas. DSc Thesis, Universidade Federal do Rio de Janeiro, Brasil

6 Diversity, Phylogeny and Classification of *Clusia*

MATS H. G. GUSTAFSSON, KLAUS WINTER, and VOLKER BITTRICH

6.1 Taxonomic Position and Delimitation

Clusia L., with over 300 species, is one of the largest genera of the Clusiaceae (Guttiferae). According to a recent classification system (Stevens 2005), the family comprises the subfamilies Clusioideae and Kielmeyeroideae. In earlier classifications it has often also included *Hypericum* L. and related genera, the Hypericoideae (Engler 1925; Thorne 1976 [using the name Hypericaceae]; Cronquist 1981). There is, however, growing evidence that the Hypericoideae do not form a monophyletic group with other Clusiaceae (Gustafsson et al. 2002; Davis et al. 2005), and in, e.g., the classification system by P. F. Stevens (Stevens 2006), they are treated as a separate family, Hypericaceae. In the following, the name Clusiaceae is therefore used in the narrow sense, excluding *Hypericum* and its relatives.

Clusiaceae comprises two well-defined monophyletic subfamilies, Kielmeyeroideae and Clusioideae. The type genus *Clusia* itself obviously belongs to the subfamily Clusioideae, and tribe Clusieae. The latter is a well-supported, strictly Neotropical group characterized by opposite entire leaves (like the whole subfamily Clusioideae), absence of bud-scales, predominant dioecy, non-fasciculate androecia and fleshy capsules with arillate seeds (Engler 1925; Stevens 2006). Apart from *Clusia*, the largest genus in the tribe, generic delimitation is somewhat problematic. Commonly recognized are *Tovomita* Aubl. (ca. 25 species), *Chrysochlamys* Poepp. (ca. 55 spp.), *Dystovomita* D'Arcy (4 spp.), and sometimes also *Tovomitopsis* Planch. et Triana (2–3 spp.). The sister-group of *Clusia* is to be found among these genera, but still has not been identified with certainty.

Recently, several small genera have been found to have phylogenetic positions nested inside *Clusia*, i.e. *Decaphalangium* Melchior (1 sp.), *Havetia* Kunth. (1 sp.), *Havetiopsis* Planch. et Triana (7 spp. or, more likely, less), *Oedematopus* Planch. et Triana (ca. 15 spp.), *Pilosperma* Planch. et Triana (1 sp.) and *Quapoya* Aubl. (7 spp.). This is based both on morphological cladistic

Ecological Studies, Vol. 194
U. Lüttge (Ed.)
Clusia: A Woody Neotropical Genus
of Remarkable Plasticity and Diversity

analyses (Stevens 2006) and molecular studies (all genera except *Pilosperma* sampled; Gustafsson and Bittrich 2003 and unpublished). Most of the necessary recombinations have already been made (Pipoly 1997; Pipoly and Cogollo 1998). The resulting *Clusia* sensu lato is a well supported taxonomic unit, both based on molecular evidence (Gustafsson and Bittrich 2003) and morphology, although it is difficult to point at an uncontradicted morphological synapomorphy for the genus. Most consistent is probably seed length, which is 5 mm or less in *Clusia* while mostly 10 mm or more in other genera of Clusieae. The aril of the seed in these genera is usually vascularized, a condition not seen in *Clusia*. Another synapomorphy with very few, if any reversals, is the presence of a two- to many-layered hypodermis in the leaves (Vesque 1892). Characters occurring in most but not all *Clusia* species, while very rare or absent from the related genera, include hemi-epiphytic life form, long aerial adventitious roots (*Tovomita* has stilt roots, but not long aerial roots), coriaceous-subcarnose leaves with obscure venation, and fruits with more than one seed per carpel.

6.2 Molecular Phylogenetics

In recent years, several widely sampled phylogenetic studies of *Clusia* have been published (Vaasen et al. 2002; Gehrig et al. 2003; Gustafsson and Bittrich 2003). All of these studies have used ITS (Internal Transcribed Spacer, in nuclear ribosomal DNA) sequences for the phylogenetic reconstruction. The number of published ITS sequences of this region for *Clusia* now exceeds 120. Plastid DNA variation has also been investigated (Hale et al. 2004), but all four regions sequenced showed very low levels of variation. ITS is more variable and has proved to be rather informative regarding the interrelationships within *Clusia*.

6.2.1 Combining Published ITS Sequences in a New Analysis

For the present review, ITS sequences published by Gehrig et al. (2003) and Gustafsson and Bittrich (2003) were analyzed together. Outgroups from outside of tribe Clusieae (10 terminals in all), as well as one poor, incomplete sequence of *Chrysochlamys* (GenBank number AY145241), and a highly divergent sequence of *C. sipapoana* (Maguire) Pipoly (AJ312562), were excluded, and the number of samples per species was reduced to one. The remaining 99 sequences were aligned using the program DIALIGN (Morgenstern 1999). The resulting data matrix (indels were not coded) was analyzed cladistically, using PAUP 4b10 (Swofford 2002). A heuristic tree search was performed, and aborted when 130,000 most parsimonious trees (1135 steps, CI=0.47, RI=0.78) had been found. A jackknife analysis was also performed, as described by Gustafsson and Bittrich (2003). Figure 6.1 shows a single most parsimonious tree, indicating which branches are absent in the strict consensus tree and which have jackknife support.

6.2.2 Character Optimizations

Two characters were optimized onto the selected tree (Fig. 6.1) using MacClade 4.02 (Maddison and Maddison 2001) and PAUP 4b10: resin production in flowers, and CAM photosynthesis.

In the case of resin production the knowledge is fairly complete, as the character is easily observed even on herbarium specimens, and can sometimes be deduced from species descriptions. For most taxa not included by Gustafsson and Bittrich (2003), descriptions in Flora of Panama (D'Arcy 1980) and herbarium specimens were consulted.

Knowledge about the type of photosynthesis is far more patchy, and for many species a question mark had to be entered. None of the outgroup species have been investigated in this respect. In one optimization they were honestly coded as unknown, in another (shown in Fig. 6.1) they were assumed to lack CAM. Information was extracted from Tinoco Ojanguren and Vàzquez-Yanes (1983), Ting et al. (1985), Franco et al. (1990, 1992, 1994, 1996), Borland et al. (1992, 1998), Winter et al. (1992), Roberts et al. (1997), Lüttge (1999), Herzog et al. (1999), and Holtum et al. (2004). Species coded as "CAM present" are those with $\delta^{13}C$ values (see Sect. 8.2) less negative than –20‰, which is indicative of strongly expressed CAM (as seen in, e.g., *C. flava* Jacq., *C. hilariana* Schltdl., *C. fluminensis* Planch. et Triana, *C. major* L., *C. rosea* Jacq., and *C. uvitana* Pittier), and species with $\delta^{13}C$ more negative than –20‰ yet exhibiting nocturnal increases in tissue acidity indicative of weakly expressed CAM (*C. aripoensis* Britton, *C. croatii* D'Arcy, *C. cylindrica* Hammel, *Havetiopsis flexilis* Planch. et Triana=*C. flavida* (Benth.) Pipoly, *C. fructiangusta* Cuatrec., *C. lanceolata* Cambess., *C. lineata* Planch. et Triana, *C. minor* L., *C. odorata* Seem., *C. pratensis* Seem., *C. quadrangula* Bartlett, *C. valerioi* Standl., *C.* sp. D). Species coded as "CAM absent" have $\delta^{13}C$ values normally associated with C_3 photosynthesis (more negative than –20‰) and no evidence of nocturnal increases in tissue acidity (*C. coclensis* Standl., *C. cupulata* Maguire, *C. divaricata* Maguire, *C. liesneri* Maguire, *C. longipetiolata* Schery, *C.* aff. *multiflora*, *C. osseocarpa* Kunth, *C. palmana* Standl., *C. stenophylla* Standl., *C. torresii* Standl., *C.* sp. A, *C.* sp. E), or in which nocturnal levels of tissue acidity have not yet been studied (*C. amazonica* Planch. et Triana, *C. columnaris* Engl., *C. congestiflora* Cuatrec., *C. grandiflora* Splitg., *C. nemorosa* G. Mey., *C. triflora* Cuatrec., *C. salvinii* Donn. Sm., *C.* sp. B, C, F). Although the latter category is currently considered to be C_3, it cannot be excluded that future measurements of tissue acidity may reveal weakly expressed CAM in some of these species.

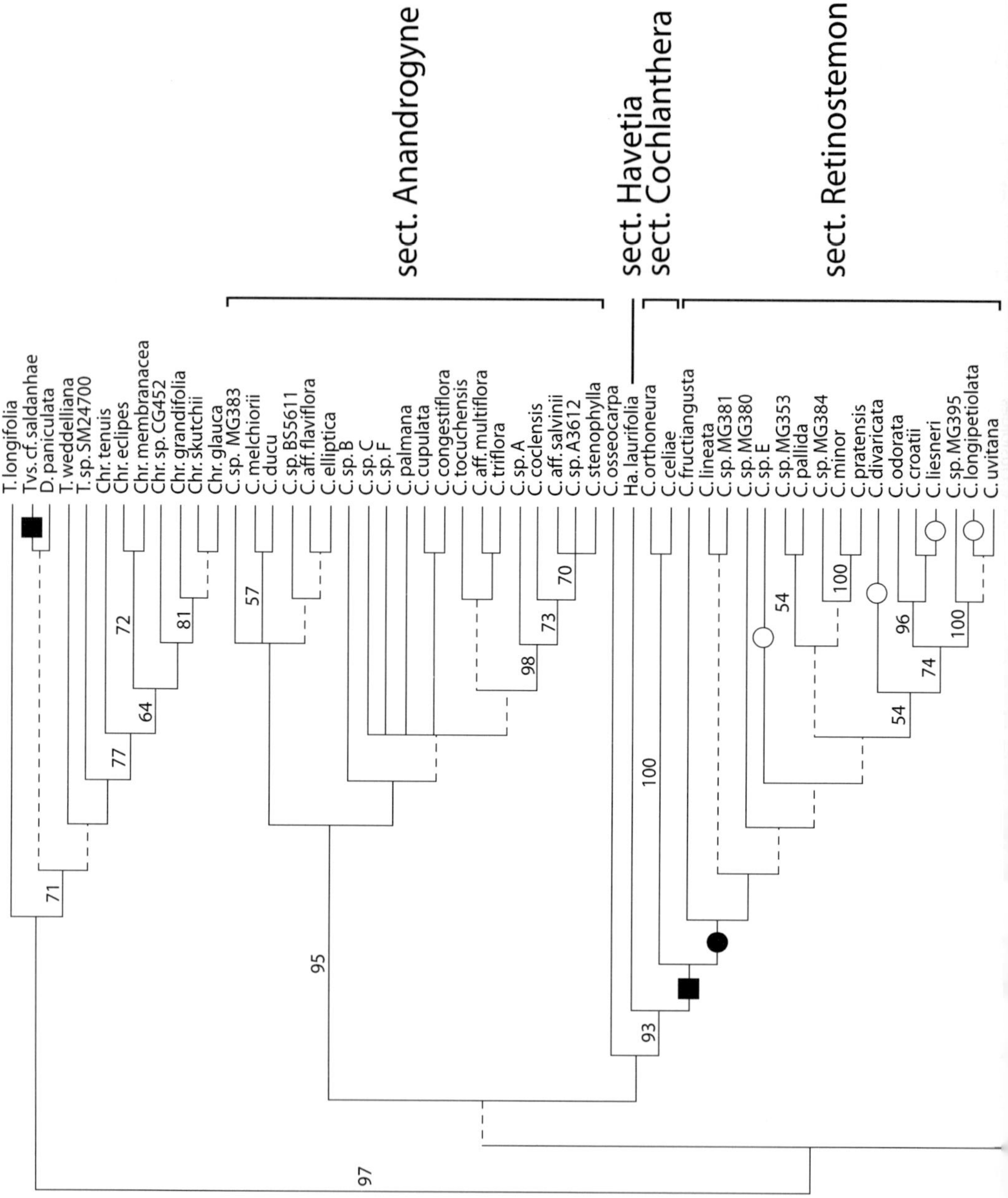
sect. Anandrogyne
sect. Havetia
sect. Cochlanthera
sect. Retinostemon
T. longifolia
Tvs. cf. saldanhae
D. paniculata
T. weddelliana
T. sp. SM24700
Chr. tenuis
Chr. eclipes
Chr. membranacea
Chr. sp. CG452
Chr. grandifolia
Chr. skutchii
Chr. glauca
C. sp. MG383
C. melchiorii
C. ducu
C. sp. BS5611
C. aff. flaviflora
C. elliptica
C. sp. B
C. sp. C
C. sp. F
C. palmana
C. cupulata
C. congestiflora
C. tocuchensis
C. aff. multiflora
C. triflora
C. sp. A
C. coclensis
C. aff. salvinii
C. sp. A3612
C. stenophylla
C. osseocarpa
Ha. laurifolia
C. orthoneura
C. celiae
C. fructiangusta
C. lineata
C. sp. MG381
C. sp. MG380
C. sp. E
C. sp. MG353
C. pallida
C. sp. MG384
C. minor
C. pratensis
C. divaricata
C. odorata
C. croatii
C. liesneri
C. sp. MG395
C. longipetiolata
C. uvitana
71
72
64
77
81
57
70
73
98
95
100
93
54
100
96
74
54
100
97

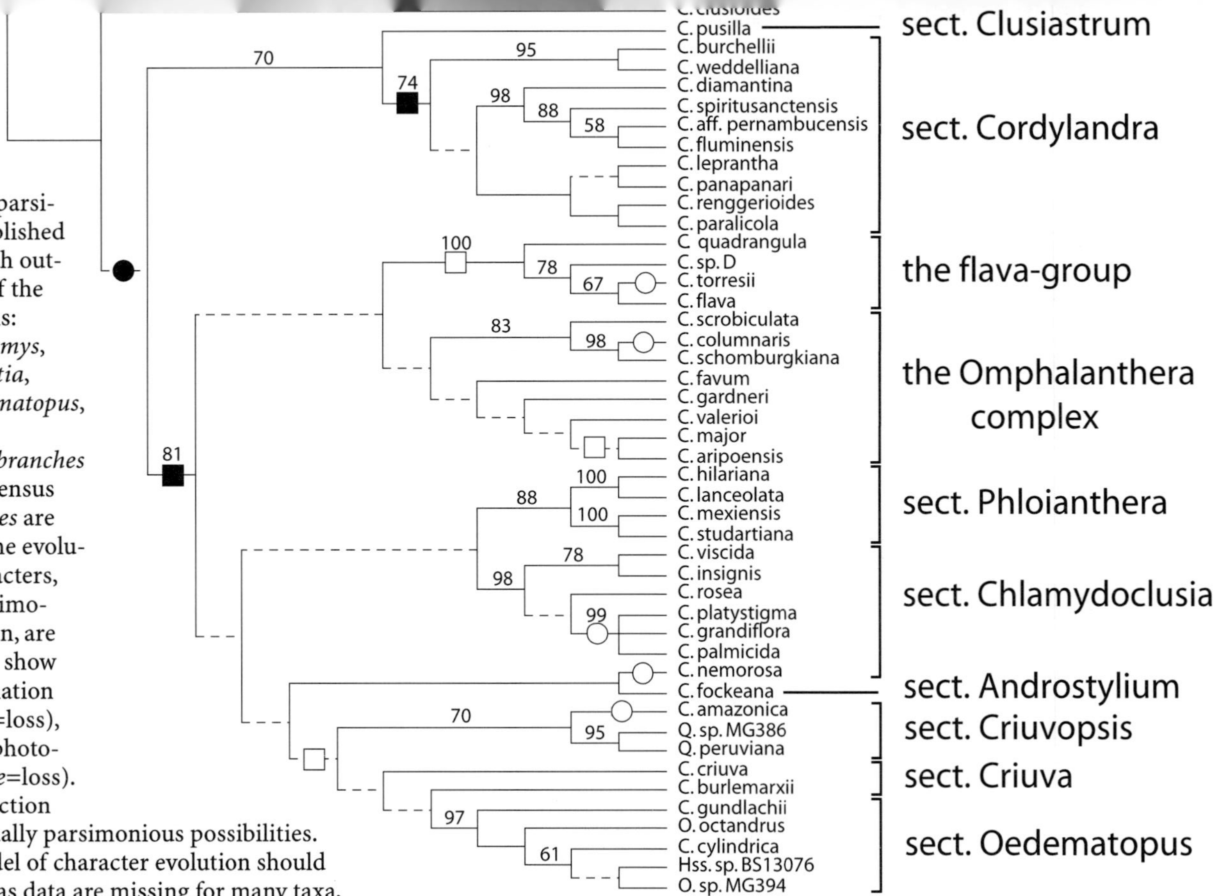

Fig. 6.1. One of many most parsimonious trees based on published ITS sequences of *Clusia*, with outgroups from other genera of the tribe Clusieae. Abbreviations: C.=*Clusia*, Chr.=*Chrysochlamys*, D.=*Dystovomita*, Ha.=*Havetia*, Hss.=*Havetiopsis*, O.=*Oedematopus*, Q.=*Quapoya*, T.=*Tovomita*, Tvs.=*Tovomitopsis*. *Dashed branches* are absent in the strict consensus tree. *Numbers above branches* are jackknife support values. The evolutionary history of two characters, reconstructed through parsimonious character optimization, are shown on the trees. *Squares* show the evolution of resin pollination (*black box*=gain, *white box*=loss), and *circles* represent CAM photosynthesis (*black*=gain, *white*=loss). In both cases, the reconstruction shown is one of several, equally parsimonious possibilities. In the case of CAM the model of character evolution should be considered preliminary, as data are missing for many taxa, and given that for many of the species coded as "CAM absent" this is based solely on $^{13}C/^{12}C$ ratios (meaning that weak CAM cannot be excluded with certainty; see text)

For each of the two characters, one optimization was selected from several, equally parsimonious possibilities (Fig. 6.1). For the resin-optimization, a scenario intermediate between ACCTRAN and DELTRAN (Swofford 2002) was selected. A "pure" ACCTRAN optimization is very similar, but shows only two origins (with a common origin for *Cordylandra* and subg. *Clusia*; with reversal in *C. pusilla*). Independent origin of resin in *Cordylandra* was considered more plausible given its chemical and physical peculiarities. A DELTRAN model would entail five origins (independently in the *Omphalanthera* complex, *Phloianthera* plus *Chlamydoclusia*, and the *C. fockeana* – *C. nemorosa* clade). The close morphological similarity between *C. nemorosa* and sect. *Chlamydoclusia* makes an independent origin of resin in these taxa seem unlikely.

In the case of CAM photosynthesis, a DELTRAN optimization was chosen. The ACCTRAN alternative is similar, but the whole clade *C. scrobiculata-columnaris-schomburgkiana* (Planch. et Triana) Benth. ex Engl. shows a reversal, and the clade *C. nemorosa* through *Havetiopsis* shows reversal with regain (considered a less plausible complexity) in the clade *C. criuva* through *Havetiopsis*. In Fig. 6.1, outgroups were assumed to lack CAM. If outgroups are coded as unknown with respect to CAM photosynthesis, the number of ambiguous branches increases, but the optimization shown in Fig. 6.1 is still valid (representing one of many maximally parsimonious solutions).

6.3 Species Diversity and Distribution

At the species level *Clusia* must be considered poorly known, at least in parts of its distribution area. Around 300 species are recognized presently (Pipoly et al. 1998), but as many as 100 species may remain to be described, many of which occur in the Andes.

The total distribution for the genus only slightly extends north and south of the tropics (Fig. 6.2). The northern limit is the Bahamas (*Clusia rosea* Jacq.), and to the south it reaches the state of Rio Grande do Sul, Brazil (*Clusia criuva* Cambess.). Species diversity is very unevenly distributed (Fig. 6.2), being highest in areas such as the Northern Andes, the Amazonian lowlands east of the Andes, the Pacific lowlands from NW Ecuador to Panama, and the Guayana Highlands. Least diversity is seen in the peripheral parts of the distribution area, in particular the Caribbean Islands (four species in Puerto Rico), Mexico and Southern Brazil (only *C. criuva* reaches south of the Tropic of Capricorn). The altitudinal range is from near sea level (e.g. *Clusia rosea* and *C. hilariana* Schltdl., in the Caribbean and SE Brazil respectively) to at least 3500 m altitude (*Clusia frigida* Cuatrec. from the Andes of Colombia).

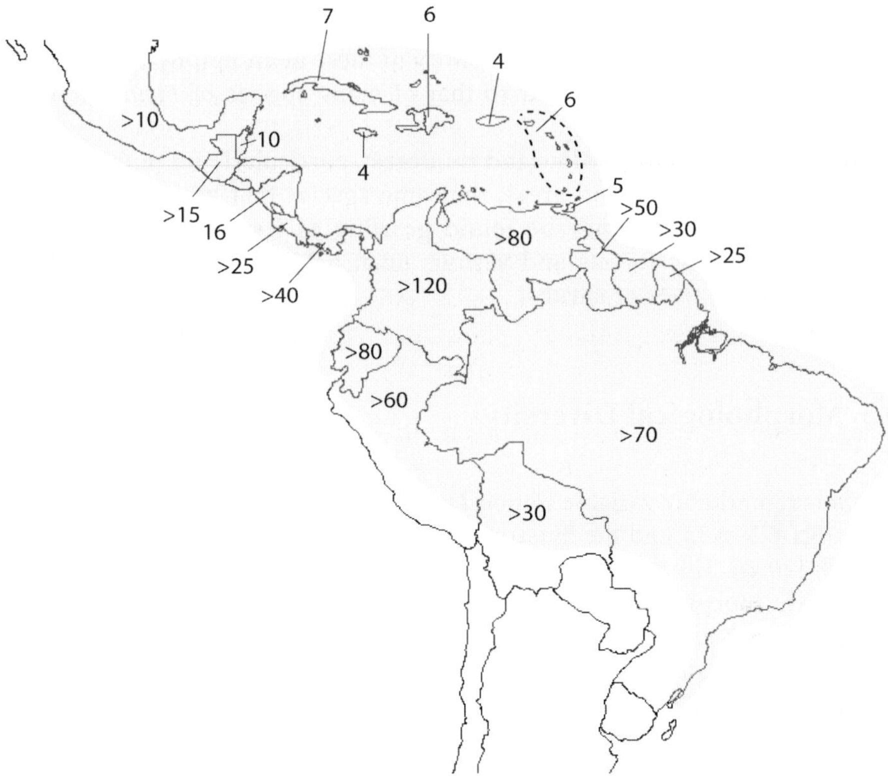

Fig. 6.2. Distribution map for *Clusia*. Estimates of species number are given for individual countries (and for Hispañola and the Lesser Antilles). Numbers are based on field- and herbarium studies and information from flora treatments, taxonomic revisions and checklists such as those by D'Arcy (1980), Liogier (1983), Pipoly et al. (1998), Jørgensen and León-Yánez (1999), and Hammel (2001). The map shows the approximate limits of the distribution; *Clusia* is absent from some regions within the *shaded area*, such as parts of the high Andes

6.4 Habitats and Habits

Habitats vary considerably, and include wet lowland forests, montane forests, paramo, open low forests on coastal sands ("restingas" of Brazil), dry scrub in interandean valleys, and sandstone and granite rocks (Sect. 9.1). *Clusia* species may be early colonizers (and more or less weedy), particularly on various nutrient poor soils, such as laterite or leached sand.

In lowland forests, most species grow predominantly as hemi-epiphytes in the form of shrubs or less commonly trees or climbers (Sect. 2.4). Hemi-epiphytes begin their life as true epiphytes up in a tree, and eventually establish contact with the ground by means of adventitious roots. Eventually, the adventitious roots may completely surround and encircle the "host" tree, which may die as a result. The life-history of these hemi-epiphytes becoming stranglers is remarkably similar to that of many species of *Ficus*, the well-known strangler figs.

In disturbed habitats in lowland rainforest, particularly on nutrient-poor soils such as in white-sand areas, the same species that grow as hemi-epiphytes in intact forest may be found growing as free-standing trees and shrubs. In montane forests and various open habitats, the growth form is commonly small trees or shrubs.

6.5 Morphological Diversity

Clusia is remarkably variable in floral morphology, as can be seen at a glance from Figs. 6.3–6.32, and the illustrations in Engler (1925) and Gustafsson and Bittrich (2003). The variation concerns size, degree of fusion and number of floral parts, morphology of individual organs, and the overall organization of the flower, such as the position of staminodes (central or peripheral) relative to functional stamens, and presence or absence of a pistillode in the male flowers.

Some species have a highly peculiar stamen morphology, with many unique traits. Often, stamens are fused into synandria of varying shape (Figs. 6.5, 6.7, 6.10, 6.12, and 6.13). This is particularly common in species that offer resin as a pollinator reward (see below), in which case the resin is produced in the androecium. In staminate flowers of some species, there is a clear division of labour between resiniferous staminodes and fertile, pollen-producing stamens (Figs. 6.10 and 6.11), while in other cases resin is produced by the fertile stamens (Figs. 6.5, 6.7, and 6.8). Anther morphology is often modi-

Figs. 6.3–6.17. Flowers of *Clusia*. ∅=diameter of flower. **6.3–6.14** Staminate flowers of *Clusia*: **6.3** *C. ducu* Benth. (sect. *Anandrogyne*), ∅ 1.2 cm; **6.4** *C. orthoneura* Standl. (sect. *Cochlanthera*; photo by G. Gehrlach), ∅ 5 cm; **6.5** *C. fructiangusta* Cuatrec. (sect. *Retinostemon*), ∅ 2.5 cm; **6.6** *C. pusilla* Steyermark (sect. *Clusiastrum*, photo by H. Balslev), ∅ 5 cm; **6.7** *C. leprantha* Mart. (sect. *Cordylandra*), ∅ 5.5 cm; **6.8** *C. schomburgkiana* (Planch. & Triana) Benth. ex Engl. (the "*Omphalanthera* complex"; photo by H. Balslev), ∅ 4 cm; **6.9** *C. major* L., sect. *Clusia* (part of the "*Omphalanthera* complex"; photo by A. Sloth), ∅ 2.5 cm; **6.10** *C. lanceolata* Cambess. (sect. *Phloianthera*), ∅ 3.5 cm; **6.11** *C. grandiflora* Splitg. (sect. *Chlamydoclusia*), ∅ 13 cm; **6.12** *C. fockeana* Miq. (sect. *Androstylium*), ∅ 2.7 cm; **6.13** *Quapoya* sp. (sect. *Criuvopsis*), ∅ 1 cm; **6.14** *Oedematopus* sp. (sect. *Oedematopus*), ∅ 7 mm. **6.15–6.17** Pistillate flowers of *Clusia*: **6.15** *C. ducu* (sect. *Anandrogyne*), ∅ 8 mm; **6.16** *C. fructiangusta* (sect. *Retinostemon*), ∅ 2.5 cm; **6.17** *C. amabilis* Maguire (sect. *Clusiastrum*), ∅ 5 cm

6.3
6.4
6.5
6.6
6.7
6.8
6.9
6.10
6.11
6.12
6.13
6.14
6.15
6.16
6.17

6.18
6.19
6.20
6.21
6.22
6.23
6.24
6.25
6.26
6.27
6.28
6.29
6.30
6.31
6.32

fied to a degree that it is difficult to recognize what represents an individual anther or theca (e.g., Figs. 6.5 and 6.12). Good examples of this are *C. schomburgkiana* and *C. columnaris* Engl., whose bizarre staminal morphology and pollination mechanisms were described by Bittrich and Amaral (1996). In contrast to the apparently fundamental differences in staminal morphology, molecular studies (Gustafsson and Bittrich 2003) have shown that the two species are almost identical in investigated DNA sequences (ITS), and it can be concluded that the morphological differences must have evolved rapidly.

In other floral traits, differences between *Clusia* species are less extreme, but still considerable. Flower diameter varies from ca. 0.5 to around 15 cm. Sepal number is variable and sometimes difficult to determine, as sepals often intergrade with bracteoles. Petals number from four to nine, and their colour can be white to yellow or green, pink to dark red or almost black. The ovary consists of 4–21 carpels, with sessile stigmata (Figs. 6.24, 6.26–6.32), or rarely with short styles (as long as the ovary in one unnamed species from the Andes of Ecuador; Fig. 6.25).

The fruit is invariably a fleshy, thick-walled, septifragal capsule, varying in size from pea to orange (Figs. 6.24–6.32). The number of seeds varies from one (Fig. 6.24) to many (Figs. 6.27 and 6.28) per carpel. Seeds are surrounded by an orange-coloured aril, which is very variable in morphology (Engler 1925). Endosperm is absent at maturity. The embryo is green and consists almost entirely of hypocotyl. Germination begins with elongation of the hypocotyl, and only later do the minute cotyledons expand.

In vegetative characters variation must be said to be moderate, at least compared to the situation for floral morphology, but leaf shape, presence of petiole, latex colour, distribution and morphology of laticiferous ducts in leaves are all variable characters useful for species identification, and sometimes for recognizing larger clades.

Figs. 6.18–6.32 Flowers and fruits of *Clusia*. ∅=diameter of flower or fruit. **6.18–6.23** Pistillate flowers of *Clusia*: **6.18** *C. leprantha* (sect. *Cordylandra*; one petal bent down), ∅ 6 cm; **6.19** *C. major* (sect. *Clusia*, part of the "*Omphalanthera* complex"; photo by A. Sloth), ∅ 2.5 cm; **6.20** *C. columnaris* Engl. ("*Omphalanthera* complex"), ∅ 3.5 cm; **6.21** *C. studartiana* C.M. Vieira & A.G. da Silva (sect. *Phloianthera*), ∅ 3 cm; **6.22** *C. rosea* Jacq. (sect. *Chlamydoclusia*), ∅ 9 cm; **6.23** *Havetiopsis flexilis* Planch. & Triana (sect. *Oedematopus*), ∅ 4 mm. **6.24–6.32** Fruits of *Clusia*: **6.24** *C. ducu* (sect. *Anandrogyne*; open fruits on the left), ∅ 7 mm; **6.25** *C.* sp. (sect. *Anandrogyne*), ∅ 1.5 mm; **6.26** *C.* sp. (sect. *Retinostemon*; immature fruit), ∅ 2.2 cm; **6.27** *C.* sp. (sect. *Retinostemon*; open fruit), ∅ 5.5 cm; **6.28** *C. leprantha* (sect. *Cordylandra*; open fruit; photo by S. Mori), ∅ 4.5 cm (before opening); **6.29** *C. paralicola* (sect. *Cordylandra*), ∅ 3.5 cm; **6.30** *C. major* (sect. *Clusia*, part of the "*Omphalanthera* complex"; photo by A. Sloth), ∅ 3.5 cm; **6.31** *C. grandiflora* (sect. *Chlamydoclusia*), ∅ 7 cm; **6.32** *C. criuva* Cambess. (sect. *Criuva*), ∅ 1.2 cm

6.6 Ecophysiological Variation

As is thoroughly described in Chaps. 8 and 9 of this volume, the genus *Clusia* shows a remarkable diversity in photosynthetic physiology. It is the only truly arborescent genus for which CAM photosynthesis is known, and CAM, C_3-CAM intermediate and C_3 species have been described (Tinoco Ojanguren and Vásquez-Yanes 1983; Franco et al. 1990; Borland et al. 1992; Lüttge 1996, 1999). CAM may be strongly (e.g. *Clusia rosea*) or weakly expressed (e.g. *Clusia minor*), and some species show considerable plasticity in the expression of CAM in relation to plant developmental stage and environmental stress (Sects. 8.1, 9.2, and 9.4; Winter et al. 1992; Lüttge et al. 1993). CAM would appear to be highly adaptive for a rainforest hemi-epiphyte, which begins its life in a periodically very dry micro-habitat on a tree branch, and only later establishes contact with the soil, which ensures a reliable supply of water (Ting et al. 1987).

From the distribution of CAM photosynthesis within *Clusia*, it can be concluded that it has arisen several times independently (Gehrig et al. 2003). If the present optimization (Fig. 6.1) is accepted, there were two independent origins of CAM, one in sect. *Retinostemon*, and another in the ancestor of a group that comprises sect. *Cordylandra* and subgenus *Clusia*. The optimization shown in Fig. 6.1 suggests that there has been no less than nine reversals, which would mean that CAM photosynthesis is a strongly homoplasious character. In some of the species involved, photosynthetic pathway classification is based solely on $\delta^{13}C$ values. Hence, the number of reversals could markedly decrease once detailed studies on tissue acidity are available for all species.

With the present state of knowledge it appears that the occurrences of CAM in *Clusia* are phylogenetically isolated. CAM photosynthesis is absent from other genera of Clusiaceae, as far as is known. This is not surprising given that these comprise mostly terrestrial trees or shrubs in wet to mesic forest. Exceptions are *Clusiella* Planch. et Triana (hemi-epiphytic like *Clusia*) and *Kielmeyera* Mart. (xeromorphic terrestrial of open, dry habitats), both belonging to the subfamily Kielmeyeroideae, and species of *Rheedia* L. (subfamily Clusioideae, tribe Garcinieae) such as the highly xeromorphic *R. ruscifolia* Griseb. from Cuba. Representatives of the two latter genera have been investigated and were found not to have CAM (U. Lüttge et al., unpublished). No examples of CAM are known from the families most closely related to the Clusiaceae, i.e., the Hypericaceae, Podostemaceae and Bonnetiaceae, although the last two families have probably never been investigated in this respect, being aquatics and difficult to access, respectively.

Apart from CAM, *Clusia* also shows other adaptations to (temporarily) dry conditions that are not seen in related genera. Leaves are generally much thicker (Sect. , Tables 2.1 and 2.2), and the hypodermis is two- to many-layered (as opposed to single-layered) in almost all investigated *Clusia* species (Sect. 2.2, Table 2.2; Vesque 1892).

6.7 Variation in Biological Interactions

Pollination mechanisms, pollinator rewards and pollinators, are all remarkably diverse in *Clusia* (see Chap. 5). Pollinator rewards include nectar, pollen and resins. Floral resin is a rare type of pollinator reward, known only from a few angiosperm genera (Armbruster 1984; Gustafsson et al. 2002; Gustafsson and Bittrich 2003). Of these, three belong to tribe Clusieae: *Clusia* (around half of the species resiniferous, with three independent origins and three reversals according to Fig. 6.1), *Tovomitopsis* (at least 1 species out of 2 or 3), *Chrysochlamys* (at least 1 species, *C. tenuifolia* Cuatrec., out of 55; Hammel 1999). Floral resin is also found in all species of *Clusiella* (7 spp.) of subfamily Kielmeyeroideae. Outside of Clusiaceae it is only known from *Dalechampia* of the Euphorbiaceae. Chemically, the resins of *Clusia* and *Tovomitopsis* flowers consist largely of polyisoprenylated benzophenones, compounds also found in latex (Oliveira et al. 1996) and present in most if not all Clusiaceae. The resin is collected by bees that use it in nest construction (Bittrich and Amaral 1996, 1997) and in the defense of the nest. Flowers offering pollen may be pollinated by beetles (Rodrigues et al. 1993) or bees (Bittrich and Amaral 1996), and nectariferous flowers are visited also by flies and wasps (Gustafsson and Bittrich 2003), lepidoptera (W. S. Armbruster, personal communication) and hummingbirds (Dziedzioch et al. 2003).

The fleshy, brightly orange-coloured aril of *Clusia* is assumed to be an adaptation to dispersal by birds. In *Clusia criuva*, birds are primary dispersers, and seeds are secondarily dispersed by ants (Passos and Oliveira 2002). Arboreal and perhaps other mammals may also act as seed-dispersers (spider monkeys observed by M. v. Roosmalen, personal communication).

6.8 Phylogeny and Sectional Classification

Groups supported by the ITS data are in many cases easily identified based on morphological characters and have sometimes been recognized at the sectional level in previous classifications (Planchon and Triana 1860a, b; Engler 1925; Gustafsson and Bittrich 2003). Around ten such groups are fairly well supported (Fig. 6.1). Morphological synapomorphies for these groups typically concern the androecium, the stigma morphology, and sometimes also characters of the fruit and leaves (often venation and course of latex canals). At higher hierarchical levels, relationships are generally less well supported.

There are, however, a few exceptions. One is the moderately supported sister-group relationship between sects. *Clusiastrum* and *Cordylandra*. Another example is the fairly well-supported group referred to as subgenus *Clusia* (Gustafsson and Bittrich 2003) or "Group I" (Gehrig et al. 2003) (*C. quadrangula* and down in Fig. 6.1). There may also be some weak support (seen in, e.g., the analysis by Gehrig et al. 2003) for a largely Andean and Central American clade comprising the species-rich sections *Anandrogyne* Planch. et Triana and *Retinostemon* Planch. et Triana and a few additional small groups. This "Cordillera-clade" is present in the single mp tree shown in Fig. 6.1.

In the following, the more important subgeneric groups, mostly recognized at the level of section, will be briefly presented.

6.8.1 *Clusia* sect. *Anandrogyne* Planch. et Triana

This may be the most species-rich section in the genus (Pipoly 1995) (Figs. 6.3, 6.13, 6.24, and 6.25). Based on molecular data, it is well supported and distinct from other groups, but the differences in DNA sequence between the species are small (Gustafsson and Bittrich 2003; Frederiksen 2005). The same result was obtained by Gehrig et al. (2003) who referred to them as "Group III". Diversity thus appears to result from a fairly recent radiation. In spite of this recent origin, the group is quite variable, especially in size and shape of flowers, fruits and leaves (Pipoly 1995; Gustafsson and Bittrich 2003; Frederiksen 2005). Stamens are numerous and almost always free (Fig. 6.3). There are usually 4–6 carpels with thick and angular stigmata that are sometimes borne on short styles (Fig. 6.25). Latex canals are often evident on the lower (abaxial) side of the leaves, as darker lines. Pollination biology is very poorly known, but based on scattered field observations, it appears that most species are either nectariferous, or produce copious amounts of pollen, indicating that these may be the pollinator rewards. Floral resin is, as far as is known, absent.

The distribution is predominantly Andean and Central American, with a few species in the Guayana Highlands and in Eastern Brazil. Most grow in montane forests, where they often form a significant component of the vegetation. The altitude record for the genus is held by a member of this section, and above ca. 2500 m no other *Clusia* (or Clusiaceae) occur. Most species are terrestrial and grow in moderately to very wet habitats. Although many species in this section have been investigated (Gehrig et al. 2003; Holtum et al. 2004; K. Winter, unpublished), in none of them has the presence of CAM photosynthesis been conclusively demonstrated. There are indications of CAM activity in *Clusia tocuchensis* Britton (Borland et al. 1992) but further studies are needed to confirm CAM in this species. In the scenario of CAM evolution presented here, the absence of CAM in sect. *Anandrogyne* is plesiomorphic. In the study by Gehrig et al. (2003), CAM seems to have been secondarily lost in this group, but this is likely an effect of the smaller and geographically biased taxon sample, which primarily comprised Panamanian species.

Clusia clusioides (Griseb.) D'Arcy, a species from the Greater Antilles that has been included in sect. *Anandrogyne* based on morphological characters (Engler 1925), has been shown to have a more basal and isolated position in molecular phylogenies (Fig. 6.1; Gustafsson and Bittrich 2003). It would be interesting to investigate photosynthesis physiology of this species, as its near-basal position in the genus is probably of importance for the optimization of CAM on the basal node of the genus, i.e., for determining which photosynthesis mechanism was present in the most recent common ancestor of *Clusia*. Given that *C. clusioides* occurs as a terrestrial tree in wet montane forests, it seems unlikely that it would exhibit CAM.

6.8.2 *Clusia* sect. *Retinostemon* Planch. et Triana and Relatives

Like sect. *Anandrogyne*, *Retinostemon* (referred to as "Group II" by Gehrig et al. 2003) is a species-rich group centered in the Andes and Central America (Gustafsson and Bittrich 2003) (Figs. 6.5, 6.16, 6.26, and 6.27). It is notably absent from eastern and southern Brazil. It reaches around 2000 m altitude in the Andes and is quite frequent in montane forests, but also has many species in the lowlands. Representatives of this section are found in some of the driest habitats known for *Clusia*, interandean valleys in rainshadow. Both terrestrial and hemi-epiphytic life-forms are common. Laticifers are usually evident as thin, white lines on the adaxial surface of the leaves. Flowers offer resin as a pollinator reward, and male flowers usually have a fungiform, resin-producing synandrium, with scattered anthers on the distal surface. Female flowers secrete resin from a ring-shaped organ surrounding the ovary, as is also seen in several other sections (e.g. *Chlamydoclusia*). There are usually around eight carpels, and the stigmata can be described as flap-like (Fig. 6.26). The widespread, apomictic *C. minor* belongs to this section. Many species investigated have been shown to have CAM photosynthesis.

Closely related to sect. *Retinostemon* are the small N. Andean sections *Havetia* (Kunth) Pipoly (one species) and *Cochlanthera* (Choisy) Engl. with seven species (Maguire 1977). *Havetia* appears as a miniaturized (in terms of flowers and fruits) member of sect. *Retinostemon*, while sect. *Cochlantera* is quite different in floral morphology (Fig. 6.4), especially the staminate flowers, which have a resiniferous disk and large, curved anthers (Maguire 1977).

6.8.3 *Clusia* sect. *Clusiastrum* Planch. et Triana

This small (ca. 10 spp) and distinctive section is restricted to the Guayana shield, where the species grow as shrubs in open habitats (Figs. 6.6 and 6.17).

The group can easily be recognized by the bent, short styles and numerous carpels in female flowers (Fig. 6.17), and the dehiscence of the anthers in male flowers, which is by short slits or pores. The pollinators are large bees that vibrate the androecium with their indirect flight muscles which causes the pollen to be released (buzz pollination; Bittrich and Amaral 1996). Representation in molecular studies is poor, but available evidence indicates that it may be the sister-group of sect. *Cordylandra*, from which it differs strongly morphologically. It has so far not been sampled in studies of photosynthesis, but would be highly desirable to include given its phylogenetic position.

6.8.4 *Clusia* sect. *Cordylandra* Planch. et Triana

Most species of sect. *Cordylandra* occur in Eastern Brazil and the Amazon, and the group is, as far as known, absent from the Andes and Central America (Figs. 6.7, 6.18, 6.28 and 6.29). The total number of species known is 17 (P. F. Stevens, V. Bittrich and M. H. G. Gustafsson, revision in preparation). The section is well supported both by molecular and morphological synapomorphies. The sister-group relationship to subgenus *Clusia* (sensu Gustafsson and Bittrich 2003) is not well supported, and the group should perhaps be treated in a subgenus of its own, maybe together with sect. *Clusiastrum*. The mostly five stigmata are covered with acute papillae (unique in the genus), and form a "crown" on the fruit (Fig. 6.29; Bittrich and Amaral 1997; P. F. Stevens in unpublished manuscript). Bracts, bracteoles and sepals are caducous. The endocarp of the fruit is bony, with transversal ridges. The stamens are short and stout (Fig. 6.7) and invariably produce resin, which in male flowers is more fluid than in other *Clusia* species and mixed with pollen.

Clusia fluminensis Planch. et Triana, from the restingas in southeastern Brazil, is reported to have CAM photosynthesis, and recently CAM has been found to occur also in *C. spiritu-sanctensis* G. Mariz et B. Weinberg (Sect. 9.4.2.10).

6.8.5 *Clusia* sections *Clusia* and *Omphalanthera* Planch. et Triana, and Related Groups

The type species of *Clusia* is *C. major* L., a species from the Lesser Antilles (Bittrich and Stevens 1998; Figs. 6.8, 6.9, 6.19, 6.20, 6.30). It belongs to a small group of Caribbean and Central American species, which should be referred to as sect. *Clusia*. This is part of a diverse and poorly supported complex with somewhat uncertain interrelationships that comprises also sect. *Omphalanthera* Planch. et Triana, with peculiar ring-shaped anthers, which is closely related to sect. *Polythecandra* (Planch. et Triana) Engl., with equally peculiar but rather different anthers, divided into numerous minute pollen sacs (Bittrich and Amaral 1996). Both groups occur mostly in the Guayana region. Here also belong some additional species with very varying floral morphology and geographical distribution, e. g. *C. gardneri* Planch. et Triana, from Central Brazil, which was placed in its own section, *Gomphanthera* Planch. et Triana, on account of its unique, irregularly dehiscent anthers (Engler 1925), which in other respects are similar to those of *C. schomburgkiana*. Like the floral morphology, the pollination mechanisms are quite variable. Three species in the complex, *C. aripoensis*, *C. major* and *C. valerioi*, are known to have CAM photosynthesis.

Related to the *Omphalanthera* complex is what could be referred to as the flava-group, comprising a few species from Central America and Jamaica. These species all lack resin and are similar in floral morphology to sect. *Oedematopus* (Planch. et Triana) Pipoly, having bud-like flowers with four thick petals. There are some indications that *C. flava* Jacq. (a CAM species) is beetle pollinated (Hammel 1986). It should be pointed out that the "flava-group" of Hammel (1986) also included species here assigned to sect. *Oedematopus*.

6.8.6 *Clusia* sect. *Phloianthera* Planch. et Triana

This is a mostly Eastern-Brazilian group comprising around 10 species (Figs. 6.10 and 6.21). Several North Andean and Amazonian species have also been ascribed to it, although they differ in aspects of androecial morphology. So far, only Eastern Brazilian species have been sampled in molecular studies. There is some weak evidence from molecular studies that the sister group is sect. *Chlamydoclusia*, but morphologically the two groups are strikingly different. In "typical" (Eastern Brazilian) members of the section, the androecium is dome-shaped, covered with numerous minute anthers, and surrounded by a ring of resiniferous staminodes (Fig. 6.10; Bittrich and Amaral 1997). Staminodes of pistillate flowers consist of several series of short, truncate, mostly anantherous, resiniferous staminodes (Fig. 6.21).

One member of this section, *C. hilariana*, is particularly well studied from the point of view of autecology and ecophysiology (Sect. 9.4.2.1.2). It exhibits CAM, and its role in the succession in the restingas of coastal South Brazil is addressed in Chap. 3.

Clusia fockeana Miq. is similar to sect. *Phloianthera* in its floral morphology, differing basically in that the fertile part of the androecium is borne on a "stalk" (Fig. 6.12). Although not supported as member of the section in the ITS phylogenies, *C. fockeana* is part of a clade comprising sects. *Phloianthera* and *Chlamydoclusia* in the most parsimonious tree shown in Fig. 6.1. Historically, it has been placed in its own section, *Androstylium* (Miq.) Engl.

6.8.7 *Clusia* sect. *Chlamydoclusia* Engl.

The most well-known member of sect. *Chlamydoclusia* is *Clusia rosea* (Figs. 6.11, 6.22, 6.31), a frequently cultivated species and a well-known CAM-plant. It is distributed around the Caribbean and throughout Central America, where it is particularly frequent in dry habitats near the sea. At least in most parts of its distribution area it is apomictic, a trait that has evolved repeatedly in the genus (also known in *Clusia minor* L. of sect. *Retinostemon*; Maguire 1976). Apart from *Clusia rosea*, the section comprises a handful of mostly Amazonian and Guayanan species, some of which are very widespread.

The flowers are some of the largest found in the genus, up to 15 cm across in *C. grandiflora* Splitg (Fig. 6.11). Staminate flowers are readily recognized on the androecium, with a central mass of resiniferous staminodes surrounded by basally fused fertile stamens with an apically prolonged connective that secretes tiny oil droplets. Fruits are also characteristic in having connivent stigmata forming a (in most cases) raised ring on the fruit (Fig. 6.31).

The delimitation of sect. *Chlamydoclusia* is somewhat problematic as one species, *C. nemorosa*, shares with sect. *Chlamydoclusia* both the characteristic organization of the androecium, the typical connective and the fruit shape, while there is no support in the ITS data for its inclusion in the section (neither is it strongly contradicted).

6.8.8 *Clusia* sect. *Oedematopus* (Planch. et Triana) Pipoly, and Relatives

Sections *Oedematopus* and *Criuvopsis* are closely related (Figs 6.13, 6.14, 6.23 and 6.32) according to molecular studies, and sections *Criuva* and *Brachystemon* may also be part of this clade (Fig. 6.1). Parts of these sections were previously treated as separate genera, viz. *Quapoya* (Fig. 6.13), *Havetiopsis* (Fig. 6.23) and *Oedematopus* (Fig. 6.14); the old names are used in the figures. Shared characters include small flowers with few stamens (4 to ca. 20; more in sects. *Criuva* and *Brachystemon*), absence of floral resin and often thick, rubber-like, yellow to white petals. Sometimes (in sect. *Oedematopus*) the flowers, during anthesis, only open with a small apical pore and look rather bud-like (Fig. 6.23). The staminodes in the female flowers have sterile anthers and the filaments are dilated at base. The stigmas are densely papillose. Fruits are small with minute, widely spaced, round, ovate or elliptic stigmata (Fig. 6.32). The epidermis on the branchlets disrupts often in characteristic annular segments and the upper leaf epidermis shows a fine scrobiculate pattern in herbarium specimens. Most leaves are distinctively discolorous in vivo, the lower side being greenish-white due to air-filled achlorophyllous subepidermal layers. Life form is usually hemi-epiphytic (at least in *Oedematopus* and *Criuvopsis*) and there are several lianescent species, particularly in sect. *Criuvopsis*. Most species occur in the Amazon region and Guayana.

6.9 Key Innovations in the Diversification of *Clusia*

If one were to point at the most characteristic and evolutionarily important traits of *Clusia*, they would most probably include resiniferous flowers, hemi-epiphytism, CAM photosynthesis and invasion of montane habitats.

Production of resin in the flowers has evolved no less than three times in *Clusia*, if the evolutionary reconstruction in Fig. 6.1 is accepted. All three origins optimize on deep branches, indicating that resin production is a relatively old phenomenon. In three of the four groups the trait is evolutionarily fixed. The relationship between presence of floral resin and speciation rate is not clear, as there are also species-rich groups (sect. *Anandrogyne*) that lack resin. It is very likely that it has been important to (at least it is correlated with) the floral morphological diversification, as resin-producing flowers are morphologically the most diverse, and show a number of odd and often unique traits. Some of these traits must, judging from the molecular phylogenies, have evolved relatively rapidly, as there may be radical differences between genetically very closely related species.

Clusia, together with *Clusiella* of Kielmeyeroideae (which curiously shares also the resin production in flowers), are the only Clusiaceae that have adopted a hemiepiphytic or lianescent growth form. The exact distribution in *Clusia* is difficult to map, as so many species are plastic in this respect, and may grow terrestrially opportunistically. The predominance of the hemi-epiphytic life form is particularly strong in wet lowland forests, and has evidently been crucial for the success (manifest in numbers of individuals) of *Clusia* in this habitat. Other genera of tribe Clusieae, among which is the sister-group of *Clusia*, are basically from the lowland tropical rainforests, a fact that, together with the widespread occurrence in this habitat by various clades of *Clusia*, suggests that this may have been the habitat of the *Clusia* ancestor, which may or may not have been hemi-epiphytic. *Clusia* may be somehow "predisposed" for the hemi-epiphytic lifestyle, through, e.g., the possession of numerous and smaller seeds than in the outgroups, and a double or multiple hypodermis that serves as an enlarged water storage tissue. Alternatively, these characters, with an evident adaptive value for an epiphyte, may have followed the evolution of hemi-epiphytism. Only careful studies, using character-optimization on well-sampled and well resolved phylogenies can reveal in which order these characters evolved.

Photosynthesis mechanisms have been studied only in a small fraction of *Clusia* species, and the role of CAM in the diversification of the genus is therefore hard to establish with certainty. The present optimization experiment provides only a very preliminary model. There is no doubt, however, that the ability to perform CAM photosynthesis has been an important factor behind the proliferation of *Clusia* in (periodically) dry habitats (including microhabitats in tree crowns where hemi-epiphytes often begin their lives), very different from the rainforest understorey where species of most other genera of tribe Clusieae are found.

Final among the assumed key innovations of *Clusia* is the invasion of the tropical montane habitat. No other Clusiaceae come close to the altitudinal limit of *Clusia* at over 3500 m (the closest is probably *Tovomita weddelliana* Planch. et Triana at around 1800 m). Two clades within *Clusia* have invaded high (over 2000 m) habitats, sect. *Anandrogyne* and sect. *Retinostemon*. Both, but particularly *Anandrogyne*, have undergone extensive speciation in the high altitude habitat. This could be referred to as a radiation, given the small genetic differences in combination with the large number of species.

The presence of all of these rare or unique innovations in a single genus, in combination with the evidently exceptionally variable floral morphology, makes it justified to speak of an extraordinary inherent evolutionary plasticity in *Clusia*.

References

Armbruster WS (1984) The role of resin in angiosperm pollination: ecological and chemical considerations. Am J Bot 71:1149–1160

Bittrich V, Amaral MCE (1996) Flower morphology and pollination biology of some *Clusia* species from the Gran Sabana (Venezuela). Kew Bull 51:681–694

Bittrich V, Amaral MCE (1997) Flower biology of some *Clusia* species from Central Amazonia. Kew Bull 52:617–635

Bittrich V, Stevens PF (1998) Lectotypification of the names *Clusia major* and *C. minor* (Clusiaceae). Taxon 47:117–122

Borland AM, Griffiths H, Maxwell C, Broadmeadow MSJ, Griffiths NM, Barnes JD (1992) On the ecophysiology of the Clusiaceae in Trinidad: expression of CAM in *Clusia minor* L. during the transition from wet to dry season and characterization of three endemic species. New Phytol 122:349–357

Borland AM, Técsi LI, Leegood RC, Walker RP (1998) Inducibility of crassulacean acid metabolism (CAM) in *Clusia* species; physiological/biochemical characterization and intercellular localization of carboxylation and decarboxylation processes in three species which exhibit different degrees of CAM. Planta 205:342–351

Cronquist A (1981) An integrated system of classification of flowering plants. Columbia University Press, New York

D'Arcy WG (1980) Flora of Panama. Family 123. Guttiferae. Ann Missouri Bot Gard 67:969–1043

Davis CC, Webb CO, Wurdack KJ, Jaramillo CA, Donoghue MJ (2005) Explosive radiation of Malpighiales supports a mid-cretaceous origin of modern tropical rain forests. Am Nat 156:E36–E65

Dziedzioch C, Stevens A-D, Gottsberger G (2003) The hummingbird plant community of a tropical rain forest in Southern Ecuador. Plant Biol 5:331–337

Engler A (1925) Guttiferae. In: Engler A, Prantl K (eds) Die natürlichen Pflanzenfamilien, edn 2, vol 21A. Wilhelm Engelmann, Leipzig, pp 154–237

Franco AC, Ball E, Lüttge U (1990) Patterns of gas exchange and organic acid oscillations in tropical trees of the genus *Clusia*. Oecologia 85:108–114

Franco AC, Ball E, Lüttge U (1992) Differential effects of drought and light levels on accumulation of citric and malic acids during CAM in *Clusia*. Plant Cell Environ 15:821–829

Franco AC, Olivares E, Ball E, Lüttge U, Haag-Kerwer A (1994) In situ studies of crassulacean acid metabolism in several sympatric species of tropical trees of the genus *Clusia*. New Phytol 126:203–211

Franco AC, Haag-Kerwer A, Herzog B, Grams TEE, Ball E, de Mattos EA, Scarano FR, Barreto S, Garcia MA, Mantovani A, Lüttge U (1996) The effect of light levels on daily patterns of chlorophyll fluorescence and organic acid accumulation in the tropical tree *Clusia hilariana*. Trees 10:359–365

Frederiksen RF (2005) Phylogeny and floral morphological diversity of *Clusia* L. section Anandrogyne Planch. & Triana (Clusiaceae). Master's Thesis, University of Aarhus

Gehrig HH, Aranda J, Cushman MA, Virgo A, Cushman JC, Hammel BE, Winter K (2003) Cladogram of Panamanian *Clusia* based on nuclear DNA: implications for the origins of crassulacean acid metabolism. Plant Biol 5:59–70

Gustafsson MHG, Bittrich V (2003) Evolution of morphological diversity and resin secretion in flowers of *Clusia* L. (Clusiaceae): insights from ITS sequence variation. Nord J Bot 22:183–203

Gustafsson MHG, Bittrich V, Stevens PF (2002) Phylogeny of Clusiaceae based on *rbcL* sequences. Int J Plant Sci 163:1045–1054

Hale ML, Borland AM, Gustafsson MHG, Wolff K (2004) Causes of size homoplasy among chloroplast microsatellites in closely related *Clusia* species. J Mol Evol 58:182–190

Hammel BE (1986) New species of Clusiaceae from Central America with notes on *Clusia* and synonymy in the tribe Clusieae. Selbyana 9:112–120

Hammel BE (1999) Synopsis of *Chrysochlamys* (Clusiaceae: Clusoideae: Clusieae) in Mesoamerica. Novon 9:360–374

Hammel BE (2001) *Clusia* L. In: Stevens WD, Ulloa Ulloa C, Pool M, Montiel OM (eds) Flora de Nicaragua. Missouri Botanical Garden Press, St. Louis, pp 620–625

Herzog B, Grams TEE, Haag-Kerwer A, Ball E, Franco AC, Lüttge U (1999) Expression of modes of photosynthesis (C-3, CAM) in *Clusia criuva* Camb. in a cerrado gallery forest transect. Plant Biol 1:357–364

Holtum JAM, Aranda J, Virgo A, Gehrig HH, Winter K (2004) $\delta^{13}C$ values and crassulacean acid metabolism in *Clusia* species from Panama. Trees 18:658–668

Jørgensen PM, León-Yánez S (1999) Catalogue of the vascular plants of Ecuador. Missouri Botanical Garden Press, St. Louis

Liogier AH (1983) La flora de la Española II. Ediciones de la UCE, San Pedro de Macorís

Lüttge U (1996) *Clusia*: plasticity and diversity in a genus of C_3/CAM intermediate tropical trees. In: Winter K, Smith JAC (eds) Crassulacean acid metabolism. Springer, Berlin Heidelberg New York, pp 296–311

Lüttge U (1999) One morphotype, three physiotypes: sympatric species of *Clusia* with obligate C_3 photosynthesis, obligate CAM and C_3 -CAM intermediate behaviour. Plant Biol 1:138–148

Lüttge U, Ziegler H, Ting IP (1993) Ecophysiological comportment of the tropical C_3/CAM intermediate tree *Clusia rosea* in the field as assessed by analyses of stable carbon- and hydrogen-isotope ratios. J Plant Physiol 142:497–501

Maddison DR, Maddison WP (2001) MacClade 4: analysis of phylogeny and character evolution. Version 4.02. Sinauer Associates, Sunderland

Maguire B (1976) Apomixis in the genus *Clusia* – a preliminary report. Taxon 25:241–244
Maguire B (1977) A revision of *Clusia* L. Section Cochlanthera (Choisy) Engler. Caldasia 11:129–146
Morgenstern B (1999) Dialign 2: improvement of the segment-to-segment approach to multiple sequence alignment. Bioinformatics 15:211–218
Oliveira CMA, Porto ALM., Bittrich V, Vencato I, Marsaioli AJ (1996) Floral resins of *Clusia* spp.: chemical composition and biological function. Tetrahedron Lett 37:6427–6430
Passos L, Oliveira PS (2002) Ants affect the distribution and performance of seedlings of *Clusia criuva*, a primarily bird-dispersed rain forest tree. J Ecol 90:517–528
Pipoly JJ (1995) *Clusia fabiolae*, a new species, with a synopsis of *Clusia* section *Anandrogyne* (Clusiaceae) in Guayana. Sida 16:737–755
Pipoly JJ (1997) Nomenclatural notes on neotropical Clusieae (Clusiaceae). Sida 17:765–767
Pipoly JJ, Cogollo PA (1998) New species and nomenclatural notes in *Clusia* (Clusiaceae) from andean Colombia and Venezuela. Sida 18:401–409
Pipoly JJ, Kearns DE, Berry PE (1998) *Clusia*. In: Steyermark JA, Berry PE, Holst BK (eds) Flora of the Venezuelan Guayana, vol. 4. Timber Press, Portland, pp 260–294
Planchon JE, Triana J (1860a) Mémoire sur la famille de Guttifères. Ann Sci Nat Bot Ser IV 13:303–376
Planchon JE, Triana J (1860b) Mémoire sur la famille de Guttifères. Ann Sci Nat Bot Ser IV Mémoire sur la famille de Guttifères. Ann Sci Nat Bot sér IV 14:226–367
Roberts A, Borland AM, Griffiths H (1997) Discrimination processes and shifts in carboxylation during the phases of crassulacean acid metabolism. Plant Physiol 113:1283–1292
Rodrigues CMC, Ormond WT, Pinheiro MCB, Alves de Lima H (1993) Estudo da biologia floral de *Clusia criuva* Camb. Um caso de mimetismo. Bradea 6:209–219
Stevens PF (2005) Angiosperm phylogeny website. Version 6, May 2005. http://www.mobot.org/MOBOT/research/APweb/
Stevens PF (2006) Clusiaceae. In: Kubitzki K (ed) The families and genera of vascular plants. Vol. III. Flowering plants, dicotyledons, dilleniid families. Springer, Berlin Heidelberg New York
Swofford DL (2002) PAUP* version 4.0b10. Sinauer Associates, Sunderland
Thorne RF (1976) A phylogenetic classification of the Angiospermae. Evol Biol 9:35–106
Ting IP, Lord EM, da SL Sternberg L, DeNiro MJ (1985) Crassulacean acid metabolism in the strangler *Clusia rosea* Jacq. Science 229:969–971
Ting IP, Hann J, Holbrook NM, Putz FE, da SL Sternberg, Price D, Goldstein G (1987) Photosynthesis in hemiepiphytic species of *Clusia* and *Ficus*. Oecologia 74:339–346
Tinoco Ojanguren C, Vázquez-Yanes C (1983) Especies CAM en la selva húmeda tropical de los tuxtlas, Veracruz. Bol Soc Bot Méx 45:150–153
Vaasen A, Begerow D, Lüttge U, Hampp R (2002) The genus *Clusia* L.: molecular evidence for independent evolution of photosynthetic flexibility. Plant Biol 4:86–93
Vesque J (1892) Epharmosis, sive materiae ad instruendam anatomiam systematis naturalis. 3. Genitalia foliaque Clusiearum et Moronobearum. Delapierre, Vincennes
Winter K, Zotz G, Baur B, Dietz K-J (1992) Light and dark CO_2 fixation in *Clusia uvitana* and the effects of plant water status and CO_2 availability. Oecologia 91:47–51

7 Population Biology of Different *Clusia* Species in the State of Rio de Janeiro

ANJA VAASEN, FABIO RUBIO SCARANO, and RÜDIGER HAMPP

7.1 Introduction

"Simple-Sequence-Repeats" (SSR) or "Microsatellites" are repeats of short sequence motifs with a length of 1 to 6 bp, which can be replicated up to 100 times (Tautz 1993) and occur in the non-coding regions of eucaryotic and chloroplast genomes in a very high diversity (Gupta et al. 1994). These sequence repeats are the main cause for the length polymorphisms of microsatellites in populations (Schlötterer and Tautz 1992) and are very useful for molecular taxonomy and population genetics (Zhivotovsky and Feldman 1995). Due to the large number of microsatellites in eukaryotes, and a high diversity within species or populations, they are the most important markers for genomic mapping and relation studies (Kashi et al. 1997; Queller et al. 1993). Trimer primers are widespread, give the most suitable results (Hughes and Queller 1993), and were used in many different studies before (Becher et al. 2000; Beyermann et al. 1992; Echt and May-Marquardt 1997; Echt et al. 1996; Jarret et al. 1997). Successful amplification of polymorphic banding pattern was possible with trinucleotide primers such as AAC_7, AAG_8, and GTG_5 (Hale et al. 2001; Poulsen et al. 1993; Squirrell and Wolff 2001). Using AAC_7 we were able to identify ecotypes of Brazil pine [*Araucaria angustifolia* (Bert.) O. Ktze.; Hampp et al. 2000].

The genus *Clusia* comprises about 350 species (Willis 1973; Pipoly et al. 1998), which occur mainly in the tropical part of South America (Lüttge 1991). These plants are the only known dicotyledonous tree species that are able to perform crassulacean acid metabolism (CAM) and which possess the ability to switch between the C_3 and the CAM mode of photosynthesis. This enables them to conserve water, depending on environmental conditions (see also Chaps. 6 and 9). Certain *Clusia* species occur at quite different habitats with regard to subsoil and climatic conditions. It could thus be possible that this has resulted in site-specific diversity. In order to investigate this hypothesis,

Ecological Studies, Vol. 194
U. Lüttge (Ed.)
Clusia: A Woody Neotropical Genus
of Remarkable Plasticity and Diversity

we performed an SSR analysis of genetic variability at the population level of three *Clusia* species, *C. parviflora* Engl., *C. fluminensis* Pl. et Tr., and *C. hilariana* Schlecht., growing at different sites along the coast of the state of Rio de Janeiro, Brazil. This was especially promising as the selected sites showed characteristic differences and thus a possible loss of genetic exchange due to habitat fragmentation. The sites mostly represent restingas with open vegetation on coastal sand dunes. *Clusia* is the main shrub of this vegetation structure and grows in the centres of the vegetation islands. Due to the severe conditions along the coast (e.g., salt stress), the vegetation is often destroyed, and the vegetation islands are dispersed, making genetic exchange difficult.

7.2 Population Studies on the Basis of Single Sequence Repeats

For our studies we collected leaf samples from nine locations along the coastline east to northeast of Rio de Janeiro (Fig. 7.1; Table 7.1). We took two leaves each from all *Clusia* plants we could find at a given site. DNA extraction,

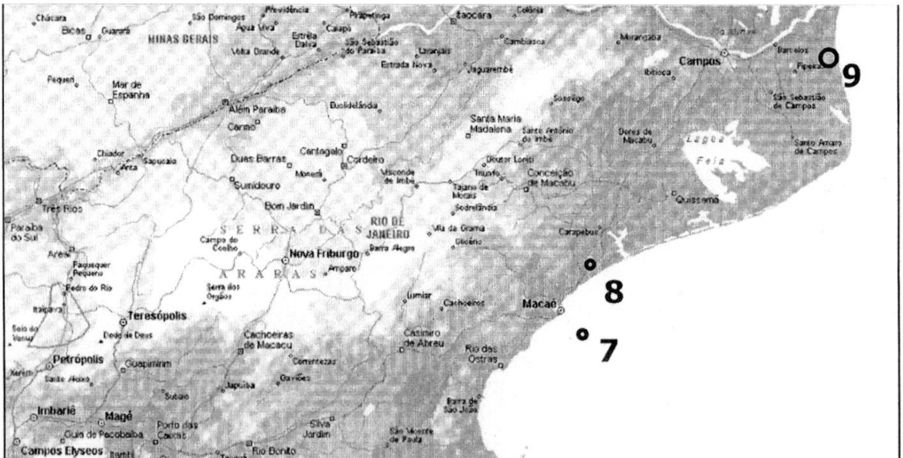

Fig. 7.1. Map of the state of Rio de Janeiro. Collection sites are marked with *circles*: 1, sugar loaf, city of Rio de Janeiro; 2, Maricá; 3, Jacarepia Nationalpark near Saquarema; 4, Figueira; 5, Arraial do Cabo; 6, Buzios; 7, Archipelago de Santana; 8, Jurubatiba Nationalpark near Macaé; 9, Iquipari near Campos de Goitacazes

Table 7.1. *Clusia* species and some properties of their habitats

Site		Vegetation (rainfall)	*Clusia* species
1 (S)	Sugar loaf	Rocky outcrops (1500 mm/a)	*Clusia parviflora*
1 (C)	Corcovado	Rocky outcrops (1500 mm/a)	*Clusia fluminensis*
2 (M)	Maricá	Wet restinga (1230 mm/a)	*Clusia fluminensis*
3 (S)	Jacarepia/Saquarema	Wet restinga (1000 mm/a)	*Clusia fluminensis*
4 (F)	Figueira	Dry restinga (800 mm/a)	*Clusia fluminensis*
5 (Ar)	Arraial do Cabo	Dry restinga (800 mm/a)	*Clusia fluminensis* *Clusia hilariana*
6 (B)	Buzios	Dry Forest (800 mm/a)	*Clusia fluminensis*
7 (A)	Archipelago de Santana	Rocky outcrops (~800 mm/a)	*Clusia fluminensis*
8 (J)	Jurubatiba NP/Macaé	Change of dry and wet Restinga (1164 mm/a)	*Clusia hilariana* *Clusia parviflora*
9 (I)	Iquipari/ Campos de Goitacazes	Transition from restinga into dry forest (800–1200 mm/a)	*Clusia hilariana* *Clusia spiritu-sanctensis*

analysis by PCR, and determination of banding patterns on polyacrylamide gels was as in Hampp et al. (2000).

7.2.1 *Clusia parviflora*

C. parviflora Engl. occurs at the coast only at two different sites, at the rocky vegetation of the sugar loaf and at the Restinga de Jurubatiba National Park, near Macaé (Fig. 7.1). This species showed the most obvious differences by use of the primer GTG_5 (Fig. 7.2).

The samples from the restinga site display clearly two different fragments with a size of approximately 1500 bp, while those of the sugar loaf have only one or even no fragment of this size. These differences are more conspicuous in the samples of the year 2001 (S1, J1) than in those of 2003 (S2, J2). Furthermore, a fragment at a size of approximately 800 bp which shows up in the samples of Macaé is extremely weak or absent in the sugar loaf samples. Likewise, a clear banding pattern appeared when using the primer AAG_8 (Fig. 7.3).

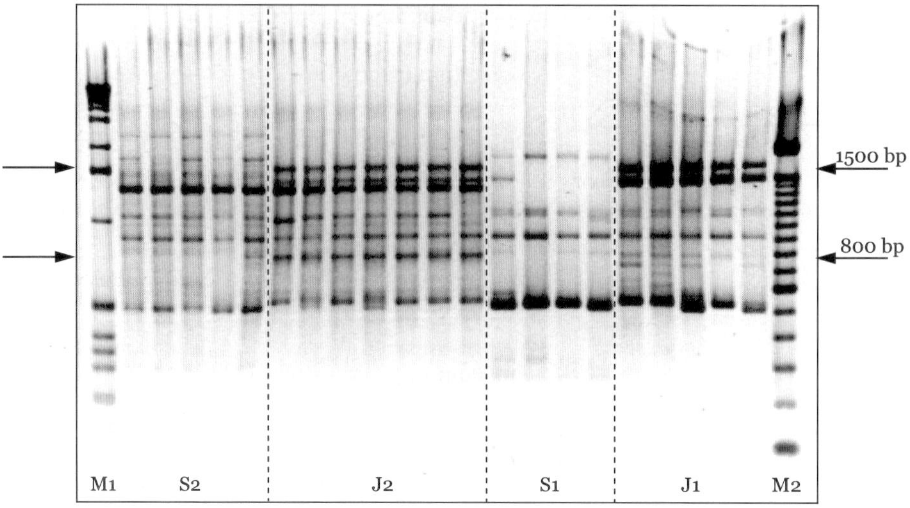

Fig. 7.2. Banding pattern of different samples of *C. parviflora* using the primer GTG_5. (Sample sites: S1, sugar loaf; J1, Restinga de Jurubatiba National Park, taken in 2001; S2, sugar loaf; J2, Restinga de Jurubatiba National Park, taken in 2003). M1 shows a 1-kb, M2 a 100-bp size marker ladder. Differences in the banding pattern are marked by *arrows*

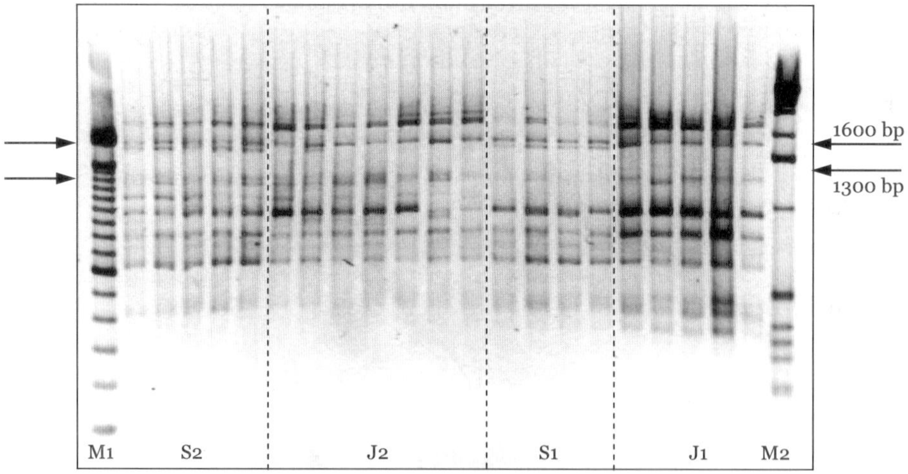

Fig. 7.3. Banding pattern of different samples of *C. parviflora* using the primer AAG_8 (for abbreviations see Fig. 7.2)

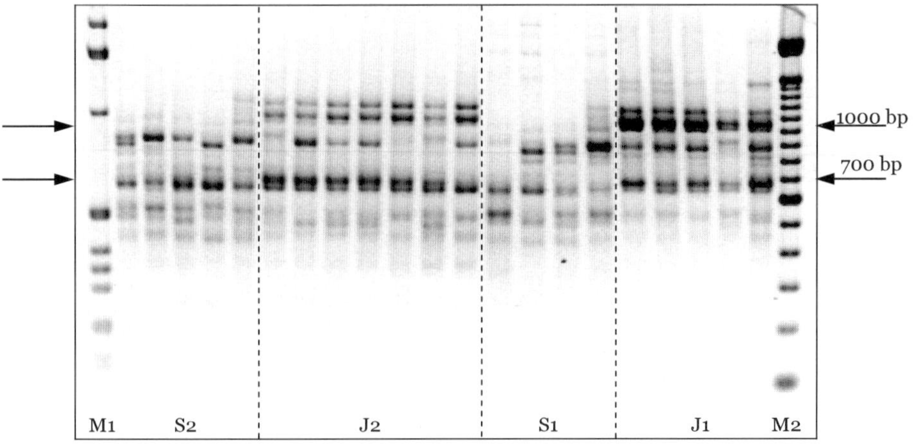

Fig. 7.4. Banding pattern of samples of *C. parviflora* using the primer AAC_7 (for abbreviations see Fig. 7.2)

At a size of ca. 1600 bp, the sugar loaf samples form a double band while the samples from the restinga display only a single band. By contrast, the samples of the sugar loaf exhibit only weak signals at a size of ca. 1300 bp while those from the restinga show one or two distinct signals. Finally, the third microsatellite primer, AAC_7, gave another polymorphic banding pattern (Fig. 7.4), which again clearly distinguished DNA samples from both sites. In a size range of between 1000 bp and 1100 bp, the samples from the restinga showed three distinct fragments while those from the sugar loaf site displayed only very weak or no bands. Similar differences were found at around 700 bp.

The banding pattern obtained with each of the three primers yielded differences between the two sites. Given that these two sites are located at a distance of ca. 150 km from each other, the lack of genetical exchange between these populations is not surprising. Furthermore, the different habitats could be responsible for these differences in the non-coding DNA regions.

7.2.2 *Clusia fluminensis*

C. fluminensis Pl. et Tr. is widely spread along the coast east of Rio de Janeiro, and this species is not found in the more northern parts of the state of Rio de Janeiro. *C. fluminensis* grows in all habitats of this region such as restingas (Maricá, Saquarema, Figueira), dry forests (Buzios), and rocky outcrops (Corcovado, Archipelago de Santana). It was only absent from the sugar loaf. In spite of this variety of habitats, there were relatively small differences in the banding patterns. Using the primer AAG_8, some variation at around 1700 bp was visible only in the Corcovado sample (C) (Fig. 7.5). At

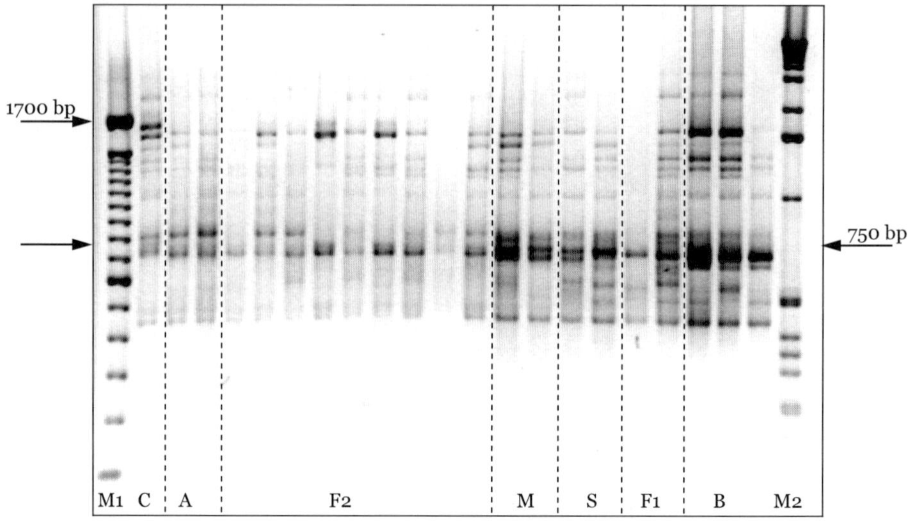

Fig. 7.5. Banding pattern of samples of *C. fluminensis* using the primer AAG_8 (M, Maricá; S, Saquarema; F1, Figueira; B, Buzios, taken in 2001; C, Corcovado; A, Archipelago de Santana; F2, Figueira, taken in 2003). M1 shows a 100-bp, M2 a 1-kb size marker ladder. Differences in the banding pattern are marked by *arrows*

about 750 bp other differences in the banding pattern occurred. The samples of Corcovado (C), Archipelago de Santana (A), and Figueira (F1, F2) yielded a single band while the samples of Maricá (M) and Saquarema (S) display an additional longer fragment, and the sample of Buzios (B) an additional smaller fragment.

The observed differences in fragment polymorphisms could be due to the climatic conditions of the different restingas. Maricá and Saquarema have more than 1000 mm rainfall per year and form a rather wet restinga, while the dry restinga in Figueira has only about 800 mm rainfall per year and thus the behaviour is more similar to the rocky sites of Corcovado and Archipelago de Santana with their increased water drain off. Buzios represents a special site because, after the separation of the tectonic plates some fragments of the African plate remained attached to the South American continent. This geological anomaly together with different subsoil, resulted in a dry forest vegetation instead of restinga and is the reason for numerous endemic plant species (Scarano et al. 2001). Owing to the different habitat, *C. fluminensis* grows in this area as a strangler on other trees, compared to a freestanding shrub in the open restinga vegetation. With this background the specific banding pattern of this provenience is not surprising.

Location-dependent differences in the banding pattern of the non-coding DNA regions of this species could also be identified with the primer AAC_7 (Fig. 7.6). The samples taken from Maricá (M), Buzios (B), and Corcovado (C)

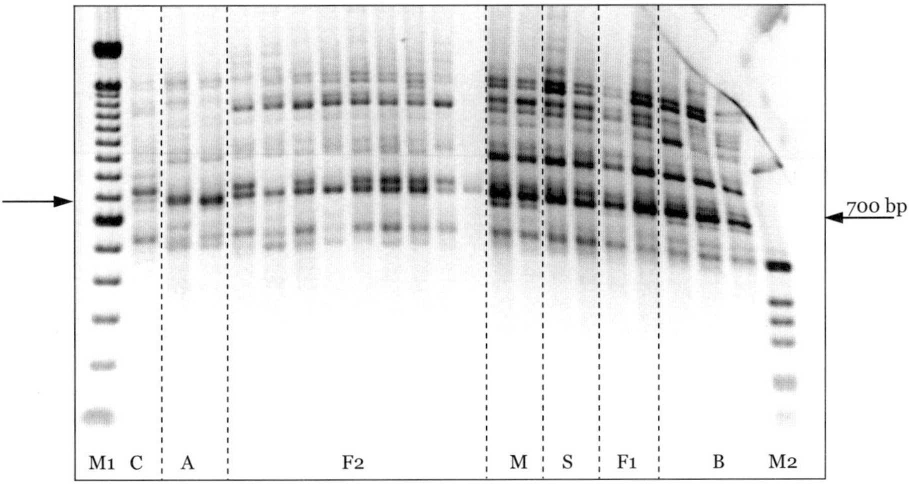

Fig. 7.6. Banding pattern of samples of *C. fluminensis* using the primer AAC_7 (for abbreviations see Fig. 7.5)

clearly show three different fragments at a size of around 700 bp, while those from Archipelago de Santana (A) exhibit only one fragment, and those from Saquarema two fragments at the respective size. The samples of Figueira (F1, F2) display only a weak third fragment. The distinctly different pattern of the sample of the Archipelago de Santana could be due to the island habitat. The distance to the coast renders genetic exchange for this location possibly more difficult. In addition, the sites of *C. fluminensis* at the coast are more restricted to the south. The banding pattern of the sample taken from Saquarema could be related to the site-specific subsoil properties, characterized by a higher moisture compared to the other restinga sites and, in consequence, a close-fitting marsh vegetation (Cirne and Scarano 2001). Furthermore, at this site *C. fluminensis* is not a dominating shrub but grows in the underwood, because the humid soil favours trees and bigger shrubs.

7.2.3 *Clusia hilariana*

C. hilariana Schlecht. occurs in the Restinga de Jurubatiba National Park near Macaé, at Iquipari near Campos de Goitacazes, and at Arraial do Cabo (Fig. 7.1). These sites are located in the eastern part of the state of Rio de Janeiro and include mostly regions where *C. fluminensis* does not occur. Both species demonstrate a spatial separation with an overlapping area at Arraial do Cabo. *C. hilariana* is the highest shrub in the restinga, forms the crown of the vegetation islands, and delivers the shade for germination and growth of other plants. The area at Macaé consists of dune walls with valleys in between.

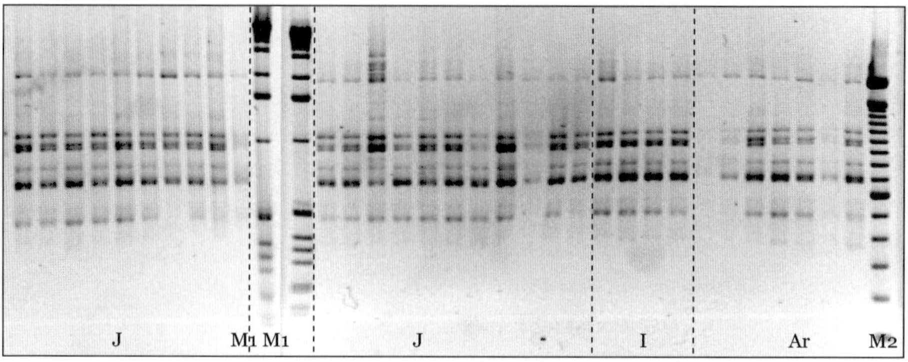

Fig. 7.7. Banding pattern of samples of *C. hilariana* using the primer AAG_8 (J, Restinga de Jurubatiba National Park; I, Iquipari near Campos de Goitacazes; Ar, Arraial do Cabo, taken in 2001). M1 shows a 1-kb, M2 a 100-bp size marker ladder

The CAM performing plant, *C. hilariana*, grows at the dry and hot dunes, while the C_3-CAM-intermediate plant, *C. parviflora*, which uses C_3 photosynthesis in this area, occurs in the valleys, providing better water supply. Again, we wanted to find out, whether the diverse habitats could have created ecotypes, which could be identified by SSR. With regard to dune walls with variable distances to the ocean, samples were taken from the first, second and fourth dune (on the third dune, *Clusia* did not occur). Using the primer AAG_8, the banding pattern was identical for all samples (Fig. 7.7). The same was true for the other primers used in this investigation. We thus conclude that the sites were of such high similarity (all represent dry restinga on a sandy soil) that only one ecotype has evolved.

7.3 Relationship of *Clusia spiritu-sanctensis* to other *Clusia* Species

The species *C. spiritu-sanctensis* G. Mariz et Weinberg was discovered not long ago and thus the phylogenetic classification is not yet completely clear. This species has a strong morphological similarity to *C. hilariana*. A comparison of the banding pattern of *C. spiritu-sanctensis* with that of the other species (primer AAG_8) indicated specific relationships. Compared to *C. hilariana*, there were many differences in the banding pattern (Fig. 7.8), and only one fragment looked identical while each of the other fragments were found only in one species. In contrast, *C. spiritu-sanctensis* and *C. fluminensis* displayed largely homologous banding patterns (Fig. 7.9). The dominating fragments with a size of 450, 700 and approximately 2000 bp existed in each of the samples, as well as a weaker fragment with a size of 1000 bp and two fragments of 1300 bp. Similar data were obtained with the other primers. This could be

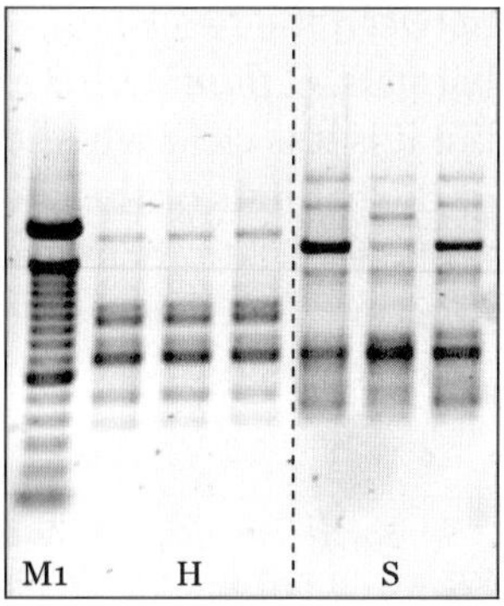

Fig. 7.8. Banding pattern of samples of *C. hilariana* and *C. spiritu-sanctensis* using the primer AAG_8 (H, *C. hilariana*; S, *C. spiritu-sanctensis* from Iquipari near Campos de Goitacazes). M1 shows a 100-bp size marker ladder

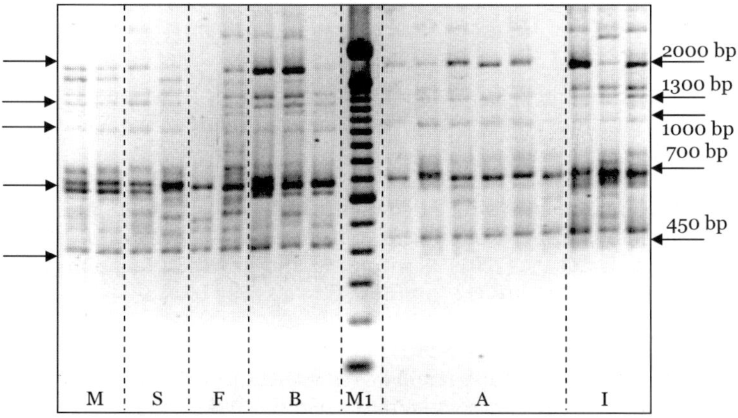

Fig. 7.9. Banding pattern of samples of *C. fluminensis* and *C. spiritu-sanctensis* using the primer AAG_8 (*C. fluminensis*: M, Maricá; S, Saquarema; F, Figueira; B, Buzios; A, Arraial do Cabo; I, *C. spiritu-sanctensis* from Iquipari near Campos de Goitacazes). M1 shows a 100-bp size marker ladder. Differences in the banding pattern are marked by *arrows*

taken as evidence that both species are closely related or even identical, which is in accordance with conclusions from a comparative sequence analysis of the ITS region as well as other molecular studies (Gustafsson and Bittrich 1999, 2002).

7.4 Conclusion

Population studies of different *Clusia* species along the coast of the state of Rio de Janeiro using short tandem repeat primers indicate a fragmentation of the *C. parviflora* populations, which may have been caused by the distance between the two sites as well as by the ecological differences between the two habitats. *C. fluminensis* exhibited differences between the diverse populations, but it is less clear whether this can be assigned to either the spatial separation of the habitats or the site specific conditions. *C. hilariana* in contrast, showed no SSR variations between the populations. This could be the result of still functional exchange of genetic information, or to an only recent disruption of this exchange.

Acknowledgements. A.V. is thankful to the Landesgraduierten-Förderung Baden-Württemberg and the Reinhold-und-Maria-Teufel-Stiftung for financial support.

References

Becher SA, Steinmetz K, Weising K, Boury S, Peltier D, Renou J-P, Kahl G, Wolff K (2000) Microsatellites for cultivar identification in *Pelargonium*. Theor Appl Genet 101:643–651

Beyermann B, Nürnberg P, Weihe A, Meixner M, Epplen JT, Börner T (1992) Fingerprinting plant genomes with oligonucleotide probes specific for simple repetitive DNA sequences. Theor Appl Genet 83 691–694

Cirne P, Scarano FR (2001) Resprouting and growth dynamics after fire of the clonal shrub *Andira legalis* (Leguminosae) in a sandy coastal plain in south-eastern Brazil. J Ecol 89:351–357

Echt CS, May-Marquardt P (1997) Survey of microsatellite DNA in pine. Genome 40:9–17

Echt CS, May-Marquardt P, Hseih M, Zahorchak R (1996) Characterization of microsatellite markers in eastern white pine. Genome 39:1102–1108

Gupta M, Chyi Y-S, Romero-Severson J, Owen JL (1994) Amplification of DNA markers from evolutionarily diverse genomes using single primers of simple-sequence repeats. Theor Appl Genet 89:998–1006

Gustafsson MHG, Bittrich V (1999) Phylogeny and floral diversity of *Clusia* L. (Clusiaceae). In: XVI International Botanical Congress. Edited by X.I.b.c. Abstracts vol, poster 529, p 381

Gustafsson MHG, Bittrich V (2002) Evolution of morphological diversity and resin secretion in flowers of *Clusia* (Clusiaceae): insights from ITS sequence variation. Nord J Bot 22:183–203

Hale ML, Bevan R, Wolff K (2001) New polymorphic microsatellite markers for the red squirrel (*Sciurus vulgaris*) and their applicability to the grey squirrel (*S. carolinensis*). Mol Ecol Notes 1:47–49

Hampp R, Mertz A, Schaible R, Schwaigerer M, Nehls U (2000) Distinction of *Araucaria angustifolia* seeds from different locations in Brazil by a specific DNA sequence. Trees 14:429–434

Hughes CR, Queller DC (1993) Detection of highly polymorphic microsatellite loci in a species with little allozyme polymorphism. Mol Ecol 2:131–137

Jarret RL, Merrick LC, Holms T, Evans J, Aradhya MK (1997) Simple sequence repeats in watermelon (*Citrullus lanatus* (Thunb.) Matsum. & Nakai). Genome 40:433–441

Kashi Y, King D, Soller M (1997) Simple sequence repeats as a source of quantitative genetic variation. Trends Genet 13:74–78

Lüttge U (1991) *Clusia* – Morphogenetische, physiologische und biochemische Strategien von Baumwürgern im tropischen Wald. Naturwissenschaften 78:49–58

Pipoly JJ, Kearns DM, Berry PE (1998) *Clusia* L. In: Berry PE, Holst BK, Yatskievych K (eds) Flora of the Venezuelan Guayana. Missouri Botanical Garden Press, St. Louis, Missouri, pp 260–294

Poulsen GB, Kahl G, Weising K (1993) Abundance and polymorphism of simple repetitive DNA sequences in *Brassica napus* L. Theor Appl Genet 85:994–1000

Queller DC, Strassmann JE, Hughes CR (1993) Microsatellites and kinship. TREE 8:285–288

Scarano FR, Duarte HM, Ribeiro KT, Rodrigues PJFP, Barcellos EMB (2001) Four sites with contrasting environmental stress in south eastern Brazil: relations of species, life form diversity, and geographic distribution to ecophysiological parameters. Bot J Linnean Soc 136:345–364

Schlötterer C, Tautz D (1992) Slippage synthesis of simple sequence DNA. Nucleic Acids Res 20:211–215

Squirrell J, Wolff K (2001) Isolation of polymorphic microsatellite loci in *Plantago major* and *P. intermedia*. Mol Ecol Notes 1:179–181

Tautz D (1993) Notes on the definition and nomenclature of tandemly repetitive DNA sequences. DNA fingerprinting: state of science. Birkhäuser Verlag, Basel, Boston, Berlin. pp 21–28

Willis JC (1973) A dictionary of the flowering plants and ferns, 8th edn. University Press, Cambridge

Zhivotovsky LA, Feldman MW (1995) Microsatellite variability and genetic distances. Proc Natl Acad Sci USA 92:11549–11552

Section III Functions and Physiological Ecology

Introduction

ULRICH LÜTTGE

Clusia minor L. is the most astonishing plant I ever had in my hands, and it offers fascinations much beyond the fact that *Clusia* is the only dicotyledonous tree genus with crassulacean acid metabolism (CAM) (Chap.1). *C. minor* shows all possible variations of CAM as well as full C_3-photosynthesis. Only C_4-photosynthesis is missing in this species as well as in the whole genus as far as this can be seen. However, the plasticity and flexibility of *C. minor* is so large that different modes of photosynthesis, C_3 and CAM, respectively, can be performed by two different opposite leaves at the same node and even by different parts of one leaf simultaneously (Chap. 8). *C. minor* displays all conceivable facets of CAM. It shows versatile dynamics of shifts between C_3-photosynthesis and CAM. Due to the longevity of its leaves for more than one season in *C. minor* these switches are reversible and can occur frequently and repeatedly, in contrast to the annual therophyte *Mesembryanthemum crystallinum* L. which is now very intensively studied and frequently considered the preferable model plant for C_3/CAM intermediate behaviour.

In addition, *C. minor* and the CAM-species of the genus *Clusia* in general show some peculiarities of CAM which suggest a re-interpretation of the ecophysiological advantages of CAM (Lüttge 2006). It appears that the important function of CAM for ecological niche acquisition does not primarily reside in the specific capacity for a particular stress response, e.g. to stressors such as drought or high irradiance. It rather is the very plasticity which is basically inherent in CAM and still more so in potential C_3/CAM-intermediate behaviour which is the basis of the ecological success of *Clusia*s, where *C. minor* alone can cover a wide range of ecologically different habitats extending from very dry savannas and shrub forests to moist montane rain forests in the neotropics. *M. crystallinum* is currently styled as a model plant for stress and CAM research, but should not *C. minor* be a paramount choice? Evidently, as a tree it is more difficult to handle than an annual plant. However, beyond the annual species it offers many additional relevant facets. Among *Clusia*s, in fact *C. minor* is the most widely and intensely studies species and clearly develops to kind of a *Clusia* model plant. However, the model plant concept can be dangerous, and dialectically I might immediately retract again from such considerations.

Some of the deeper insights that *C. minor* provided are based on the fact that it has a photosynthetically much less plastic counterpart in *Clusia multiflora* H.B.K., which offers itself for comparative studies. *C. multiflora* is an obligate C_3-species. However, its ecological amplitude is as large as that of *C. minor*, perhaps even somewhat larger. The pair of *C. minor*/*C. multiflora* has been frequently studied both in phytotrons and in the field and these comparative analyses are much adding to our understanding of the roles of C_3-photosynthesis and CAM, respectively, in performance at the community level. A model pair? It is really schizophrenic in current biology, however, that on the one hand research policy strongly and widely favours the fashion of model organisms, while on the other hand simultaneously bio-diversity enjoys very high popularity.

Considering *Clusia* here, we might perhaps switch our primary attention from *C. minor*/*C. multiflora* to *C. hilariana* Schlechtendal, the dominant species of Brazilian restingas where in the combination of phytogeographical/ecological and physiological/ecological studies (Sects. II and III, respectively) astonishing new insights are unravelled. To date from the circa 300–400 species of *Clusia* only circa 20 have been studied physiologically and/or ecophysiologically and we may only guess about the wealth of surprises which are still inherent in a broader and deeper assessment of physiological diversity of the whole genus. It is promising to note in this respect the observation that the community of researchers strongly engaged in studies of *Clusia* form a gang of people who are always equally interested in work both in the laboratory and in the field (Chaps. 8 and 9).

Notwithstanding successful applications of molecular approaches in studying phylogeny and population genetics of *Clusia* (Sect. II, Chaps. 6 and 7) the "-omics" have not yet arrived at *Clusia*-research. It would be highly desirably that the interest of researchers and the motivation of funding organizations to support them forcefully turn to the -omics. In view of the great flexibility of *Clusia*, which includes plasticity of biochemical pathways and metabolites involved in CAM (Chap. 8), it is not hazardous to predict a wealth of novel findings if systems analysis with transcriptomics and metabolomics were applied in ecophysiological studies of *Clusia*.

Section III of this book explores the physiological characteristics of *Clusia*, where by far the major body of work available is on photosynthesis (Chap. 8). Mineral nutrition is far less studied (Sect. 9.3). Nevertheless, although Wanek et al. (2002) did not detect mycorrhiza in hemi-epiphytic *Clusia* species, the short Chap. 10 can now at least provide a background and present some new findings on the functions of mycorrhiza in *Clusia*. The imbalance of the volume covered on above and below ground physiology of *Clusia*, respectively, reveals another deficit of current research and knowledge on *Clusia* bearing in mind the richness of discoveries which is always made when pedospheric relations are included in ecophysiological studies.

The basic functions of photosynthesis in *Clusias* treated in Chap. 8 are related to ecophysiological performance in an aut-ecological vein. Based on Chap. 8 the following Chap. 9 then fathoms the realization of options of photosynthesis in the actual behaviour in the field. In the complement of different traits which are involved in fitness we consider photosynthesis, the path of primary energy input into and productivity within ecosystems, as fundamentally important (Lüttge and Scarano 2006). In Chap. 9 we move on to physiological syn-ecology by comparatively evaluating the performance of *Clusias* with different modes of photosynthesis and other species of similar life form. This supports the notion of CAM-flexibility rather than CAM per se being the more important advantage at the community level.

The high speciation rate in the large genus of *Clusia* (Chap. 6) together with the enormous ecological amplitude of the genus as a whole as well as individual species and the flexible ecological performance raises thoughts about the relationships between physiological phenotype (physiotype) plasticity and bio-diversity of *Clusias* (Sect. 9.5).

References

Lüttge U (2006) Photosynthetic flexibility and ecophysiological plasticity: questions and lessons from *Clusia*, the only CAM tree, in the neotropics. New Phytol 171:7–25

Lüttge U, Scarano FR (2006) Synecological comparisons sustained by ecophysiological fingerprinting of intrinsic photosynthetic capacity of plants as assessed by measurements of light response curves. Bazilian J Bot (in press)

Wanek W, Arndt SK, Huber W, Popp M (2002) Nitrogen nutrition during ontogeny of hemiepiphytic *Clusia* species. Funct Plant Biol 29:733–740

8 Photosynthesis

ULRICH LÜTTGE

8.1 Photosynthetic Physiotypes

One of the great fascinations of the genus *Clusia* is that its single leaf morphotype (Sect. 2.1) expresses different photosynthetic physiotypes. Often different photosynthetic types are even expressed by the same species and even in clones of vegetatively propagated plants or in different leaves of single individual plants depending on environmental conditions. The photosynthetic types observed among *Clusia*s are based on the modes of C_3-photosynthesis and of crassulacean acid metabolism (CAM) and its variants. Figure 8.1 presents a schematic overview of the basic features of the three to four different photosynthetic physiotypes found among *Clusia*s. Figures 8.2 and 8.3 show typical patterns of photosynthetic CO_2-gas exchange for different modes of photosynthesis expressed among four different species under identical conditions in a phytotron (Fig. 8.2) and for different modes of photosynthesis expressed in one species, *Clusia minor* L., under different conditions (Fig. 8.3), respectively.

In C_3-photosynthesis primary fixation of CO_2 by ribulose-bis-phosphate carboxylase/oxygenase (RubisCO) occurs simultaneously with the energy providing light reactions of photosynthesis. Stomata are open during the light period for CO_2-uptake. In the dark period stomatal opening is much reduced or stomata are closed. Some respiratory CO_2 is released. Expression of C_3-type CO_2 exchange is shown by *Clusia venosa* Jacq. in Fig. 8.2A and by well watered *C. minor* under high irradiance (Fig. 8.3A).

In CAM primary fixation of CO_2 occurs in the dark period and is mediated by phospho*enol*pyruvate carboxylase (PEPC). Stomata are open during the dark period for CO_2 uptake. The CO_2 fixed is stored nocturnally in the cell vacuoles in the form of organic acids (Sects. 8.3.2 and 8.3.3). This is called phase I of CAM (Osmond 1978). Organic acids are remobilized from the vacuoles, decarboxylated and the CO_2 regenerated is refixed and assimilated via RubisCO and the Calvin cycle in the light period. This is called phase III of

Ecological Studies, Vol. 194
U. Lüttge (Ed.)
Clusia: A Woody Neotropical Genus
of Remarkable Plasticity and Diversity

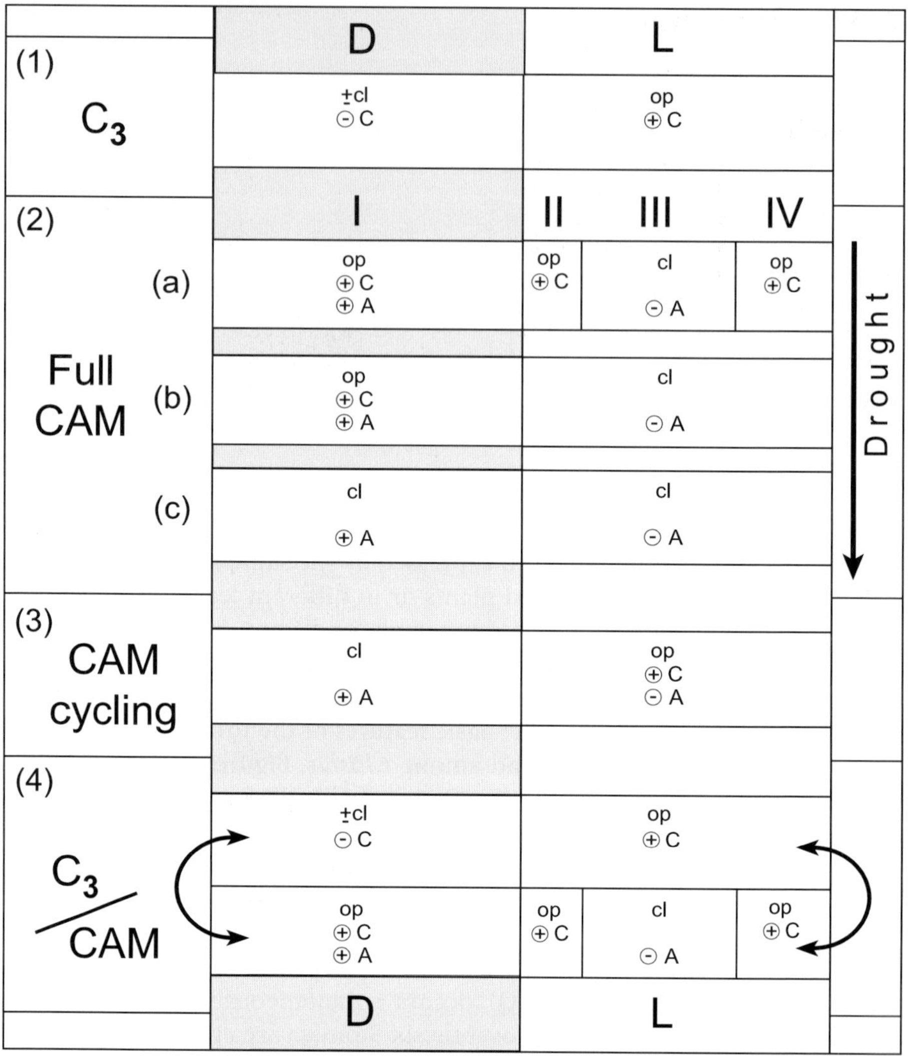

Fig. 8.1. Schematic overview of the photosynthetic physiotypes found among *Clusia* species. **(1)** C_3-photosynthesis; **(2)** full CAM, **(a)** with the four CAM-phases I to IV, **(b)** with phases II and IV suppressed, and **(c)** with CAM-idling as drought increases; **(3)** CAM-cycling; **(4)** C_3/CAM intermediate behaviour. D=dark period, L=light period, cl or op=stomata closed or open, C=net CO_2-exchange (+=uptake, -=release), A net change of organic acid levels (+=accumulation, -=remobilisation), roman numbers=CAM phases

CAM during which stomata remain closed. Since stomata are closed in phase III, its activity cannot be studied by measurements of gas exchange. It can be measured, however, by analyses of declining levels of organic acids, which requires destructive leaf-tissue sampling. In addition phase III activity can be assessed experimentally by non-destructive measurements of chlorophyll flu-

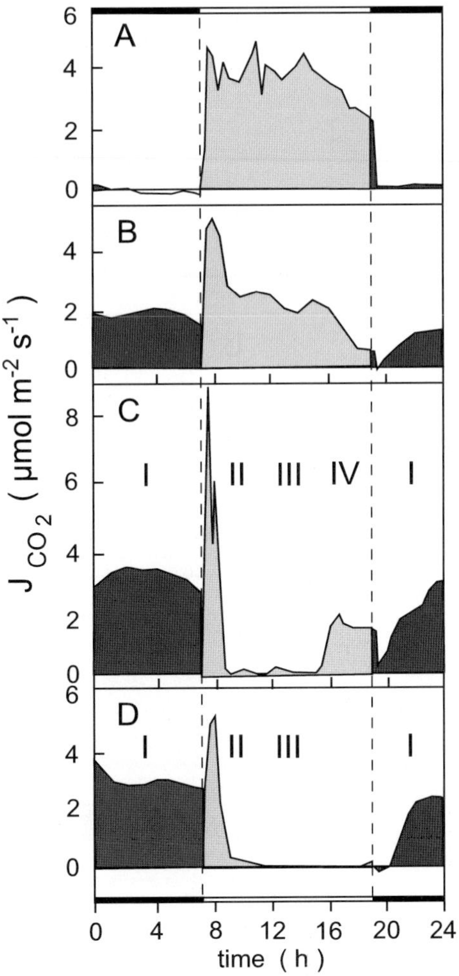

Fig. 8.2A-D. Photosynthetic types expressed by four different species of *Clusia* kept under identical conditions in a phytotron. Net CO_2-exchange, J_{CO2}, of: **A** C_3-photosynthesis: *C. venosa*; **B** CO_2-uptake around the clock: *C. minor*; **C** CAM: *C. major* with phases I to IV; **D** CAM with suppressed phase IV: *C. alata*. *Dark bars* on the abscissa indicate dark periods, *Roman numbers* refer to CAM phases explained in the text (from Lüttge 1991, after data of Franco et al. 1990)

orescence parameters from which apparent photosynthetic electron transport rate, ETR, can be deduced. ETR is determined as

$$ETR = 0.5 \times 0.86 \times PPFD \times (F_m'-F)/F_m' \quad (8.1)$$

(Genty et al. 1990), where F is the ground fluorescence and F_m' the maximum fluorescence of chlorophyll *a* of photosystem II of a light adapted leaf, and

$$\Delta F/F_m' = (F_m'-F)/F_m' \quad (8.2)$$

is effective quantum yield; PPFD is photosynthetic photon flux density; the factor 0.5 accounts for equal distribution of irradiance energy between photosystems II and I, and the factor 0.86 accounts for an estimated average reflec-

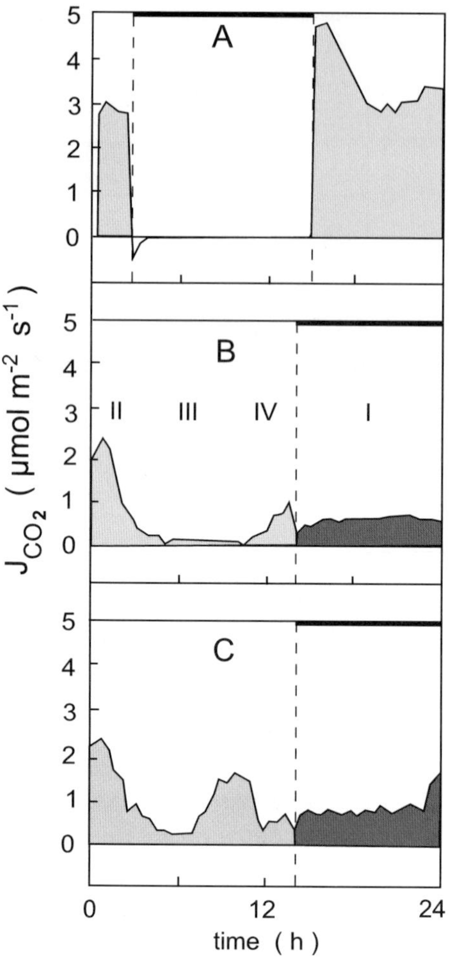

Fig. 8.3A–C. Three different photosynthetic types (net CO_2-exchange, J_{CO2}) expressed by *C. minor* under different conditions in a phytotron: **A** C_3-photosynthesis under well watered conditions with a PPFD of 1700 µmol photons m^{-2} s^{-1} and a leaf/atmosphere VPD of 6.6 mbar bar^{-1}; **B** CAM under drought stress with a PPFD of 400 µmol m^{-2} s^{-1} and a VPD of 13.5 mbar bar^{-1}; **C** CO_2 uptake around the clock under well watered conditions with a PPFD of 400 µmol m^{-2} s^{-1}and a VPD of 3.4 mbar bar^{-1}. *Dark bars* on the abscissa indicate dark periods, *Roman numbers* refer to CAM phases. PPFD=photosynthetic photon flux density, VPD=vapour pressure difference. (From Lüttge 1991, after data of Lee et al. 1989, where *C. minor* is wrongly called *C. rosea*)

tion of PPFD of 14%. In Fig 8.4 the ETR is plotted vs decarboxylation rates calculated from the decline of malate levels at various times during phase III. This example is from *C. minor* growing in the shade of a deciduous dry forest in Venezuela in the dry season and performing weak CAM. The linear correlation between ETR and decarboxylation shows that reduction of CO_2 derived from malate remobilization drives photosynthetic electron transport in phase III.

In a transition phase between phases I and III in the early morning (phase II) stomata are still open and both PEPC and RubisCO may contribute to CO_2 fixation in *Clusia*s in this phase. It appeared particularly in the obligate CAM species *Clusia fluminensis* Planch. et Triana that both carboxylases were active in phase II, while in the C_3/CAM-intermediate species *C. minor* RubisCO made a greater contribution to CO_2-fixation in phase II than PEPC

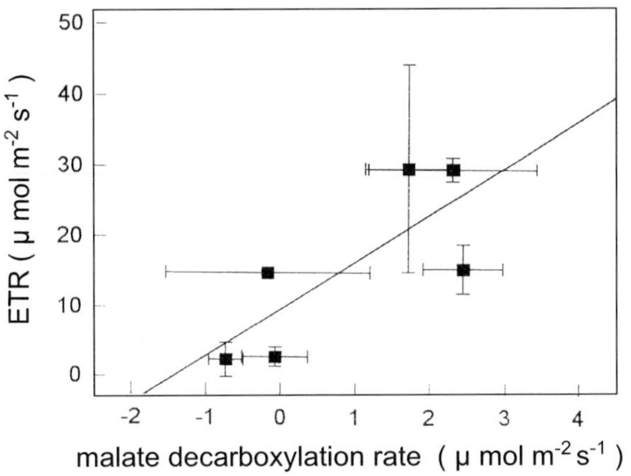

Fig. 8.4. Correlation of apparent electron transport rate of photosystem II, ETR, and internal CO_2 production by decarboxylation of malate in *C. minor* in a semideciduous dry forest in the field in Venezuela in the dry period (Grams et al. 1997)

(Roberts et al. 1997). In most CAM plants phase II is rather short in the early light period and perhaps it is really not much more than a transition phase between the dark period and the light period, where PEPC is gradually down-regulated and RubisCO is activated. However, in some *Clusia* species it has been observed that phase II can be much more extended. In *Clusia rosea* Jacq. in the field on St. John Island, Lesser Antilles, an extension of phase II to almost mid-day was observed (Fig. 8.5). The phenomenon has been studied in *Clusia* in more detail by A. Borland, H. Griffiths and their collaborators. In a comparison with the CAM-plant *Kalanchoë daigremontiana* Hamet et Perrier, in the C_3/CAM-intermediate *C. minor* phase II was more extended and PEPC was much more slowly down-regulated, i.e., it remained active for 4 h after the start of the light period in contrast to *K. daigremontiana* where PEPC was inactivated within the first 30 min (Borland et al. 1993; Borland and Griffiths 1997; Roberts et al. 1998). Roberts et al. (1997) suggest that the extended phase II may have important functions. Extended activation of PEPC in phase II increases overall organic acid accumulation and thus, increases the amount of CO_2 available from organic acid remobilization later in the light period when external conditions such as temperature and irradiance are maximal at midday. Thus, the extended phase II that is typical of several *Clusia* species may have a cardinal role in terms of photosynthetic efficiency.

In the afternoon, when the nocturnally accumulated organic acids are largely consumed depending on conditions, stomata may open and CO_2 is taken up from the atmosphere and fixed directly by RubisCO and assimilated via the Calvin cycle (Phase IV).

Expression of this typical CO_2-exchange pattern of CAM with the four phases is shown by *Clusia major* L. (Fig. 8.2C) and by *C. minor* under drought stress and at medium irradiance (Fig. 8.3B).

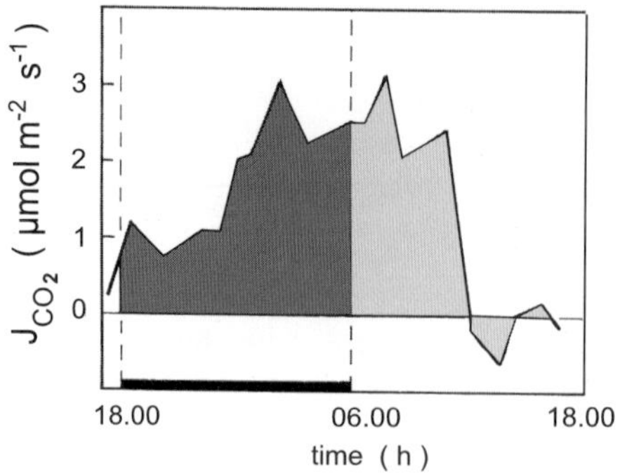

Fig. 8.5. Extended phase II of CAM (net CO_2-exchange, J_{CO2}) observed in *C. rosea* in the field on St. John Island, Lesser Antilles. The *dark bar* on the abscissa indicates dark period (from Lüttge 1991, after data of Ball et al. 1991)

Full CAM with its four phases can be modified particularly in response to drought (Smith and Lüttge 1985). As drought increases, phases II and IV start to be suppressed. A CAM CO_2 exchange pattern without expression of phase IV is shown by *Clusia alata* Planch. et Triana in Fig. 8.2D. With more severe drought, stomatal opening and CO_2 uptake may also be reduced in the dark period (phase I). This may culminate in total stomatal closure night and day. In this condition respiratory CO_2 is recycled via PEPC, organic acids are accumulated in the dark period and are subsequently decarboxylated and the CO_2 is re-assimilated to carbohydrate in the light period, a process that is driven by solar irradiance. Water loss is limited to cuticular transpiration as stomata are closed. Naturally under these conditions there is no carbon gain, but plants can use such respiratory recycling to overcome periods of drought until water is available again. This variation of full CAM is called CAM-idling (Sipes and Ting 1985). It is often observed in succulents of arid habitats during dry periods. *Clusia*s also exploit the photosynthetic flexibility conferred by the plasticity of expressing the different CAM phases to varying degrees (e.g. *C. alata*, Fig. 8.2D). While full CAM idling involves complete closure of stomata in the dark period, partial stomatal closure in phase I will not eliminate but reduce uptake of atmospheric CO_2 and at the same time internal respiratory CO_2 can be re-fixed.

The extent to which this occurs can be measured by comparing CO_2-gas exchange and nocturnal malic acid accumulation. In *Clusia*s in addition to malic acid citric acid may also be accumulated. Citrate does not count in the context of CO_2 recycling, because no net fixation of CO_2 is involved in the formation of citrate, while one CO_2 is fixed per malate accumulated (see Sect. 8.3). If uptake of atmospheric CO_2 accounts for all the malic acid accumulated, recycling of respiratory CO_2 is zero, and conversely, in full CAM-idling recycling is 100 %. Any values in between may be observed. In Table 8.1 examples for a recycling of 28–89 % are given for four species of *Clusia* studied in a phytotron. In *C. minor* 100 % recycling, i.e. full CAM-idling, was observed in plants grown at an irradiance of 260–300 µmol photons m^{-2} s^{-1}

Table 8.1. Recycling of respiratory CO_2 in phase I of CAM in four species of *Clusia* in a phytotron

	CO_2	mal	citr	RCR	RCR %
C. venosa	2.2	19.7	6.9	17.5	89
C. minor	54.1	85.2	37.0	31.1	37
C. major	109.4	151.9	61.0	42.5	28
C. alata	91.2	192.4	103.0	101.2	53

CO_2=uptake of atmospheric CO_2 integrated for the whole dark period
mal=accumulation of malate
citr=accumulation of citrate
RCR=respiratory CO_2 recycling equal to mal minus CO_2
All data are mmol m^{-2} leaf surface, RCR is also given in % of total malate accumulated (Franco et al 1990)

and studied at various temperatures in the day and in the night, i.e. in many cases when day and night temperatures were identical or similar there was nocturnal malate accumulation without a net uptake of CO_2 (Haag- Kerwer et al. 1992). In this study citrate accumulation also occurred in the absence of net CO_2 uptake, i.e. at day/night temperatures of 20/20 and 25/25 °C. Although no net recycling of respiratory CO_2 is involved in citrate accumulation we may call this CAM-idling with a recycling of whole carbon skeletons from hexose to citrate (see also Sect. 8.3). (More detailed results of this elaborate experiment are presented in Fig. 8.29 and Sect. 8.8.3 when the role of temperature as an external control parameter is discussed.)

Another mode of CO_2 assimilation is CAM-cycling (Sipes and Ting 1985). Here stomata are open during the light period and plants perform normal C_3-photosynthesis. Stomata close in the dark period and plants fix respiratory CO_2 via PEPC and store it in the form of organic acids. In the light period CO_2 released from the organic acids supplements CO_2 taken up from the atmosphere for photosynthesis via RubisCO. CAM-cycling has been considered as a kind of an incipient CAM, i.e. in evolution as a first step towards development of full CAM (Lüttge 2004). So far we do not have much information about typical CAM-cycling among *Clusias*. It appears, however, that *Clusia arrudae* Planch. et Triana under drought conditions in the field is performing gas exchange and organic acid accumulation close to the features of CAM-cycling. We have unpublished data showing that at the end of the dry season *C. arrudae* behaved neither as a typical CAM-cycling plant – since photosynthesis only took place during a couple of hours in the early morning and not throughout the whole day – nor as a typical CAM-idling plant – since there was nocturnal acid accumulation but stomata were not continuously closed.

However, Veste et al. (2001) found a similar photosynthetic pattern in *Monilaria moniliformis* and classified it as a CAM-cycling plant, and Martin (1996) argued that CAM-cycling plants seldom perform CAM sensu stricto, but rather exhibit C_3 gas exchange patterns concomitant with overnight acid accumulation, which also fits the pattern of *C. arrudae*. *C. minor* under certain conditions is able to take up CO_2 from the atmosphere day and night, around the clock (Figs. 8.2B and 8.3C). This is not typical CAM-cycling as this would largely dwell on internal recycling of respiratory CO_2. On the other hand, the combination of night time and day time CO_2 fixation is the characteristic feature of CAM cycling. Thus, we might possibly discuss the CO_2 exchange pattern of *C. minor* shown in Fig. 8.2B in relation to CAM-cycling. Conversely, the pattern of Fig. 8.3C may be closer to a full CAM with strongly expressed phases II and IV. A strong expression of phase II is often observed among *Clusias* (see above and Fig. 8.5). However, real CAM-cycling was seen in some other experiments with *C. minor*, where the effects of nitrogen nutrition on CAM expression were studied, and which are therefore presented below in Sect. 8.8.4 (Table 8.10; Franco et al. 1991). Plants grown with and without nitrogen at low PPFD and plants grown at higher PPFD without nitrogen showed no CO_2 uptake and a small loss of CO_2 in the dark period. Nevertheless they nocturnally accumulated appreciable amounts of citrate plus a small amount of malate in the case of high PPFD minus nitrogen while daytime CO_2 uptake was substantial (Table 8.10). These are clear indications of CAM-cycling with nocturnal accumulation of mainly citrate and some malate. At the higher PPFD plus nitrogen daytime CO_2 uptake was highly increased and there was only a very low night time CO_2 uptake which, however, was accompanied by quite large accumulations of both malate and citrate, a situation that also comes very close to CAM-cycling, i.e. high C_3-photosynthesis in the light period plus acid accumulation in the dark period with very little CO_2 uptake from the atmosphere.

Finally, there are many species of *Clusia* which are C_3/CAM intermediate. This means that they can readily switch between these two major modes of photosynthesis. We know many C_3/CAM-intermediate species in different phyla of the angiosperms. In well studied cases, such as *Mesembryanthemum crystallinum* L. and *Kalanchoë blossfeldiana* cv. Tom Thumb Poelln., the switch is only in one direction, i.e. from C_3-photosynthesis to CAM, and in addition to environmental factors intrinsic developmental programmes are involved in generation of the switch (see Lüttge 2004). For hemi-epiphytic *Clusia* species it has been assumed initially that CAM would be a particular property of the epiphytic stage and not so much expressed in free standing trees (Ting et al. 1987). However, this has not been supported by much later work. Evidently CAM is expressed in both life forms and there is no developmental programme modulating CAM expression in *Clusias* (Wanek et al. 2002). Moreover, for a tropical tree like *Clusia* a one-way irreversible switch from C_3 to CAM would confer limited physiological advantages. *Clusia* leaves are persistent and often used for several vegetation periods. In *C. multiflora* H.B.K. only leaves older than two years in the furthest position relative to the apex are shed (Olivares 1997). If the two modes of photosynthesis, C_3-photosynthesis and CAM, respectively, are different options fit for different environmental conditions only repeated reversible switching between the two modes would suit leaves persistent under varying conditions. Thus, reversible C_3-CAM-C_3 changes are commonly observed among *Clusia*s (Sect. 8.8).

Hence, we can conclude that the *Clusia* morphotype and even individual species, such as *C. minor* in particular (Fig. 8.3) but many others in addition, are very flexible in expressing up to four photosynthetic physiotypes, i.e.

- C_3-photosynthesis
- Full CAM with flexible expression of the different CAM phases including CAM-idling
- C_3/CAM-intermediate behaviour
- CAM-cycling

8.2 Stable Carbon Isotope Signatures

Because of the high flexibility of *Clusia*s with respect to photosynthetic metabolism it is highly desirable to have a tool to determine the extent to which species and individual plants make use of options for C_3-photosynthesis and CAM, respectively, and when performing CAM the relative magnitude of carbon acquisition in phases I and IV. Assessing the contributions made by the different photosynthetic pathways to plant carbon balance may be achieved via stable carbon isotope analysis. The stable carbon isotope ^{13}C in nature has an overall abundance of 1.11 %, while the abundance of ^{12}C is 98.89 %. During carbon acquisition and assimilation from the inorganic CO_2 in the environment there are different mechanisms affecting discrimination against the heavy isotope ^{13}C. Isotope discrimination is influenced by stomatal conductance to CO_2, diffusion of CO_2 through the leaf mesophyll as well as other thermodynamic consequences of discrimination in metabolic reactions. However, all these effects are quantitatively over-ruled by the large differences between the $^{13}CO_2$ discrimination of the two key enzymes of primary CO_2 fixation, namely RubisCO and PEPC (Table 8.2). The discrimination of RubisCO to ^{13}C in CO_2 is +27‰. PEPC has the much lower discrimination of −5‰ to ^{13}C in CO_2, which is determined by the isotope effect of the hydration of CO_2 (−7‰) because the actual substrate of PEPC is not CO_2 but bicarbonate, HCO_3^-. Thus, relative to the actual substrate HCO_3^- PEPC has a ^{13}C discrimination of +2.0‰ (Table 8.2). Stable carbon isotope ratios of plant material are obtained by mass spectrometry of the CO_2 released from combusted samples and expressed as

Table 8.2. ^{13}C-discrimination (‰) in various steps of the CO_2 fixation process (from Ziegler 1994)

Step	^{13}C isotope discrimination
Diffusion of CO_2 in the gas phase	4.4
Dissolution of CO_2 in water	–0.9
Liquid phase diffusion of CO_2 or HCO_3^-	0.0
Hydration of CO_2	–7.0
Carboxylation of phospho*enol*pyruvate	
relative to HCO_3^-	2.0
relative to CO_2	–5.0
Carboxylation of ribulose-bisphosphate	27

$$\delta^{13}C = \left(\frac{^{13}C/^{12}C \text{ of sample}}{^{13}C/^{12}C \text{ of an international standard}} - 1 \right) \times 10^3 \,(‰) \tag{8.3}$$

Thus, the enzyme responsible for the primary fixation of CO_2 grossly determines the carbon isotope signature of the plant material analysed. Therefore, C_3 plants with primary CO_2 fixation via RubisCO have highly negative $\delta^{13}C$ values and CAM plants where PEPC dominates CO_2 fixation have less negative $\delta^{13}C$ values due to the high and low discrimination, respectively, of the two enzymes against $^{13}CO_2$. In this way C_3 and CAM plants can often be readily distinguished.

However, in CAM plants the leaf $\delta^{13}C$ signature is also determined by the extent to which the different phases of CAM contribute to carbon balance. Borland et al. (1993) have measured instantaneous ^{13}C-discrimination concurrently with CO_2 exchange during the four CAM phases in *C. minor*. In their analysis instantaneous discrimination according to Evans et al. (1986) is defined as

$$\Delta^{13}C = \frac{\xi\,(\delta_o - \delta_e)}{\Delta + \delta_o - \xi(\delta_o - \delta_e)} \;(‰) \tag{8.4}$$

where $\xi = p_e/(p_e - p_o)$ and p_e and p_o are CO_2 partial pressures of the air entering and leaving the gas exchange chamber, respectively, when a leaf is enclosed. δ_o is the carbon isotope ratio (see Eq. 8.3) of the air leaving the gas exchange chamber with a leaf enclosed and δ_e the carbon isotope ratio in control air leaving a gas exchange chamber without a leaf. While for $\delta^{13}C$ negative values indicate ^{13}C discrimination (Eq. 8.3) in the case of $\Delta^{13}C$ (Eq. 8.4) discrimination is given by positive values. Figure 8.6 shows that discrimination was lowest in phase I of CAM where PEPC is the only CO_2 fixing enzyme, higher in phase II where both PEPC and RubisCO participate in CO_2 fixation, and highest in phase IV where CO_2 fixation is dominated by RubisCO.

Hence, the range of $\delta^{13}C$ values overall obtained from CAM plants is much broader than that of C_3 plants. Moreover, $\delta^{13}C$ signatures also indicate to which extent C_3/CAM-intermediate plants made use of their option for the two different modes of photosynthesis during the life time of the plant material sampled and analysed. In CAM plants, $\delta^{13}C$ values correlate linearly with the proportions of CO_2 taken up during the light and dark (Winter and Holtum 2002). Thus, stable carbon isotope analysis is a very powerful tool for assessing the expression of photosynthetic physiotypes and their variants. The plant material required is easily sampled and transported because only small samples of dried tissue are required and as suggested below the interpretation of $\delta^{13}C$ data can supply very important information. Nonetheless, we must note that $\delta^{13}C$ values per se do not allow to decide conclusively if a plant has an intrinsic potential for CAM or not, because the ability to perform CAM may be hidden by highly negative $\delta^{13}C$ values if the plant actually makes little use of its CAM option and only expresses CAM for limited periods over the life span of the leaf. Additional physiological information is required to assess the inherent capacity for CAM.

$\delta^{13}C$-values for a number of *Clusia* species are shown in Fig. 8.7. The range is continuous from as negative as about –30‰ in *C. parviflora* Saldanha et

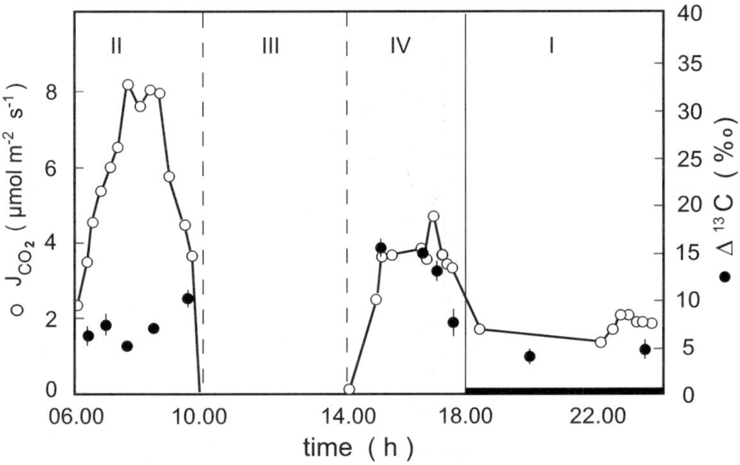

Fig. 8.6. CO_2-exchange (J_{CO2}, *open circles*) and ^{13}C-discrimination ($\Delta^{13}C$ according to Eq. (8.4), *closed circles*) by sun exposed leaves of CAM-performing *C. minor* measured in the dry season in the field in Trinidad. The *solid bar* on the abscissa indicates the dark period, *Roman numbers* give the CAM phases (Fig. 4a in Borland et al. 1993)

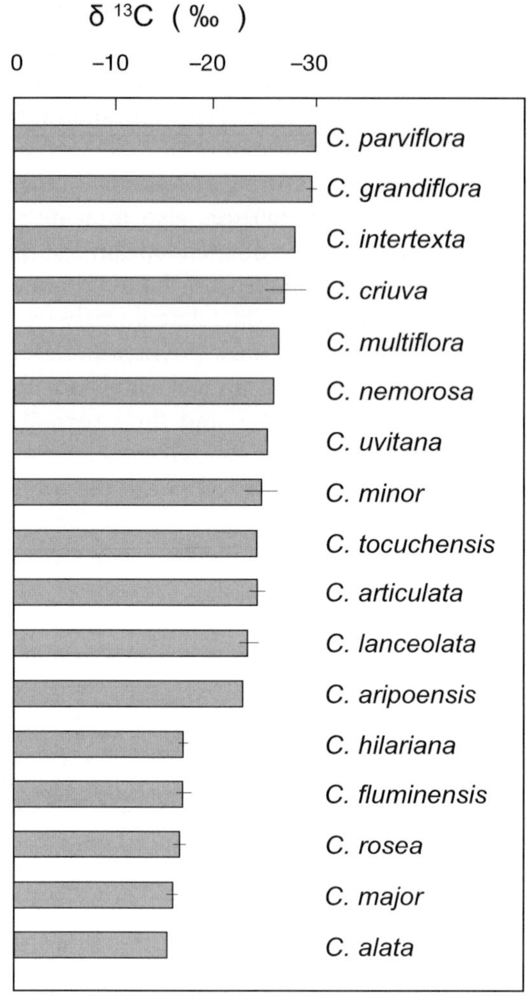

Fig. 8.7. $\delta^{13}C$ values obtained for a range of *Clusia* species (Fig. 1 in Lüttge 1999)

Engl. to the much less negative value of about −15‰ in the obligate CAM species *C. alata*. One can compare these values with results obtained from measurements of gas exchange and night/day changes of organic acid levels. The physiological measurements indicate the capability of the plants to perform C_3-photosynthesis or CAM under the given conditions of the experiments, while the $\delta^{13}C$-values provide information on the actual performance over time. Thus, it is seen in Fig. 8.7 that, for instance, the C_3/CAM-intermediate species *C. minor* under natural conditions makes more use of C_3-photosynthesis than of CAM as its $\delta^{13}C$ values are rather negative.

A more extensive survey was performed on *Clusia* species from Panamá where night/day changes of acidity (ΔH^+) were measured in 25 species and $\delta^{13}C$ values analysed for 38 species (Holtum et al. 2004). The distribution of

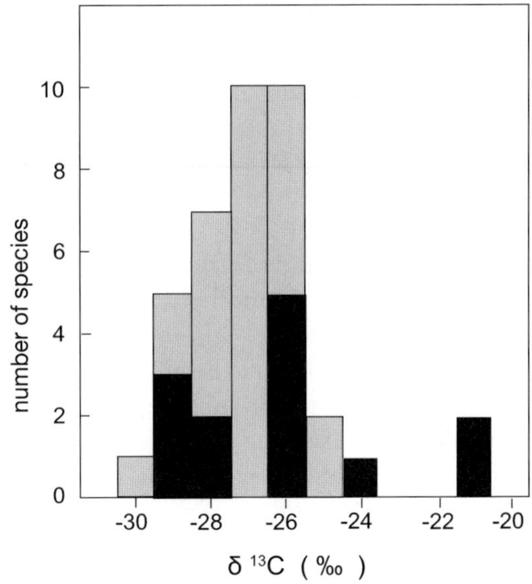

Fig. 8.8. Range of $\delta^{13}C$ values obtained from mature leaves of *Clusia* species growing in their natural environment in Panamá and the presence (*solid bars*) and absence (*shaded bars*) of an intrinsic CAM potential as derived from ΔH^+- measurements (Fig. 3 of Holtum et al. 2004)

values is shown in Fig. 8.8. The values exhibit a predominant C_3-type peak ($\delta^{13}C$ –24 to –30‰), in which the weak CAM performing species are hidden as identified by ΔH^+, and a very small more CAM-like peak at less negative $\delta^{13}C$ values. This suggests that in the field, at least in Panamá, strong CAM is the exception rather than the rule in *Clusia*. Moreover, like *C. minor* mentioned above, many other *Clusia*s which have the intrinsic potential to perform CAM make predominant use of the C_3-option in the field.

8.3 Biochemistry of Crassulacean Acid Metabolism (CAM)

The standard pathway of CAM is that CO_2 is fixed nocturnally via phospho*enol*pyruvate carboxylase (PEPC), malate (mal) is synthesized by reduction of the oxaloacetate (OAA) formed and stored in the cell vacuoles, from where it is mobilized during the day and decarboxylated, and the CO_2 regenerated is assimilated via ribulose-bis-phosphate carboxylase/oxygenase (RubisCO) in the Calvin cycle (Sect. 8.1). The precursor, phospho*enol*pyruvate (PEP) for nocturnal dark fixation of CO_2 is usually formed from starch stored during the day and broken down to PEP via glycolysis during the night. However, CAM in *Clusia* shows several deviations from this general scheme. Therefore, in this section we need to consider CAM metabolism in more detail. (The reader may note that here just once we have deliberately used the pleonasm, which is so frequently produced in the literature, as a warning: The *M* in CAM, of course, is already standing for *metabolism*.)

8.3.1 Turnover of Carbohydrates

The relevant reactions of intermediary metabolism are summarized schematically in Fig. 8.9, where major intermediates are shown so that turnover of energy equivalents (ATP and reducing potential [2H]) can also be assessed (for details see Holtum et al. 2005). The stoichiometries are based on one C_6-unit (hexose-unit).

In the dark period of the CAM cycle PEP as the acceptor of CO_2 in dark fixation via PEPC is formed glycolytically from carbohydrates. In some CAM species, e.g. pineapple (*Ananas*; Black et al. 1996), in addition to glucans (starch) mobilized from the chloroplasts the precursors for PEP may include free sugars (hexoses or sucrose) mobilized from the vacuoles. An important involvement of free sugars is also very often observed in the CAM of *Clusia* (Fig. 8.10 and Table 8.3). Hexose-phosphate is thought to be mobilized from starch by phosphorolysis. Free hexoses are assumed to be released passively

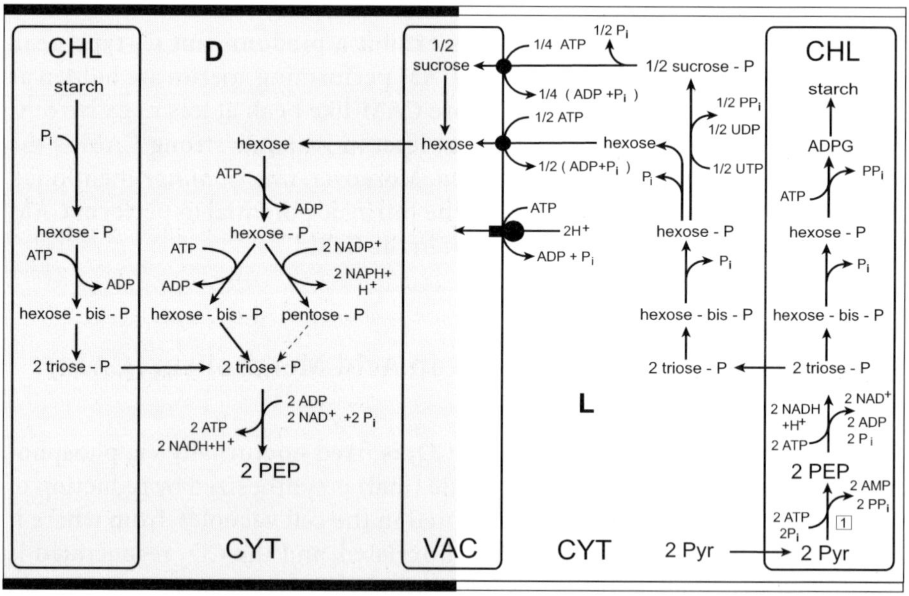

Fig. 8.9. Turnover of carbohydrates in the cycle of CAM, with nocturnal formation of phospho*enol*pyruvate and daytime regeneration of carbohydrate from pyruvate. Nocturnal formation of triose-phosphate is via glycolysis, and an alternative pathway to triose-phosphate is from pentose-phosphate via the oxidative pentose-phosphate cycle. Daytime regeneration of carbohydrates is mediated via gluconeogenesis. Reaction sequences are abbreviated. An important individual key enzyme is (1)=pyruvate-P_i-dikinase. CHL=chloroplast, CYT=cytosol, D=dark period, L=light period, P=phosphate residues, PEP=phospho*enol*pyruvate, pyr=pyruvate, P_i=free inorganic phosphate, VAC=vacuole

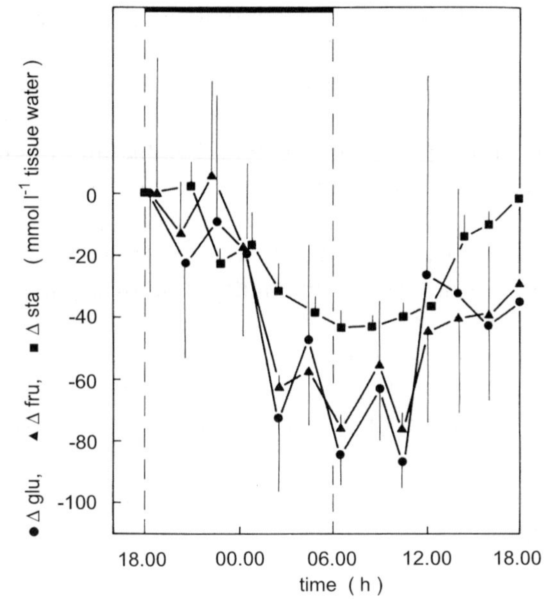

Fig. 8.10. Diurnal changes of the levels of glucose (*circles*), fructose (*triangles*) and starch (*squares*) during the CAM-cycle in *C. rosea* in the field on St. John Island, Lesser Antilles. Starch is given in hexose units. Changes of carbohydrates were related to the levels obtained at 18:30 h at the start of the measurements, namely glucose 98.0 mmol L^{-1} tissue water, fructose 84.7 mmol L^{-1} and starch 50.4 mmol L^{-1}. The *dark bar* on the abscissa indicates dark period (Fig. 4 from Ball et al. 1991)

Table 8.3. Dusk minus dawn values of the levels of free sugars (glucose: glu, fructose: fru, sucrose: su) and of starch (in hexose units) in various species of CAM-performing *Clusia* sampled in the field. Data are in mmol L^{-1} of plant water or fresh weight and were adapted from the references indicated. (*C. hilariana* is not mentioned in the text before, the authority is Schlecht.)

		glu	fru	su	starch	Referencee
C. alata	Dry season	93	94	13	n.d.	Popp et al. 1987
	End of dry season	45	38	−10	22	
C. rosea	Dry season	92	79	5	n.d.	
	End of dry season	174	165	−8	31	
C. rosea		75	85	0	43	Ball et al. 1991
C. hilariana		|→	111 ←|	0	17	Berg et al. 2004
C. minor	Exposed leaves					
	Wet season	|→	65	←|	31	Borland et al. 1994
	Exposed leaves					
	Dry season	|→	81	←|	39	
	Shaded leaves					
	Dry season	|→	49	←|	105	

from the vacuole and sucrose to be split hydrolytically into hexoses by invertase. Thus, there is a net production of one ATP when the precursor is starch but not when it is free hexose or sucrose, and in all three cases 2 [2H] are formed when one hexose unit is broken down to PEP (Table 8.4B).

During the day, the decarboxylation of organic acids formed in the previous night, generates pyruvate (pyr). Therefore, we start the scheme of carbohydrate turnover of CAM in the light period with pyr (Fig. 8.9). PEP is formed from pyr by pyr-Pi-dikinase (PPDK). Reduction equivalents ([2H]) are consumed for reduction of PEP to triose-phosphate. Formation of sucrose and starch is energetically a little more costly than formation of hexose (Table 8.4B, where UTP is taken to be equivalent to ATP). Daytime storage of free sugars, hexoses and sucrose, in the vacuole needs energy for secondary active

Table 8.4. Energy budgets based on one hexose unit of CAM-cycles with different carbohydrates, starch, free hexose or sucrose, and organic acids, malate or citrate, as shown in Figs. 8.9, 8.12 and 8.14. (A) Organic acid turnover, (B) carbohydrate turnover and overall budgets.

(A)	Dark period		Light period			
	ATP	[2H]	ATP	[2H]		
2 PEP → 2 vacuolar mal	−2	−2				
2 vacuolar mal → 2 pyr			0	+2		
2 PEP → 1 vacuolar citr	−1.5	+1				
1 vacuolar citr → 1 pyr			+1	+4		
(B)	**Starch**		**Free hexose**		**Sucrose**	
	ATP	[2H]	ATP	[2H]	ATP	[2H]
Carbohydrates						
Dark period formation of PEP	+1	+2	0	+2	0	+2
Light period recycling of pyr	−5	−2	−4.5	−2	−4.75	−2
Malate and carbohydrates						
Dark period	−1	0	−2	0	−2	0
Light period	−5	0	−4.5	0	−4.75	0
Net	−6	0	−6.5	0	−6.75	0
Citrate and carbohydrates						
Dark period	−0.5	+3	−1.5	+3	−1.5	+3
Light period[a])	−10.5	−3	−10.25	−3	−10.375	−3
Net	−11.0	0	−11.75	0	−11.875	0

[a] Only one pyr is regenerated, 3 CO_2 need to be fixed in the Calvin cycle to regenerate the second C_3 compound.

[2H]=reduction equivalent; citr=citrate, mal=malate, pyr=pyruvate; +=net production, −=net consumption.

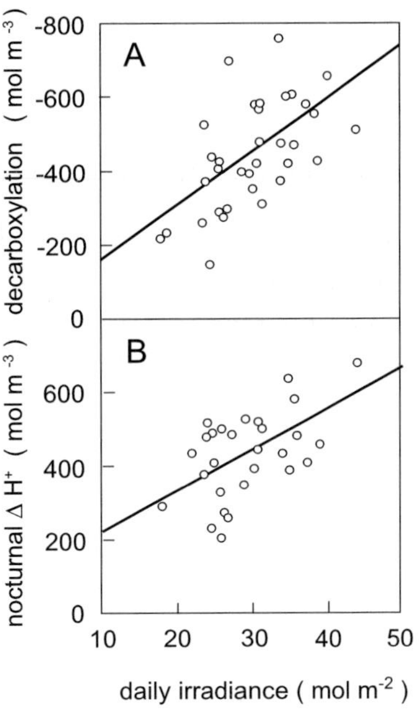

Fig. 8.11A,B. Correlations of: **A** organic acid decarboxylation in phase III of CAM to daily irradiance received; **B** organic acid accumulation in the following dark period (phase I) in *C. minor* (after Fig. 2a,b of Roberts et al. 1998)

transport via sugar translocators at the tonoplast, and it is assumed that free sugars (hexoses, sucrose) are transported into the vacuole by H^+/sugar co-transport driven by the tonoplast H^+-ATPase pumping 2 H^+ into the vacuole per ATP hydrolysed (Holtum et al. 2005).

Energetically the amount of organic acids accumulated during phase I of CAM depends on the irradiance received during the previous light period which determines both the degree of organic acid breakdown in phase III and the accumulation of photosynthetic products, viz. carbohydrates, for the generation of phospho*enol*pyruvate (PEP) as CO_2 acceptor in phase I (Kluge 1968; Nobel 1988). These relations have also been documented in leaves of *C. minor,* where positive linear correlations between decarboxylation and nocturnal acidification, respectively, and light period irradiance have been observed (Fig. 8.11; Roberts et al. 1998). In *C. uvitana* Pittier high acid levels remaining in the leaves at dusk inhibited organic acid accumulation during the following night (Zotz and Winter 1993).

8.3.2 Organic Acid Turnover with Nocturnal Storage of Malic Acid

At night, malate is formed via PEPC and malate dehydrogenase (Fig. 8.12). Nocturnal energy metabolism when starting from PEP is determined by the consumption of [2H] by malate dehydrogenase and of ATP for vacuolar accumulation, where malate is transported into the vacuoles via an inward recti-

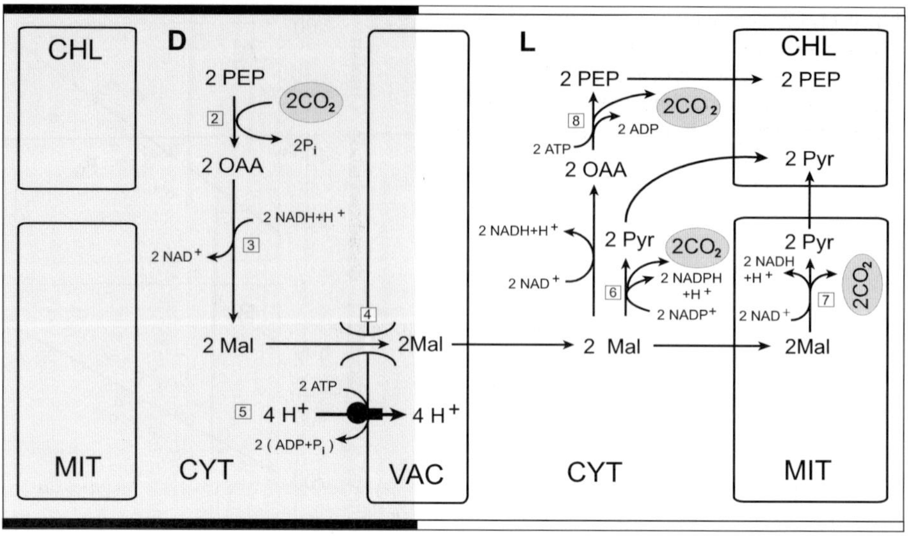

Fig. 8.12. Organic acid turnover in the cycle of CAM with nocturnal storage of malic acid. General explanation as for Fig. 8.9. Abbreviations not yet explained in Fig. 8.9 are as follows: MIT=mitochondrion, Mal=malate, OAA=oxaloacetate. Key enzymes are: (2) PEP-carboxylase (PEPC), (3) malate dehydrogenase, (4) inward rectifier vacuolar malate channel, (5) H^+-transporting ATPase of the tonoplast (V-ATPase), (6) cytosolic NADP-dependent malate decarboxylase or malic enzyme (NADP-ME), (7) mitochondrial NAD-dependent malate decarboxylase (NAD-ME), (8) PEP-carboxykinase (PEPCK)

fier malate channel (Hafke et al. 2003) energetically driven by the H^+-transporting ATPase at the tonoplast (V-ATPase), pumping 2 H^+ per ATP (Table 8.4A; Holtum et al. 2005).

During the light period malate is remobilized passively from the vacuole (Lüttge and Smith 1984). Malate can be decarboxylated to pyr by NADP-dependent malate decarboxylase or "malic enzyme" (NADP-ME) in the cytosol or by NAD-dependent malate decarboxylase (NAD-ME) in the mitochondria. Transport of malate into the mitochondria could occur via energetically neutral exchange systems (Holtum et al. 2005). A third way of malate decarboxylation is via oxidation to OAA and then by PEP-carboxy-kinase (PEPCK) to PEP. This reaction is realized in *Clusia*s (Black et al. 1996; Borland et al. 1998; Holtum et al. 2005). Energetically all three mechanisms are identical. In the scheme of Fig. 8.12 the way via PEPCK appears to be more costly, but it already produces PEP instead of pyr so that the first reaction of light period carbohydrate turnover shown in Fig. 8.9 is saved. Pyruvate is largely transferred to gluconeogenesis and not further broken down to CO_2 (Robinson et al. 1992; Holtum et al. 2005). To avoid futile recycling of CO_2 via PEPC in the light period the activity of PEPC, which has a 60-fold higher affinity for CO_2 than RubisCO, needs to be down-regulated. This is achieved by reversible

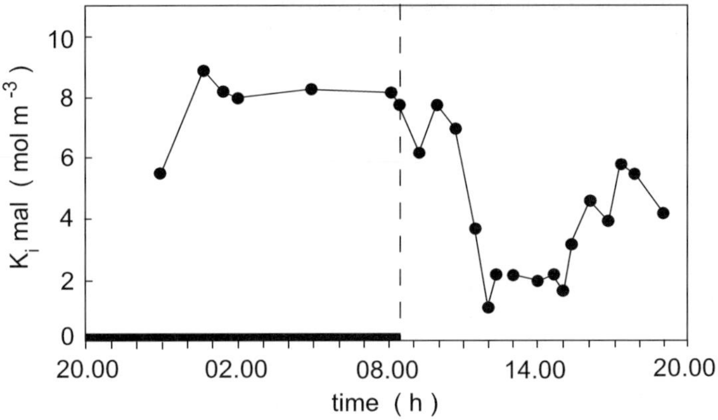

Fig. 8.13. Diurnal changes of $K_{i\text{-mal}}$ for extracts of PEPC prepared from leaves of *C. minor*. (Fig. 7a of Borland and Griffiths 1997)

phosphorylation of the enzyme (Nimmo et al. 1987; Kusumi et al. 1994; Carter et al. 1995a, b, 1996). The night-form of PEPC is phosphorylated and is much more active and less sensitive to lowered pH and to feedback inhibition by its product malate than the de-phosphorylated day form. For *C. minor*, Borland and Griffiths (1997) have demonstrated this phenomenon by analysing the malate inhibitor constant, $K_{i\text{-mal}}$, of PEPC during the diurnal CAM-cycle (Fig. 8.13), showing that inhibition was low (high $K_{i\text{-mal}}$) in the dark period and increased in the light period (lower $K_{i\text{-mal}}$).

Turnover of malate leads to a net gain of carbon. Per malate accumulated 1 CO_2 is fixed (2 CO_2 per hexose unit turned over) in the dark period, which is released in the light period and available for assimilation via RubisCO in the Calvin cycle (Fig. 8.12).

8.3.3 Organic Acid Turnover with Nocturnal Storage of Citric Acid

A number of CAM species are known from the literature, which nocturnally accumulate citric acid in addition to malic acid (Milburn et al. 1968; Lüttge 1988). Generally, citric acid turnover in the CAM cycles is much less important than that of malic acid with day/night changes of citric acid levels of up to 26 mmol kg^{-1} fresh weight in some species of *Kalanchoë*. However, in the CAM cycle of *Clusia*s diurnal changes of citric acid levels may be much higher, i.e. up to 200 mM (Franco et al. 1992), so this is a special trait of CAM in *Clusia*. Citrate metabolism of CAM (Fig. 8.14) is more complex and much less well understood than that of malate.

Starting from two molecules of PEP citrate is formed nocturnally from OAA produced by PEPC in the cytosol and via pyr and generation of acetyl-CoA in the mitochondria. We note that one CO_2 is fixed to one molecule of PEP via PEPC and at the same time one CO_2 is lost by oxidative decarboxylation of pyr obtained from the second molecule of PEP when acetyl-CoA is

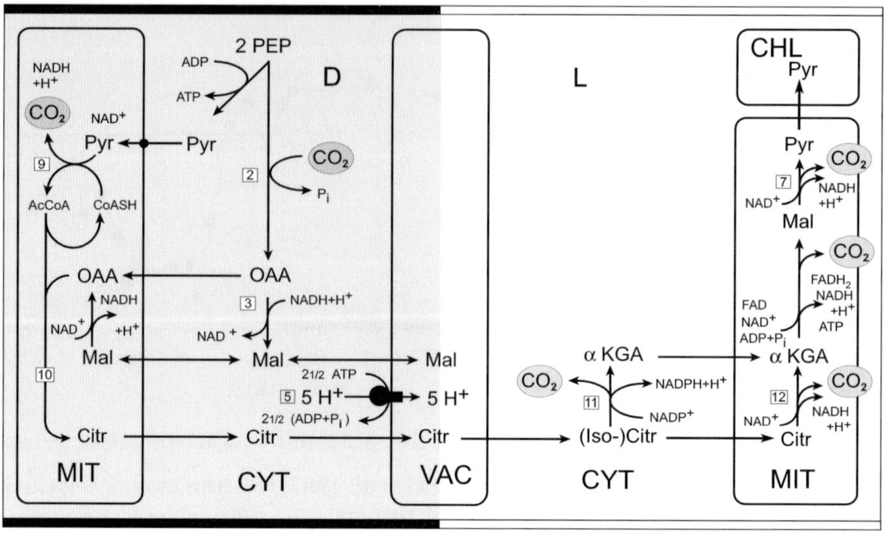

Fig. 8.14. Organic acid turnover in the cycle of CAM with nocturnal storage of citric acid. General explanation as for Fig. 8.9, abbreviations not yet explained in Figs. 8.9 and 8.12 are as follows: AcCoA=acetyl coenzyme A, Citr=citrate, CoASH=coenzyme A, αKGA=α-keto-glutaric acid. Additional key enzymes are: (9) pyruvate decarboxylase, (10) citrate synthetase, (11) cytosolic NADP-dependent iso-citrate dehydrogenase, (12) mitochondrial NAD-dependent iso-citrate dehydrogenase

formed. Hence, in contrast to malic acid, nocturnal accumulation of citric acid is not accompanied by a net gain of carbon. For energy stoichiometry (Table 8.4A) we must note that pyruvate transport into the mitochondria is consuming some of the electrochemical proton gradient at the inner membrane of mitochondria. In the energy budget we take this as being equivalent to one ATP consumed per pyruvate taken up. Moreover, we assume that vacuolar accumulation of citric acid is energized by the V-ATPase in the same way as that of malic acid but using 3 H^+ per citrate transported (see Holtum et al. 2005). Pulse-chase experiments with *C. minor* using radioactively labelled $^{14}CO_2$ have shown that in nocturnal citrate formation malate is formed first and even transported transiently into the vacuoles from where it is re-allocated to the mitochondria to regenerate OAA for citrate synthesis (Fig. 8.15; Olivares et al. 1993; see also Kalt et al. 1990). This peripatetic flow of malate may consume one extra ATP for transport into the vacuoles. The biochemical reactions (OAA to malate in the cytosol and malate to OAA in the mitochondria) together are energetically neutral. Overall, nocturnal citric acid accumulation appears to be energetically somewhat more favourable than malic acid accumulation. ATP consumption is less because one ATP is formed in the reaction from PEP to pyr (pyruvate kinase) and one ATP each is consumed for the peripatetic flow of malate and

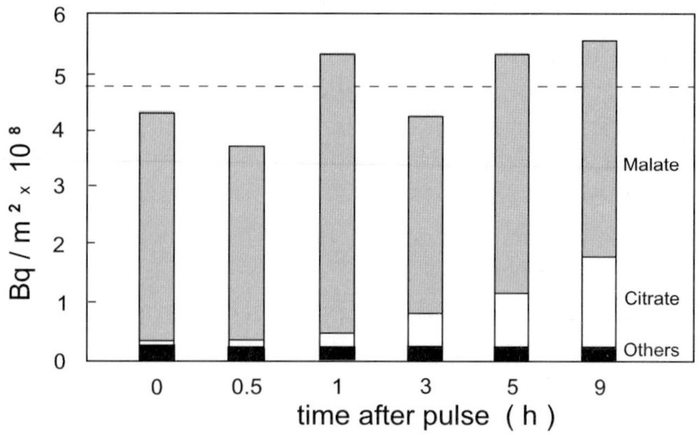

Fig. 8.15. Distribution of ^{14}C label in malate, citrate and other compounds after an exposure of detached leaves of *C. minor* to $^{14}CO_2$ during the last hour of the light period and during a subsequent chase in normal air (0.5 to 9 h). The *horizontal line* gives the average of total radioactivity in the samples taken for the times indicated. It is known that at the end of the light period in the CAM-cycle PEPC already may dominate CO_2 fixation (Kluge et al. 1982). This is corroborated here as almost all of the label of a ^{14}C-pulse at the end of the light period is found in malate. This malate must be sequestered in the vacuole. During the dark period some of the label is gradually transferred to citrate without an overall loss of label from the leaves indicating that some of the malate is transported from the vacuoles into the mitochondria for citrate synthesis and citrate transported back into the vacuoles (From data of Olivares et al. 1993)

citrate during transport into the vacuole. Moreover, starting from PEP in citrate production one [2H] is formed while in malic acid accumulation [2H] is consumed (Table 8.4A). The overall nocturnal budget including carbohydrate breakdown for citrate is a gain of 3 [2H] while for malate it is zero (Table 8.4B). ATP can be formed from the [2H] generated, and overall nocturnal citric acid accumulation remains energetically favourable. It must be noted, however, that in contrast to these theoretical evaluations of paper-biochemistry, measurements of nocturnal citrate accumulation and respiratory O_2 consumption have shown for four species of *Clusia* (*C. venosa, C. minor, C. major* L. and *C. alata*) that O_2-consumption was lower by a factor of 1.3 to 3.1 than theoretically expected from measured citrate accumulation (Franco et al. 1990), and hence, only part of the reducing power generated by the synthesis of citrate enters the respiratory chain.

No investigations are available on metabolic pathways starting with citrate in the light period. This is one of the astonishing gaps in CAM-research. Hence, we are left here with assumptions based on general biochemistry (Fig. 8.14). When released from the vacuole citrate can be decarboxylated via aconitase and iso-citrate by iso-citrate dehydrogenase in the cytosol or in the mitochondrial tricarboxylic acid cycle. Two NADP-dependent iso-citrate dehydrogenases are known in plants, one is located in chloroplasts and only occurs in ferns and dicotyledons, the other one is cytosolic and appears to be the major form (Chen 1998; Popova and Appenroth 2002; Popova et al. 2002). Another pathway for cytosolic citrate breakdown is via ATP-dependent citrate lyase (Rangasami and Ratledge 2000; Fatland et al. 2002), which generates acetyl-CoA and is involved in fatty acid metabolism (not shown in Fig. 8.14). α-Keto-glutarate generated from citrate can be further processed in the mitochondria to malate and pyr via NAD-ME which then may be used to regenerate carbohydrate (Fig. 8.9).

It has been argued that potentially citrate breakdown in the light period may generate 1, 3 or 6 CO_2. In the first case α-keto-glutarate would accumulate or would have to be used in other reactions. In the case of 6 CO_2 breakdown would have to occur in the tri-carboxylic acid cycle. This would generate a large amount of reduction equivalents, and it would be difficult to envisage how respiratory electron transport might deal with them while photosynthesis is occurring at the same time in the light (Lüttge 1988). One possibility is the cyanide resistant non-phosphorylating alternative pathway of respiration which allows mitochondrial acid breakdown to escape from the control of cellular energy charge (Rustin and Lance 1986). Cyanide-insensitive mitochondrial substrate oxidation was particularly enhanced after CAM induction in the C_3/CAM intermediate species *Kalanchoë blossfeldiana*, and an increase of cyanide resistance of leaf respiration has been observed in the early light period when acid is remobilized in the CAM cycle (Rustin and Queiroz-Claret 1985). In any case, the simplest assumption at this stage appears that 3 CO_2 are generated from each citrate as shown in Fig. 8.14. An indirect way of testing this is via apparent photosynthetic electron transport, ETR, since ETR is correlated to the rate of CO_2-production from acid decarboxylation (Sect. 8.1). In experiments with *Clusia*s in the field, in fact a linear relationship was obtained between ETR and acid decarboxylation in *C. alata* plus an unidentified species when it was assumed that citrate breakdown produces 3 CO_2 per citrate. This corroborates the stoichiometry given in Fig. 8.14. However, it needs to be mentioned that in the case of *C. rosea* a linear relationship was only obtained when assuming the production of only one CO_2 (Fig. 8.16).

With 3 CO_2 produced per citrate broken down only one pyruvate is regenerated in the light period per hexose molecule turned over (Fig. 8.14), which then may be reduced to triose-phosphate in the chloroplasts (Fig. 8.9). This means that to complete the cycle 3 CO_2 need to be fixed via RubisCO and the Calvin cycle with an energy requirement of 3 ATP and 2 [2H] per CO_2 to produce the second triose-phosphate which needs to be considered in the energy budget (Table 8.4B). Overall it is important to stress that in terms of carbon the CAM cycle using citrate is a futile cycle. There is no net gain of carbon. In the dark period the balance of CO_2 liberated and CO_2 fixed is zero. The CO_2 liberated in the light period needs to be refixed and returned into the cycle to maintain the carbon balance. Hence, the expression of a CAM cycle with cit-

Fig. 8.16A,B. Correlations between rates of CO_2 production from decarboxylation of organic acids and apparent photosynthetic electron transport rates, ETR, (A) for *C. rosea* and (B) for pooled data of *C. alata* plus an unidentified species of *Clusia* measured in the field in Venezuela. To determine decarboxylation rates from measured disappearance of organic acids in the leaf tissue during the light period, it was assumed that one malate releases one CO_2 and that one citrate releases: **A** one CO_2; **B** three CO_2. Correlation coefficients, r, are 0.93 and 0.88 in A and B, respectively (Fig. 6 of Haag-Kerwer et al. 1996)

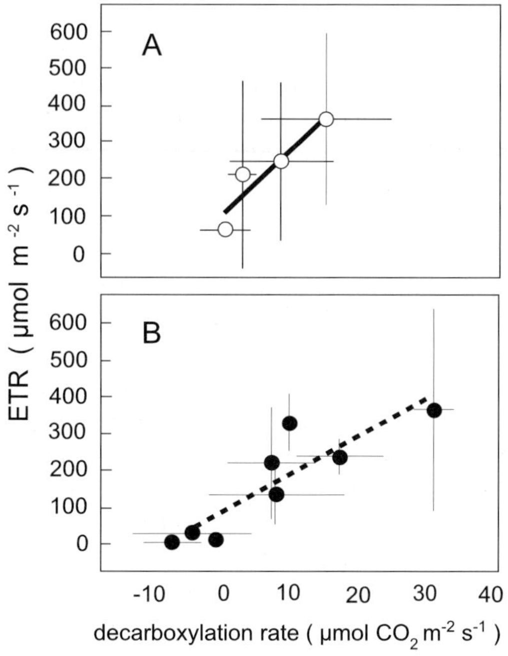

rate accumulation in *Clusias*, as large as it is under certain conditions, must have other functions than carbon acquisition, some of which will be discussed below.

8.3.4 Concluding Evaluation

1. The overall energy budgets of the CAM cycle (Table 8.4B) with both malate and citrate are only marginally different for the involvement of starch, free hexose or sucrose as carbohydrates stored during the day for nocturnal PEP production. Hence, the variation of the form of carbohydrate involved can not be well explained by energy metabolism and there may be other reasons.
2. Energetically citrate accumulation is superior to malate in the dark period. However, the recirculation of carbon from citrate to carbohydrate in the light period is much more energy demanding than in the case of malate. It is consistent with this theoretical estimate that observations with *C. minor* in the field have shown a much closer correlation of the citrate breakdown in the light period with irradiance than in the case of malate (Borland et al. 1996). The overall day and night energy budget with citrate is much more costly, i.e. almost twice that with malate (Table 8.4B).
3. Only with malate but not with citrate the CAM cycle leads to a net gain of carbon. Productivity and growth, of course, are only possible if there is car-

bon available for export from photosynthesising source leaves. The two CO_2 produced per hexose unit turned over in phase III of the CAM-cycle with malate may contribute to this. Carbon isotope signatures of exported material have shown, however, that in the C_3/CAM- intermediate species *C. minor* carbon exported from source leaves mainly originates from CO_2-fixation via RubisCO in phase IV of CAM (Borland et al. 1994).

8.4 CO_2 Concentrating Consequences of CAM

The basic function of CAM is that of a CO_2 concentrating mechanism (Lüttge 2002). This is based on the properties of PEPC which has a 60-fold higher affinity for CO_2 than RubisCO. Hence, at prevailing low atmospheric CO_2-concentrations PEPC can support carbon acquisition very effectively. However, this only leads to storage of carbon in the form of the organic acids produced in the dark period phase I of CAM. The actual CO_2-concentrating effect occurs during organic acid remobilization in the light period phase III of CAM when internal CO_2 concentrations ranging from 0.09 % to as much as 2.5 % may build up in the photosynthesising organs of CAM plants, i.e. twice to 60 times atmospheric CO_2-concentration (Lüttge 2002). CO_2 assimilation in the light at high CO_2-concentrations behind closed stomata is also accompanied by a built up of high internal oxygen concentrations of up to over 40 % (Spalding et al. 1979) in various CAM plants. Unfortunately, we have little information in this respect about CAM-performing *Clusias*. Sternberg et al. (1987) have estimated internal CO_2 concentration in *C. rosea* from gas exchange measurements when stomata were nearly closed at the time of phase III to be about 15 times that of ambient CO_2. Furthermore, we have the very early observation of Alexander von Humboldt in 1800 that *C. rosea* produced internal O_2-concentrations of up to 35 % in its leaves in the light (see Sect. 1.2 and Lüttge 2002), which remains an essential observation for us.

Although the CAM cycle with citrate is futile with respect to carbon gain (Sect. 8.3.3), citrate decarboxylation makes a somewhat larger contribution to the built up of high CO_2-concentrations in the air spaces of leaves than malate since per hexose unit turned over citrate may generate 3 CO_2 and malate only 2 CO_2 in phase III of CAM (Sect. 8.3.3). Whether elevated internal CO_2-concentration from citrate decarboxylation provides any metabolic or physiological advantage or not remains debatable (see Sect. 8.5). High internal CO_2-concentrations in phase III of CAM occurring at day times when solar irradiance is highest have often been considered to protect against high light stress because they support strong photochemical work at substrate saturation of RubisCO (Sect. 8.5). The higher energy demand of organic acid turnover with citrate during the light period as compared to malate (Sects. 8.3.3 and 8.3.4)

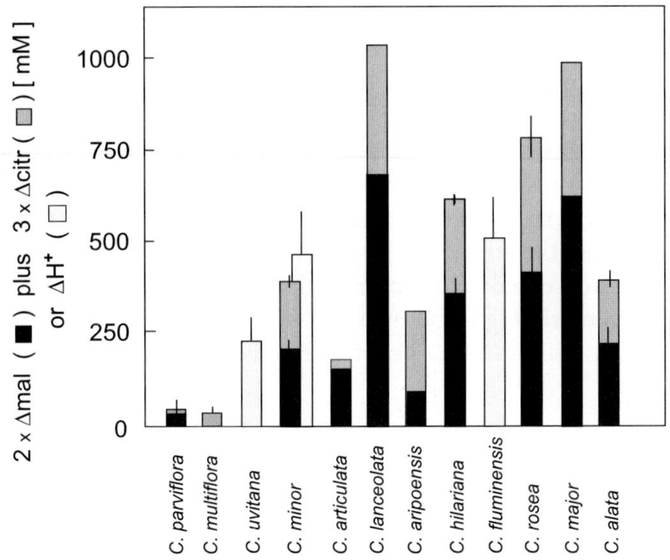

Fig. 8.17. Night/day oscillations (Δ) of organic acid accumulation and remobilisation in various species of *Clusia*. In some cases only night/day changes of titratable protons (ΔH^+) have been analysed. Values for malate (Δmal) and citrate (Δcitr) are given with a factor of 2 and 3 to account for the 2 and 3 carboxyl groups, respectively. The stoichiometry of 2 × Δmal+3 × Δcitr=ΔH^+ is realized in many analyses in different species of *Clusia* in the literature (Fig. 1 in Lüttge 1999)

may be an additional advantage for harmless use of energy at excess irradiance (Roberts et al. 1998).

Among the CAM-performing species of *Clusia* we find the strongest nocturnal acid accumulators known. The well studied model plant of CAM research, *Kalanchoë daigremontiana* Hamet et Perrier, may achieve night/day changes of malate (Δmal) of up to about 250 mM amounting to night/day changes of titratable vacuolar protons (ΔH^+) of 500 mM considering the two carboxyl groups of malic acid. With accumulation of malic acid plus citric acid (three carboxyl groups per molecule) in some *Clusia* species nocturnal acidification of a ΔH^+ of more than 1000 mM titratable protons was observed (Fig. 8.17; Franco et al. 1992; see Lüttge 1999, where a large number of analyses presented in the *Clusia* literature are summarized). These plants are also the strongest nocturnal citric acid accumulators known, with Δcitr values of up to 160 mM (Franco et al. 1992). With a nocturnal acid accumulation of ΔH^+=1 410 mM observed in *C. minor* in the field in Trinidad, Borland et al. (1992) certainly make it for the Guinness book of records with the highest nocturnal acidification ever recorded in any CAM plant.

Nocturnal vacuolar accumulation of malic and citric acid is driven by the V-ATPase at the tonoplast (Sects. 8.3.2 and 8.3.3). The capacity of accumulation is limited by the electrochemical proton gradient which is built up at the tonoplast and against which the H^+ transporting V-ATPase must pump (Lüttge and Smith 1984; Franco et al. 1992). The proton gradient is strongly affected by the buffering capacity of the vacuole regulating the concentration of free protons. Citrate is a much stronger buffer compound than malate. Therefore, citrate accumulation itself may facilitate strong accumulation of total acid in the vacuole via a positive feedback effect. Hence, it may be one of the essential functions of citrate accumulation in the *Clusias* that their very high nocturnal acid accumulation becomes possible and, of course, that such particularly high acid turnover in the CAM cycle of *Clusias* also supports particularly high CO_2-concentrating in phase III.

8.5 Photorespiration

CAM plants possess the entire biochemical complement for performance of photorespiration. However, it was long assumed that photorespiration would only occur during phase IV of CAM when plants perform C_3-like photosynthesis because the CO_2-concentrating mechanism of phase III would elevate the internal CO_2/O_2 ratio in favour of the carboxylation activity of RubisCO. In view of the concomitant O_2 concentrating effect that occurs behind closed stomata during phase III it is now realized that the situation is more complex. Studies of photorespiration online with photosynthesis are not available for CAM plants in general.

However, data have been obtained for *C. minor* in both the C_3- and the CAM-state using automatic applications of air with only 1 % O_2 for 20 min at intervals during recordings of photosynthetic parameters, such as net CO_2 exchange, J_{CO2}, leaf conductance for water vapour, g_{H2O}, and internal CO_2 partial pressure, p^i_{CO2}, and relative quantum use efficiency of photosystem II, $rel\Phi_{PSII}$ (Fig. 8.18; Duarte 2006, Duarte and Lüttge 2006). Application of 1 % O_2 elicits non-photorespiratory conditions, and therefore, J_{CO2} at 1 % O_2 minus J_{CO2} at 21 % O_2 is a measure of photorespiratory oxygen uptake, J_{O2}. In the C_3-mode a change from 21 % to 1 % O_2 had the expected effects of increased g_{H2O} and CO_2 uptake and decreased p^i_{CO2}. Photorespiration was rather constant over the light period. In the CAM mode photorespiration depended on the CAM phases. In phase II the relative contribution of oxygenase activity to total RubisCO activity was about half that observed in the C_3-mode leaves (32.1–35.7 %) at the beginning of the light period but still 15.6 % of total RubisCO activity, i.e. the onset of the CO_2 concentrating effect reduced photorespiration but did not prevent it. In phase III due to low g_{H2O} there was little effect of 1 % O_2 and the calculated values of p^i_{CO2} are not very reliable and most likely a strong underestimation. At the beginning of phase IV photorespiratory activity was highest in all measurements and even higher than in the

C_3 mode (37.9 %) perhaps due to still high internal O_2 levels. RelΦ_{PSII} is a measure of the photosynthetic irradiance and excitation use. In the C_3-mode it was constant under 21 % O_2 throughout the day but strongly reduced under 1 % O_2 reflecting the particular energy demand of photorespiration (Osmond and Grace 1995; Heber et al. 2001; Heber 2002). Since relΦ_{PSII} was obtained by chlorophyll fluorescence imaging the measure of heterogeneity of relΦ_{PSII} over the leaves could be calculated using an algorithm based on the nearest neighbour matrix concept (Hütt and Neff 2001). In the C_3-mode leaf heterogeneity was generally low and constant under 21 % O_2, but applications of 1 % O_2 dramatically increased heterogeneity. This suggests that photorespiration synchronized the energy demand of different parts of the leaf and had a stabilizing effect on overall energy use of the leaves. In the CAM-mode leaf relΦ_{PSII} increased gradually at the beginning of the light period and decreased again in the late afternoon, in relation to phases II, III and IV (see also Sect. 8.8.1 and Fig. 8.25). As in the C_3-mode reduction of relΦ_{PSII} by 1 % O_2 was also observed in phases II and IV but by contrast to the C_3-adapted plants het-

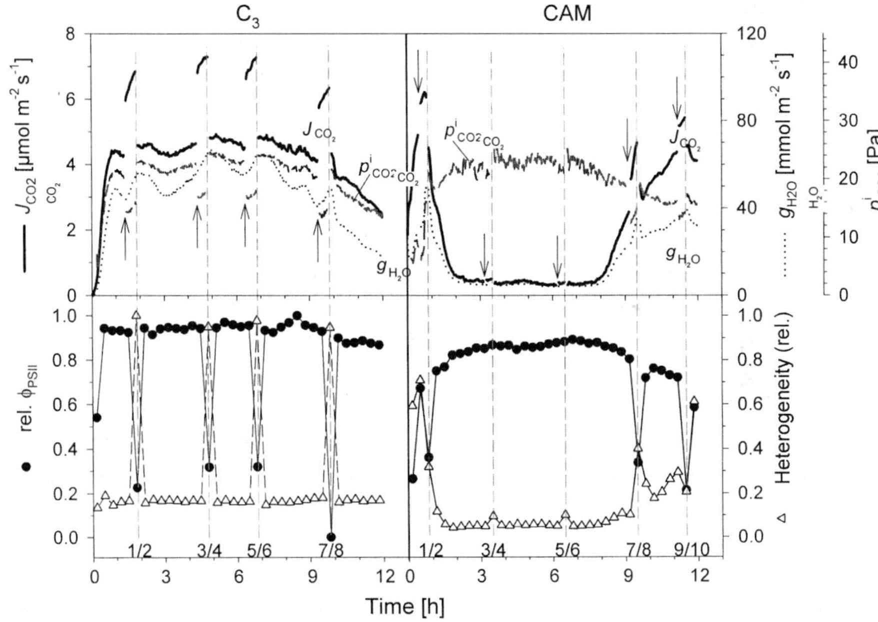

Fig. 8.18. Light period dynamics of photorespiratory activity as revealed by applications of air with only 1 % O_2 (*arrows*) causing non-photorespiratory conditions in plants of *C. minor* adapted to perform C_3-photosynthesis and CAM, respectively. J_{CO2}, net CO_2 exchange, g_{H2O}, leaf conductance for water vapour, p^i_{CO2}, internal partial pressure of CO_2, relΦ_{PSII}, relative quantum use efficiency of photosystem II and its heterogeneity over the leaves. The values obtained during 1 % O_2 (i.e. under non-photorespiratory conditions) minus the values obtained under 21 % O_2 (i.e. under photorespiratory conditions) reflect the effects of photorespiration (from Duarte 2006, Duarte and Lüttge 2006, same plants as in Fig. 8.25)

erogeneity was much more dependent on the CAM phases than on application of 1 % O_2 indicating that under 21 % O_2 heterogeneity is a particular feature of CAM and especially due to desynchronization of leaf parts occurring during CAM phase transitions.

8.6 High Light and Oxidative Stress

We have already mentioned above (Sect. 8.4) that the very high internal CO_2-concentrations prevailing in phase III of CAM at times of the day when solar irradiation tends to be particularly high were thought for a long time to protect from over-energization of the photosynthetic electron transport apparatus. This was even considered to be one of the forces driving the evolution of CAM (Gil 1986). In this hypothesis photosynthesis at CO_2 concentrations saturating RubisCO is assumed to utilize most excitation energy effectively for the photochemical work of CO_2 assimilation. Over-energization is accompanied by oxidative stress where excited electrons cause the formation of reactive oxygen species (ROS) which elicit photochemical damage. Hence, for their protection plants including CAM-species have developed various mechanisms for harmless dissipation of excitation energy not used by the photochemistry of CO_2 reduction as well as antioxidative mechanisms for detoxifying ROS (Lüttge 2000, 2002).

Based on the above hypothesis it was assumed that in CAM plants such mechanisms would be mainly relevant in phases II and IV but not so much in phase III. Indeed, this is suggested in experiments of Winter et al. (1990) on the CAM species *C. rosea*. Also in *C. uvitana* photosynthetic photon use efficiency was highest in phase III (Winter et al. 1992). One of the major mechanisms in plants for harmless dissipation of photosynthetic excitation energy is based on the xanthophyll zeaxanthin (Schindler and Lichtenthaler 1996). Zeaxanthin may divert excitation energy in photosystem II away from the central reaction centers (Horton et al. 1994). Zeaxanthin is also involved in dissipation of excitation energy in the form of heat and is the substrate for the formation of the epoxides antheraxanthin and violaxanthin by protective binding of singlet activated oxygen and consumption of reduction equivalents. Thus, the electrochemical proton gradient at the thylakoid membranes of chloroplasts is used when zeaxanthin is regenerated from the two epoxides in the so-called xanthophyll cycle (Hager 1980; Demmig-Adams 1990; Demmig-Adams and Adams 1992; Pfündel and Bilger 1994). This protective mechanism requires high zeaxanthin levels at times of high light stress. The measurements of Winter et al. (1990) show that in the early light period in phase II of CAM in *C. rosea* the level of zeaxanthin increased substantially over the much lower levels maintained in the dark period, decreased in phase III and subsequently increased again in phase IV. The levels of violaxanthin, the double epoxide of zeaxanthin, showed the opposite development during the CAM phases. This supports high energy dissipation as heat in phases II and IV and low dissipation in phase III mediated by the xanthophyll cycle.

This protective thermal energy dissipation is reflected in non-photochemical quenching of chlorophyll fluorescence. The non-photochemical quenching factor, q_N is given as

$$q_N=1-(F_m'-F_o')/(F_m-F_o) \quad (8.5)$$

and non-photochemical quenching (NPQ) is also calculated by the Stern-Volmer equation

$$NPQ=(F_m-F_m')/F_m' \quad (8.6)$$

(Schreiber and Bilger 1993), where F_o and F_o' are the minimal fluorescence of the dark adapted and light adapted sample, respectively, and F_m and F_m' the maximum fluorescence of the dark adapted and light adapted sample, respectively. NPQ values observed in species of *Clusia* can be quite high, e.g. about 3 in *C. minor* and *C. multiflora* at a PPFD of 600–800 μmol m^{-2} s^{-1} (Herzog et al. 1999) or 5 in *C. minor* at 1500 μmol m^{-2} s^{-1} (Roberts et al. 1998). Somewhat lower values were observed for *C. minor* and *C. multiflora* at 1500 μmol m^{-2} s^{-1} by Grams et al. (1997), i.e. 1.5 and 1.8, respectively. The different activities of the xanthophyll cycle in the different CAM phases noted above are also corroborated by the observation of much lower non-photochemical quenching, q_N, of chlorophyll *a* fluorescence of photosynthesis in phase III than in phases II and IV, where non-photochemical energy dissipation is much higher (Fig. 8.19).

However, the relations of high light and oxidative stress in CAM plants generally are not that simple. These stresses always cause photoinhibition, which is reflected in a reduction of the potential quantum yield of photosystem II (PS II), F_v/F_m, where F_v is the variable and F_m the maximum fluorescence of chlorophyll *a* of PS II of a dark adapted leaf (Bilger et al. 1995). The protective mechanism of harmless dissipation of energy in the form of heat described above is a type of acute photoinhibition (Thiele et al. 1998). Chronic photoinhibition is obtained after photochemical damage. Non-photoinhibited leaves show F_v/F_m ratios close to 0.83 because the maximum efficiency of the use of photosynthetically active radiation (PPFD) is generally around 83 % (Björkman and Demmig 1987). Reduced values of F_v/F_m below 0.83 indicate photoinhibition which is chronic when they are not reversible over night and acute if they are not reversible after some time of darkening during the light period but recover over night (Thiele et al. 1998). A survey of *Clusias* shows that they are often subject to both acute and chronic photoinhibition and that this is particularly prevalent in the CAM species compared to the C_3-species (Fig. 8.20). The C_3/CAM-intermediate *C. minor* allows comparisons for C_3-photosynthesis and

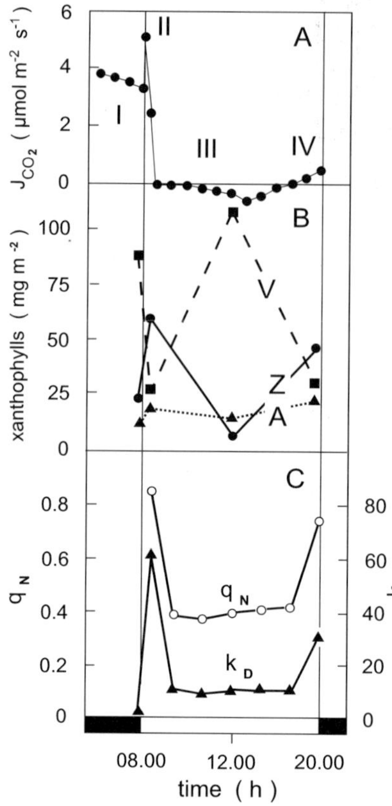

Fig. 8.19A–C. Changes in xanthophyll-cycle compounds in relation to the expression of CAM phases and non-photochemical quenching of chlorophyll *a* fluorescence of photosystem II in *C. rosea*: **A** CO_2-exchange (J_{CO2}), roman numbers refer to CAM phases; **B** xanthophyll-cycle compounds, Z=zeaxanthin, A=antheraxanthin, V=violaxanthin; **C** non-photochemical quenching, q_N, and rate constant of thermal energy dissipation k_D. *Dark bars* on the abscissa indicate dark periods (after data of Winter et al. 1990)

the CAM phases for the same individual plants or plants of clones obtained by vegetative propagation. As expected, acute photoinhibition is highest in phase IV of CAM. However, photoinhibition is not lower in phase III of CAM than in phase II of CAM or in the C_3-state (Fig. 8.21). A very important aspect of phase III that is overlooked by the hypothesis of a protection of leaves from high light and oxidative stress in phase III is the built up of very high internal O_2 concentrations in this phase as noted above.

The combined roles of CO_2-concentrating in phase III of CAM and zeaxanthin in photoprotection are well illustrated by a comparison of the behaviour of the C_3-species *C. multiflora* and the C_3/CAM intermediate species *C. minor* under light stress (Fig. 8.22; Herzog et al. 1999). The two species were grown at low irradiance of a PPFD of 4 mol m^{-2} day^{-1} in a glass house and then transferred to high irradiance of 24.5 or 33.5 mol m^{-2} day^{-1} with and without watering in a phytotron. After five days at the high irradiance of 24.5 mol m^{-2} day^{-1} the C_3-species *C. multiflora* was still capable of using the increased PPFD for increased CO_2 uptake independent of the water supply. However, at the highest irradiance of 33.5 mol m^{-2} day^{-1} it was highly inhibited. When the treatment under 24.5 mol m^{-2} day^{-1} was extended for a longer

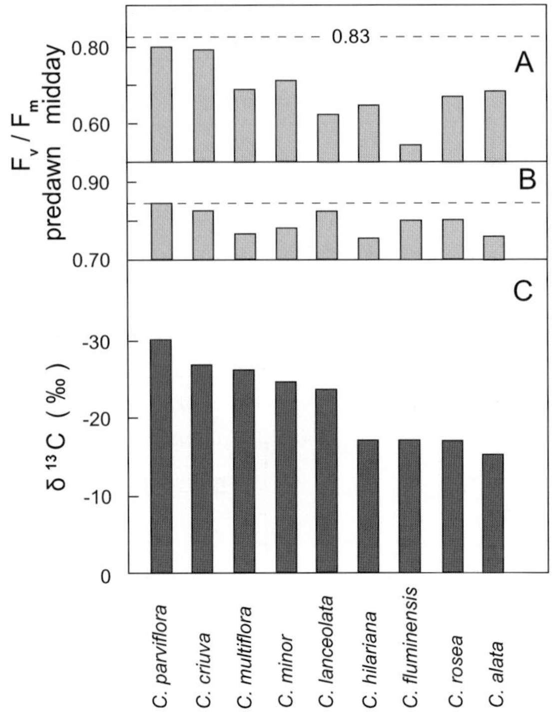

Fig. 8.20A–C. Potential quantum yield of photosystem II, F_v/F_m: **A** at midday after darkening for several tens of minutes; **B** before dawn for various species of *Clusia* with increasingly less negative $\delta^{13}C$-values (**C**) (see Fig. 8.7), i.e. C_3-photosynthesis to CAM performance, *from left to right*, where values below 0.83 indicate acute (**A**) and chronic (**B**) photoinhibition, respectively (Fig. 1B in Lüttge 1999 summarizing a large body of literature; species names not yet mentioned above with authorities are *C. criuva* Camb., *C. lanceolata* Camb.)

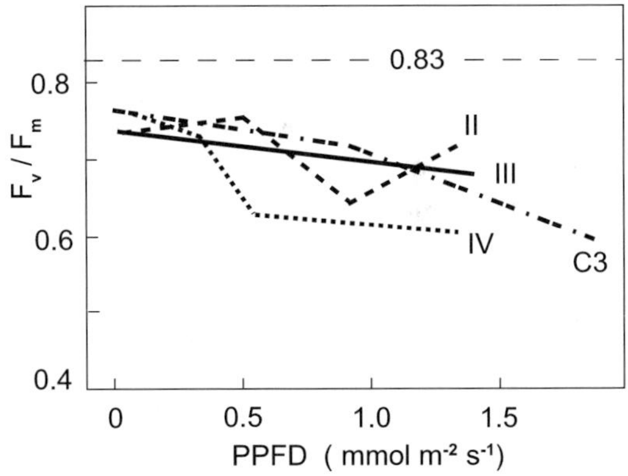

Fig. 8.21. Light response characteristics of acute photoinhibition given by measurements of potential quantum yield of photosystem II, F_v/F_m, in plants of the C_3/CAM-intermediate *C. minor* in the C_3 state (C_3) and in the CAM state during phases II, III and IV of CAM. Plants were kept at the irradiances indicated and darkened for 10 min before F_v/F_m was measured (after Lüttge 2000, with data after Haag-Kerwer 1994)

time leaves became necrotic and died. Conversely, *C. minor* switched to CAM when transferred to the high irradiance and leaves never became necrotic. As shown by F_v/F_m-values both species were not photoinhibited at the irradiance of 4 mol m^{-2} day^{-1}. Chronic photoinhibition at the beginning of the light period and acute photoinhibition building up during the light period at the higher irradiances were similar in both species. However, the responses of thermal energy dissipation via zeaxanthin were different

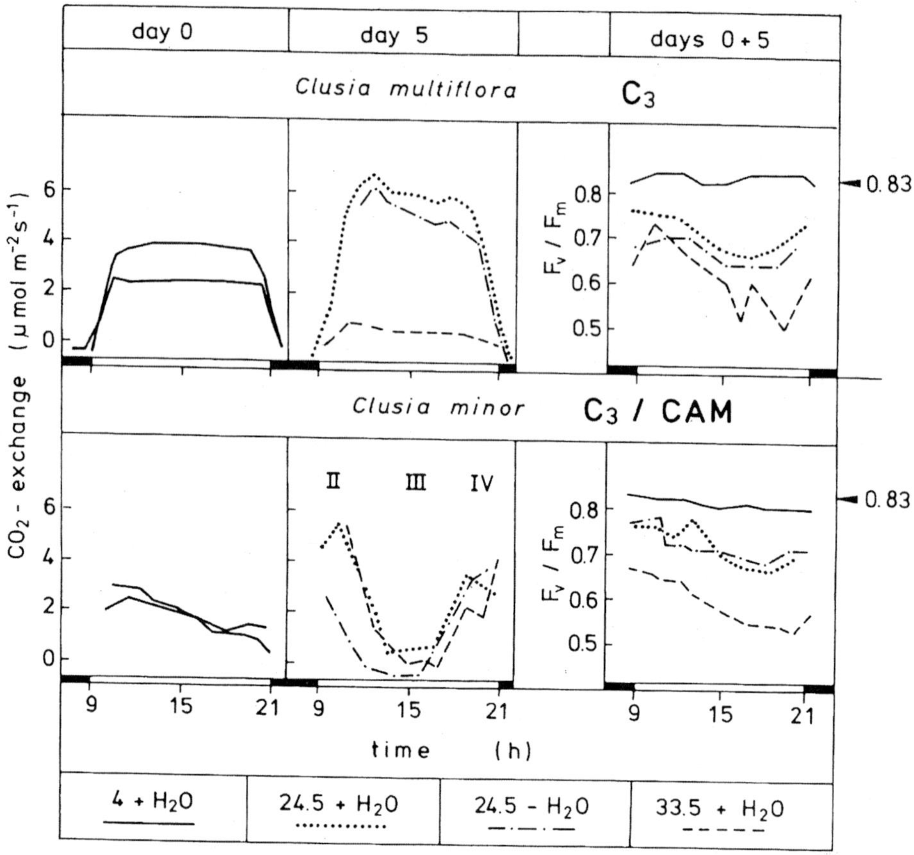

Fig. 8.22. CO_2-exchange and photoinhibition given by measurements of potential quantum yield of photosystem II, F_v/F_m, in the C_3-species *C. multiflora* and the C_3/CAM-intermediate species *C. minor*. Plants were grown at low irradiance and after day 0, the last day at low irradiance, they were transferred to two different high daily doses of irradiance applied in a bell-shaped time course (not shown) with additional stress due to withholding water in one case. The behaviour on day 0 (*solid lines*) and 5 days after the transfer (*other lines*) is shown; numbers at the bottom of the graph explaining the lines are daily doses of photosynthetic photon flux density (mol m^{-2} day^{-1}), roman numbers for gas exchange of *C. minor* on day 5 indicate CAM phases (after Lüttge 2000, with data of Herzog et al. 1999)

(Fig. 8.23). In the phytotron experiments irradiance was increased gradually towards noon to 1260 μmol m^{-2} s^{-1} and then decreased again in a bell shaped fashion over the light period. Accumulation of zeaxanthin and non-photochemical quenching of chlorophyll *a* fluorescence of PS II exactly followed irradiance and this was similar in both species. Short term exposures where different irradiances up to 1200 mol m^{-2} s^{-1} were applied for 20 min or where the irradiance of 1000 mol m^{-2} s^{-1} was applied for various times up to 80 min

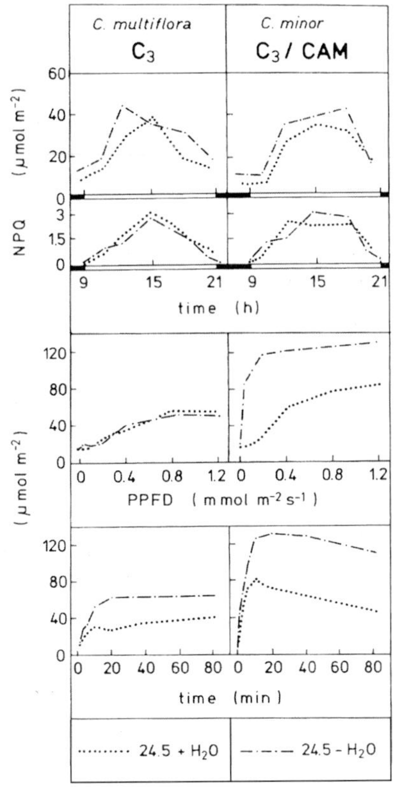

Fig. 8.23. Daily courses of zeaxanthin levels and thermal energy dissipation (non-photochemical fluorescence quenching, NPQ) in the plants of *C. multiflora* and *C. minor* also shown in Fig. 8.22 on day 5 after transfer to high daily doses of irradiance (24.5 mol m^{-2} day^{-1}) with and without drought stress as indicated (*two upper panels*). *Lower panels* show short term experiments with *C. multiflora* and with *C. minor* in the CAM-state, where photosynthetic photon flux density, PPFD, was presented for 20 min as indicated, or a PPFD of 1000 μmol m^{-2} s^{-1} was given for various times up to 80 min, as indicated, before analysis of zeaxanthin levels (after Lüttge 2000, with data of Herzog et al. 1999)

show, however, that the C_3/CAM-intermediate species *C. minor* intrinsically was able to accumulate about three times as much zeaxanthin as the C_3-species *C. multiflora*. However, it did not make use of its full capacity for zeaxanthin accumulation as shown by the day courses where it did not reach higher zeaxanthin levels than the C_3-species. Thus, we may conclude that it was the combination of the induction of CAM and thermal energy dissipation that lead to the overall superior performance of the C_3/CAM-intermediate species under the stress by increased irradiance.

8.7 Osmotic Implications of Night/Day Changes of Organic Acids and Soluble Carbohydrates

Malate/malic acid is an osmotically active solute. In the standard mode of CAM, where starch is broken down to PEP and malate is formed via PEPC and accumulated as malic acid in the vacuole this accumulation generates vacuolar osmotic pressure. Malate acts osmotically, while the accompanying proton counterions do not count due to the buffering capacity of the vacuolar cell sap. It has been well documented in various CAM plants that nocturnal malic acid accumulation increases cell sap osmolarity, π. Water follows osmotically and this leads to a nocturnal increase in turgor pressure, P. Thus, nocturnal malate accumulation also supports nocturnal water storage (Lüttge 1986; Eller and Ruess 1986; Ruess et al. 1988; Eller et al. 1992; Murphy and Smith 1998).

In *Clusias* these relationships are more complex due to the larger variety of osmotically active solutes involved. Accumulation of organic acids is only fully effective in increasing π when the precursor is osmotically inert starch, and then malate is twice as effective as citrate because two molecules of malate and only one molecule of citrate are formed per hexose unit of starch. When the precursor is sucrose one osmotically active molecule, i.e. sucrose, is lost while four molecules of malate or two molecules of citrate are produced, and in both cases the process would be increasing π, where again malate would be twice as effective as citrate. When the precursor is free hexose one osmotically active molecule, i.e. hexose, is lost while two molecules of malate or one molecule of citrate are produced, and therefore only in the case of malate the process would be increasing π, while in the case of citrate the bal-

Table 8.5. Diurnal changes of solute levels and calculated osmotic effects during the CAM cycle of various species of *Clusia* in the field. Dawn minus dusk values are presented for malate (Δmal), citrate (Δcitr) and free sugars (Δ free sugars, i.e. hexoses and sucrose) and the calculated osmotic effects ($\Delta\pi$). Values are in mmol L^{-1} tissue water or kg^{-1} fresh weight

		Δmal	Δcitr	Δ free sugars	$\Delta\pi$	Reference
C. alata	Dry season	83	77	–200	–40	Popp et al. 1987
	End of dry season	63	35	–73	25[a]	
C. rosea	Dry season	124	90	–176	38	
	End of dry season	240	112	–331	21[b]	
C. hilariana		96	74	–112	58	Berg et al. 2004
C.minor	Exposed leaves Wet season	10	85	–65	30	Borland et al. 1994
	Exposed leaves Dry season	110	61	–81	90	
	Shaded leaves Dry season	105	22	–49	78	
C. rosea		171	106	–160	117	Ball et al. 1991

Measured π was: [a]–46 mosmol L^{-1} and [b]+130 mosmol L^{-1}.

Table 8.6. Water potential, Ψ, osmotic potential, π , and calculated turgor pressure, P, (all in MPa), of *C. uvitana*. Averages of 40 measurements predawn and mid-day, obtained over a whole year with standard errors, SE. (Adopted from data of Zotz and Winter 1994b)

	Ψ	π	P
Predawn	−0.79±0.03	1.38±0.05	0.59
Mid-day	−0.71±0.03	1.26±0.04	0.55

ance is osmotically neutral. Thus, it depends on the balance of malate and citrate produced, which increases π, and free sugars consumed, which decreases π, to which extent the CAM cycle in *Clusia*s would affect osmotic activity of the vacuolar cell sap. A survey of data available in the literature is provided in Table 8.5. In all cases except one the balance is positive, i.e. the turnover of solutes would result in a nocturnal increase of π. However, in contrast to CAM-plants where only malate is accumulated and only starch is serving as precursor, the magnitude of this net osmotic effect in the *Clusia*s in Table 8.5 is much smaller than nocturnal organic acid accumulation, because besides starch free sugars also serve as precursors for acid synthesis. Zotz and Winter (1994a, b) have measured water potential, Ψ, and osmotic potential, π, in *C. uvitana* for over a year at the end of the night period (pre-dawn) and at noon and have not found significant changes of π and turgor pressure , P, calculated from $\Psi = P - \pi$ (Table 8.6). We have no simultaneous analyses of solutes, i.e. organic acids and free sugars, for this study but we may assume that their balance was neutral in this case and thus there were no diurnal changes in π and P. In two cases of the compilation of data in Table 8.5 π was not only calculated from the theoretical solute balance but actually measured (footnotes in Table 8.5). The differences between calculated and measured π, of course, are explained by transpiration and water equilibration of the cells overriding consequences of changes in solute levels. This shows that when osmotically relevant changes in solute levels do occur they also have direct effects on the water relations of the plants.

8.8 Environmental Factors Regulating Reversible Changes Between the C_3 and CAM-Mode of Photosynthesis and the Degree of CAM Expression

The most important environmental parameters affecting the performance of C_3-photosynthesis and CAM, respectively, in *Clusia*s are water, light, temperature and mineral nutrition, especially nitrogen. After we have now surveyed basic features of modes of photosynthesis and physiotypes in *Clusias* (Sects. 8.1–8.6) and before turning to the ecophysiological implications (Chap. 9) we may review the action of these factors on the engagement of CAM. Mostly the environmental parameters do not act singly but interact between each other.

8.8.1 Water

Although it is thought that the environmental pressure originally driving the evolution of CAM was acquisition and concentrating of CO_2 at low ambient CO_2 concentrations (Lüttge 2002, 2004) among extant land plants the balance of water relations is the decisive ecophysiological benefit of CAM. C_3-species may lose considerable amounts of water during the light period and respond to stress by midday closure of stomata and repression of photosynthesis ("midday depression"). CAM plants show much less transpirational loss of water during CO_2-acqusition because evaporative demand is much lower in the dark period when they open stomata for CO_2-uptake. Of course, for day-time CO_2-uptake in phases II and IV of CAM transpirational loss of water will approach that of C_3 plants. C_3/CAM-intermediate plants may switch from C_3-photosynthesis to CAM in response to drought stress.

The relations of CO_2-gain and water loss are expressed as water use efficiency (WUE), i.e. CO_2 taken up and fixed per H_2O transpired. WUE strongly correlates with $\delta^{13}C$ values, which indicate the degree of CAM expression in plants and the nocturnal CO_2 gain (Winter and Holtum 2002; Winter et al. 2005). Table 8.7 gives some examples for WUE in *Clusias*. The theoretical expectations described above are largely borne out by these experimental observations, i.e. that CAM performance leads to better WUE. However, the results compiled in Table 8.7 also indicate that fine-tuned reactions may be more complex. Night time WUE was always higher than day time WUE in the CAM-performing plants. However, in one experiment *C. minor* in the C_3-state had higher WUE than during the dark period of the CAM-state. When performing CO_2 fixation around the clock (see Fig. 8.2B), in one case *C. minor* as expected showed much higher WUE in the dark period than in the light period, but in another case WUEs were quite similar night and day. This reflects the high flexibility of *C. minor* in multi-factor responses under various experimental and environmental conditions as we shall also see several times again below.

Considering reactions in response to severe drought stress Borland et al. (1998) underlined the importance of genotypic differences in the capacity for CAM which are related to the capacities and amounts of PEPC and PEPCK increasing under stress. The *Clusia* species studied by these authors were categorized as follows: *C. aripoensis* Britt., weak CAM inducible; *C. minor*,

Table 8.7. Water use efficiencies (WUE=mmol CO_2 fixed/mol H_2O transpired[a] or g dry mass produced in growth/kg H_2O transpired[b]) in the light period (L) and the dark period (D) and both periods together for *Clusia* species with different modes of photosynthesis. (Names of species not yet mentioned above with authorities are as follows: *C. cylindrica* Hammel, *C. odorata* Seeman, *C. pratensis* Seeman, *C. valerioi* Standl.)

	L	D	L and D	Reference
C. rosea (CAM)			11.9[b]	Winter et al. 2005
C. uvitana (C3/CAM intermediate)			4.4[b]	
C. cylindrica			3.4[b]	
C. odorata			3.0[b]	
C. pratensis			2.0[b]	
C. valerioi			3.4[b]	
C. venosa (C_3 plus weak nocturnal acid accumulation)	7.1	2.8	6.9[a]	Franco et al. 1990
C. minor (CO_2 uptake around the clock)	6.4	33.8	8.8[a]	
C. major CAM	13.6	24.9	20.2[a]	
C. alata CAM	11.5	29.4	21.6[a]	
C. minor				Lee et al. 1989
C_3	38.8	–	38.8[a]	
CAM	8.0	22.9	23.6[a]	
CO_2-uptake around the clock	9.8	11.1	10.4[a]	

C_3/CAM intermediate; *C. rosea* constitutive CAM (note that these are the same plants also presented in Table 8.9 and discussed below). *C. uvitana* also showed a capacity to increase levels of PEPC under drought stress (Winter et al. 1992). Grams et al. (1998) used drought stress in order to check if even putative obligate C_3 species of *Clusia* may have at least weak capacities for CAM. *C. multiflora* emerged as truly obligate C_3, but the potential for CAM appears to be cryptically wide spread among *Clusia*s.

However, even a given genotype can show different reactions to environmental hydraulic parameters underlining flexibility and plasticity, which probably culminates in *C. minor.* Figure 8.24 shows that even the two opposite leaves at one node of a plant of *C. minor* can simultaneously perform in different ways depending on external conditions (Schmitt et al. 1988, where *C. minor* is wrongly called *C. rosea*). Each of the two leaves was kept in a separate gas exchange chamber. The plant was drought stressed for four days when both leaves showed a CAM pattern of CO_2-exchange. On the fifth day the plant was re-watered while one of the two opposite leaves was kept at a low leaf-to-air water vapour pressure difference (ΔW=6.2 mbar bar^{-1}) and the other leaf at a high ΔW (=13.1 mbar bar^{-1}), i.e. in a drier atmosphere in the gas exchange

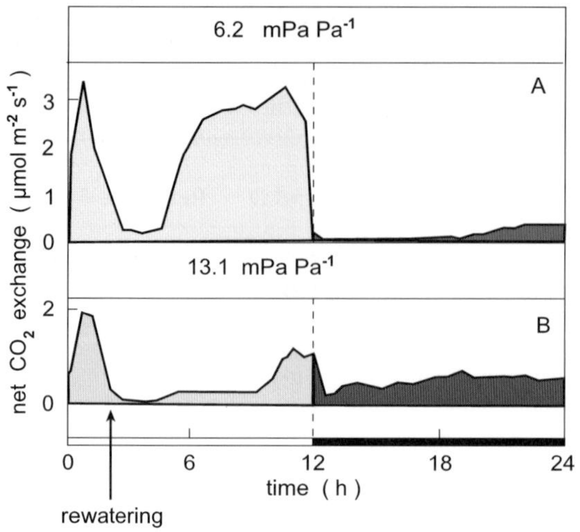

Fig. 8.24A,B. Net CO_2 exchange of two opposite leaves of *C. minor* exposed to two different leaf-to-air water vapour pressure differences, ΔW, of: **A** 6.2 mbar bar^{-1}; **B** 13.1 mbar bar^{-1}. The plant was droughted for several days and re-watered at the time indicated by the *arrow*. The *dark bar* indicates night-time (from Schmitt et al. 1988)

chamber. Only a few hours after re-watering the leaf in the wetter atmosphere changed to a very pronounced C_3-like CO_2 uptake in the light period and drastically reduced nocturnal CO_2 uptake in the following dark period, while the leaf in the drier atmosphere continued to perform full CAM.

Spatiotemporal resolution of the dynamics of photosynthesis using chlorophyll fluorescence imaging suggests that even different parts of one given leaf of *C. minor* may perform C_3-photosynthesis and CAM, respectively Duarte 2006, (Duarte and Lüttge 2006), so that *C. minor* perhaps really is the most flexible and plastic species in which photosynthesis ever has been studied. Figure 8.25 shows experiments, where plants of a clone of *C. minor* were kept at 25 °C night and day and an irradiance of 120 µmol m^{-2} s^{-1} during the day and were adapted to perform C_3-photosynthesis and CAM by watering regularly and by withholding water for three to four days, respectively. Net CO_2-exchange, J_{CO2}, conductance for water vapour, g_{H2O}, and internal partial pressure of CO_2 calculated from these parameters, p^i_{CO2}, integrated over the entire leaves showed the typical C_3- and CAM-characteristics. In the C_3-adapted plants these parameters increased rapidly at the beginning of the light period and were constant over most of the day, starting to decrease slightly towards the end of the light period. In the CAM-adapted plants J_{CO2} and g_{H2O} displayed the daytime phases of CAM with an early morning peak (phase II) followed by stomatal closure and an increase in p^i_{CO2} due to organic acid remobilisation and decarboxylation (phase III) and then again stomatal opening and net CO_2-uptake at the end of the light period (phase IV). Relative quantum yield of photosystem II, relΦ_{PSII}, as a measure of the use of irradiance and excitation energy could be calculated separately for the interveinal lamina tissue and for the major leaf vein from chlorophyll *a* fluorescence images. In the C_3-mode plants, relΦ_{PSII} in the lamina follows J_{CO2}

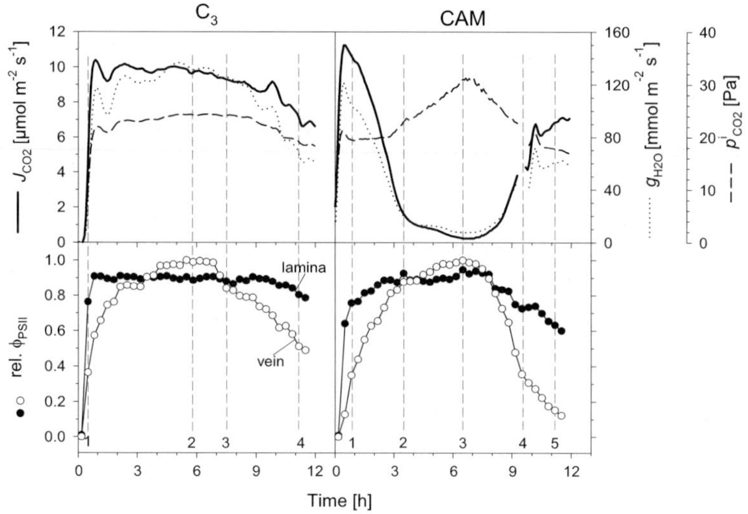

Fig. 8.25. Light period dynamics of photosynthesis parameters in plants of *C. minor* adapted to perform C_3-photosynthesis and CAM, respectively. J_{CO2}, net exchange of CO_2, g_{H2O}, leaf conductance for water vapour, and p^i_{CO2}, internal CO_2 partial pressure, integrated for the entire leaves. RelΦ_{PSII}, relative quantum use efficiency of photosystem II separated for the interveinal lamina tissue (*filled circles*) and the major leaf vein chlorenchyma (*open circles*) (from Duarte 2006, Duarte and Lüttge 2006, i.e. same plants as in Fig. 8.18)

and g_{H2O} with a rapid increase at the start of the light period. Conversely, in the CAM-mode plants relΦ_{PSII} in the lamina increases more slowly and gradually. This can be explained by the fact that in phase II of CAM both carboxylating enzymes PEPC and RubisCO are active in CO_2-fixation (Borland and Griffiths 1996), while only RubisCO carboxylation activity contributes to photosynthetic energy use. The decline of relΦ_{PSII} also is much faster in the afternoon when energy demand decreases with decreasing p^i_{CO2} in phase IV. In the CAM-mode this pattern of relΦ_{PSII} is even more pronounced in the major vein, and even higher values are reached in phase III of CAM than in the lamina tissue. Most interestingly, the vein tissue of the C_3-mode plants in contrast to their lamina tissue shows a very similar pattern to that of the CAM-mode plants. This suggests that in the C_3-mode plants a residual CAM-activity was maintained in the major vein tissue. This is also supported by anatomical observations showing a much more CAM-like anatomy of the vein chlorenchyma and lack of internal air spaces preventing a strong coupling with the lamina tissue by lateral gas diffusion (Sect. 2.2).

As an important response of CAM-species to strong drought stress we have described above internal CO_2 recycling culminating in the expression of full CAM-idling (Sect. 8.1), where CO_2 from respiration is recycled via nocturnal acid accumulation and daytime decarboxylation behind closed stomata day and night. This now brings us back to considering the role of citrate in the CAM of *Clusia*, which remained enigmatic in the above considerations (Sects. 8.3.3, 8.4 and 8.6). Internal CO_2 recycling and CAM-idling is futile with respect to carbon gain. CO_2 can be recycled via malate (Sect. 8.1, Table 8.1). While citrate accumulation in CAM is always futile with respect to the carbon budget, and there is no carbon gain even if stomata are open and CO_2 exchange with the atmosphere is possible, recycling of carbon via citrate where the entire C_6-carbon skeleton is recycled is more effective than via malate (Sect. 8.3.3, Fig. 8.14). The various potential roles of citrate in CAM in comparison to malate are surveyed in Table 8.8. If recycling in response to drought really were a major advantage of citrate, we would expect the relative importance of citrate turnover to increase in relation to malate turnover, when plants are under drought stress. Table 8.9 summarizes some experimental and analytical results from the literature. Indeed, in many cases the ratios of day/night changes of malate (Δmal) and citrate (Δcitr) decreased in CAM performing species of *Clusia* under drought stress indicating an increasing relative contribution of citrate. However, this pattern was not observed in growth chamber studies of Borland et al. (1998), including *C. minor* which showed a drought elicited increase of Δmal/Δcitr. In the experiments of Borland et al. (1998), however, the Δmal/Δcitr ratios were already rather low without drought stress. Under drought stress absolute Δcitr values increased

Table 8.8. Possible roles of night/day changes in the levels of malate (Δmal) and citrate (Δcitr) during the CAM cycle (see Franco et al. 1992; Lüttge 1996)

	Δmal	Δcitr
CO_2 acquisition	Yes	No
H_2O saving during CO_2 acquisition	Yes	No
Recycling in the night time	CO_2	Carbon skeleton
Recycling in the day time	CO_2	CO_2, more than for malate
Increase in internal CO_2 concentration in the day time	Yes	Yes, more than for malate
Energy budget: dark period	ATP-consumption	ATP-consumption and production of redox power
Energy budget: light period	ATP-consumption	Larger ATP-consumption and consumption of redox power
Osmotic changes	Yes	Yes, less than for malate
Buffer capacity	Yes	Yes, higher than for malate

Table 8.9. Ratios of night/day changes of malate and citrate levels (Δmal/Δcitr) in *Clusia* species in response to drought. (Name of species not yet mentioned above with authority is *C. lanceolata* Camb.)

	Days of drought stress			Reference
	0	10	16	
C. lanceolata	2.8	1.7	1.6	Franco et al. 1992
C. rosea	2.4	0.7		
C. minor	1.8		0.9	
C. minor	5.5		2.1	de Mattos et al. 1999
C. minor				Borland et al. 1998
Young leaves	0.8	1.3		
Mature leaves	0.5	0.5		
C. rosea				
Young leaves	0.2	1.0		
Mature leaves	0.7	0.8		
C. aripoensis				
Young leaves	0.2	0.5		
Mature leaves	0.1	0.5		

slightly in mature leaves of *C. aripoensis* and pronouncedly in mature leaves of *C. minor* and in young leaves of *C. rosea*, but Δmal values also increased and in mature leaves of *C. rosea* Δmal and Δcitr did not differ with and without drought stress. As a result there were no reductions in Δmal/Δcitr-ratios in response to drought and perhaps even some increases. These discrepancies between results in the literature may be explained by the flexible multi-parameter responses of *Clusia*s under various experimental and environmental conditions. Much more strictly comparative work is needed if we really wish to understand the phenomenon of the often quantitatively important role of citrate in night/day oscillations of organic acid levels in the CAM of *Clusia*s. However, currently there does not appear to be much motivation for funding the study of such an exciting problem of biochemical ecology.

8.8.2 Light and Water

High irradiance is often acting in relation to the water factor. One example was described above discussing high light and oxidative stress in the C_3-species *C. multiflora* and the C_3/CAM-intermediate species *C. minor* after transfer from low to high irradiance (Sect. 8.6). In this experiment drought stress was also applied but high light stress alone was the major factor under

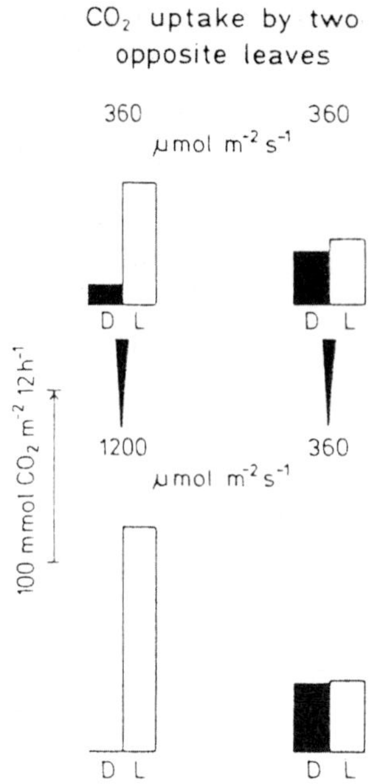

Fig. 8.26. Net CO_2 exchange of two opposite leaves on the same node of a well watered plant of *C. minor* at varied irradiance. One leaf was transferred from a PPFD of 360 µmol m^{-2} s^{-1} during the light period to a PPFD of 1200 µmol m^{-2} s^{-1}, while the other one was kept at 360 µmol m^{-2} s^{-1}. *Columns* give integrated net CO_2 uptake for the light period (L) and the dark period (D), respectively (from Lüttge 1996, after data of Schmitt et al. 1988)

the conditions applied (Figs. 8.22 and 8.23). Another example is given by two opposite leaves of a well watered plant of *C. minor* where one leaf was transferred from an irradiance of 360 µmol m^{-2} s^{-1} to a high irradiance of 1200 µmol m^{-2} s^{-1} while the other one was kept at the lower irradiance (Fig. 8.26). Nocturnal CO_2 uptake was abolished at the higher irradiance but remained at the lower irradiance. This elimination of PEPC mediated CO_2 fixation by a leaf of *Clusia* taking up CO_2 night and day by transfer to high irradiance only occurred in well watered plants. Under limited water availability the opposite effect, i.e. an increase in night time CO_2 uptake would have occurred. Blue light and UV-A radiation are involved in rapid switches from C_3-photosynthesis to CAM in *Clusia* (Grams and Thiel 2002).

8.8.3 Light and Temperature

It is widely assumed that night temperatures lower than day temperatures are favourable for the performance of CAM. This is based on the observation of effects of temperature on the overall performance of the counter acting enzymes of nocturnal carboxylation (PEPC) and daytime decarboxylation of CAM, where lower temperatures favour the former and higher temperatures the latter (Brandon 1967; Buchanan-Bollig and Kluge 1981; Buchanan-Bollig et al. 1984; Carter et al. 1995a). Stimulation of CAM expression by reduced night temperatures has been confirmed in many experimental observations (Kluge et al. 1973; Neales 1973; Medina et al. 1977; Kluge and Ting 1978; Nobel 1988; Fetene and Lüttge 1991). High night temperatures decrease night-time acid accumulation because of respiratory consumption and loss of accumulated acid (Kaplan et al. 1976) or increased organic acid efflux from the vacuoles (Friemert et al. 1988). On the other hand, in wet tropical forests many CAM species perform well at rather similar night and day temperatures.

In *C. minor* day/night temperature regimes strongly affect C_3-CAM transitions dependent on irradiance during growth. In a very elaborate comparative study Haag-Kerwer (1994, summarized by Haag-Kerwer et al. 1992) measured gas exchange and night/day oscillations of malate and citrate in two sets of regularly well watered plants grown at a low irradiance (PPFD=photosynthetic photon flux density) of 30–50 μmol m^{-2} s^{-1} and a higher irradiance of 260–300 μmol m^{-2} s^{-1}, respectively, for 13 different combinations of day/night temperatures (D/N °C). At similar D/N-temperatures all plants performed C_3-photosynthesis. Plants consistently changed to a CAM pattern in response to

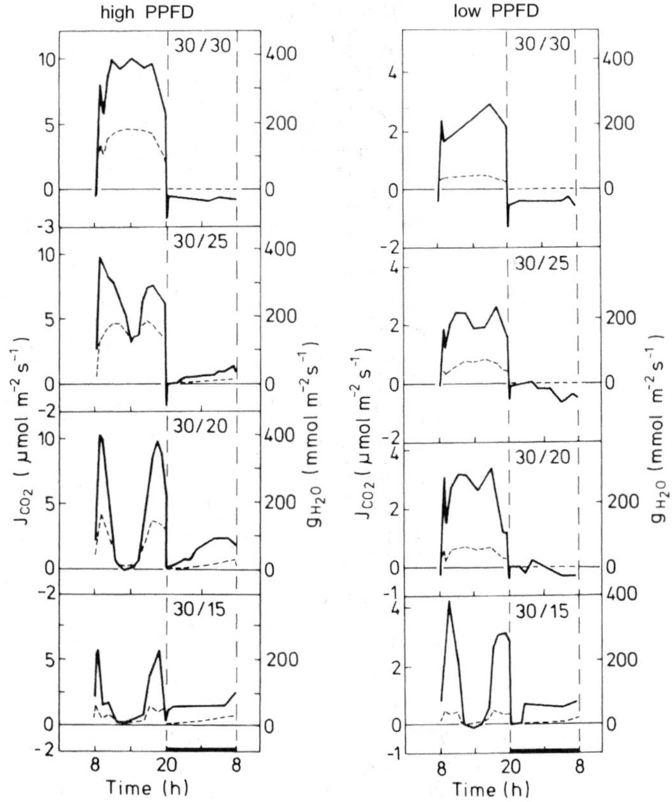

Fig. 8.27. Net CO_2 exchange (J_{CO2}, *solid lines*) and leaf conductance for water vapour (g_{H2O}, *dotted lines*) over a 24-h period for high (260–300 μmol m^{-2} s^{-1}) and low (30–50 μmol m^{-2} s^{-1}) PPFD grown *C. minor* at a day temperature of 30 °C and at various night temperatures. The *dark bars* indicate dark periods (Fig. 1 from Haag-Kerwer et al. 1992)

an increase in the difference between day and night temperatures. With a D/N temperature difference of 10 °C high-PPFD grown plants showed a substantial night-time CO_2 uptake for all temperatures tested, with a marked depression of daytime CO_2 uptake around midday. Low-PPFD grown plants on the other hand needed at least a D/N temperature difference of 15 °C to induce CAM (Fig. 8.27). These changes of the mode of photosynthesis in response to different temperature regimes were induced within a few hours and were fully expressed after three days. They were fully reversible after a return to identical D/N temperatures (Fig. 8.28). Malate and citrate accumulation in the CAM mode responded in different ways to the different temperature regimes. Evidently the temperature optimum for citrate accumulation is different from that of malate accumulation. The summary of the comparison for 13 combinations of D/N temperatures and 4 different parameters (leaf conductance for water vapour indicating stomatal width, CO_2 exchange, Δmal and Δcitr) each for two different growth temperatures of this study presented in Fig. 8.29 is so elaborate that we cannot describe and discuss it here in detail. Thus, we may leave it to the discretion of the respected reader which of the possible comparisons she or he may find particularly interesting. We may note, however, that this in fact provides the recipes for establishing plants of *C. minor* performing CAM with any combinations of nocturnal acids accumulated, e.g. of only malate accumulation (low PPFD, D/N temperatures of 30/15 °C), high malate and low citrate accumulation (high PPFD, D/N temperatures of 25/15 or 30/15 °C), equal malate and citrate accumulation (high PPFD, D/N temper-

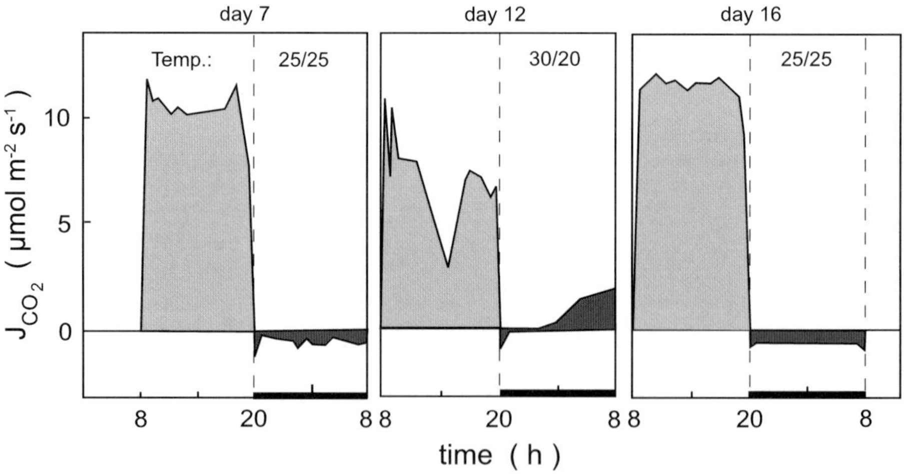

Fig. 8.28. Net CO_2 exchange measured continuously over 17 days for one given leaf of *C. minor* kept for 7 days at D/N temperatures of 25/25 °C, switched to D/N temperatures of 30/20 °C on day 7 and back to D/N temperatures of 25/25 °C on day 13. Recordings of the 7th, the 12th and the 16th day of the experiment are presented. The *dark bars* indicate dark periods (Fig. 3 from Haag-Kerwer et al. 1992)

atures of 25/20 °C), only citrate accumulation (low PPFD, D/N temperatures of 30/30 °C), and high citrate and low malate accumulation (high PPFD, D/N temperatures of 20/20 °C). If and how this may contain an explanation of the citrate enigma discussed above (Sect. 8.8.1) remains a question open to speculation.

Making use of the expression of CAM at D/N temperatures of 30/18 °C and a switch from CAM to C_3-phosotsynthesis under a D/N temperature regime of 30/30 °C, de Mattos and Lüttge (2001) studied the potential ecophysiological

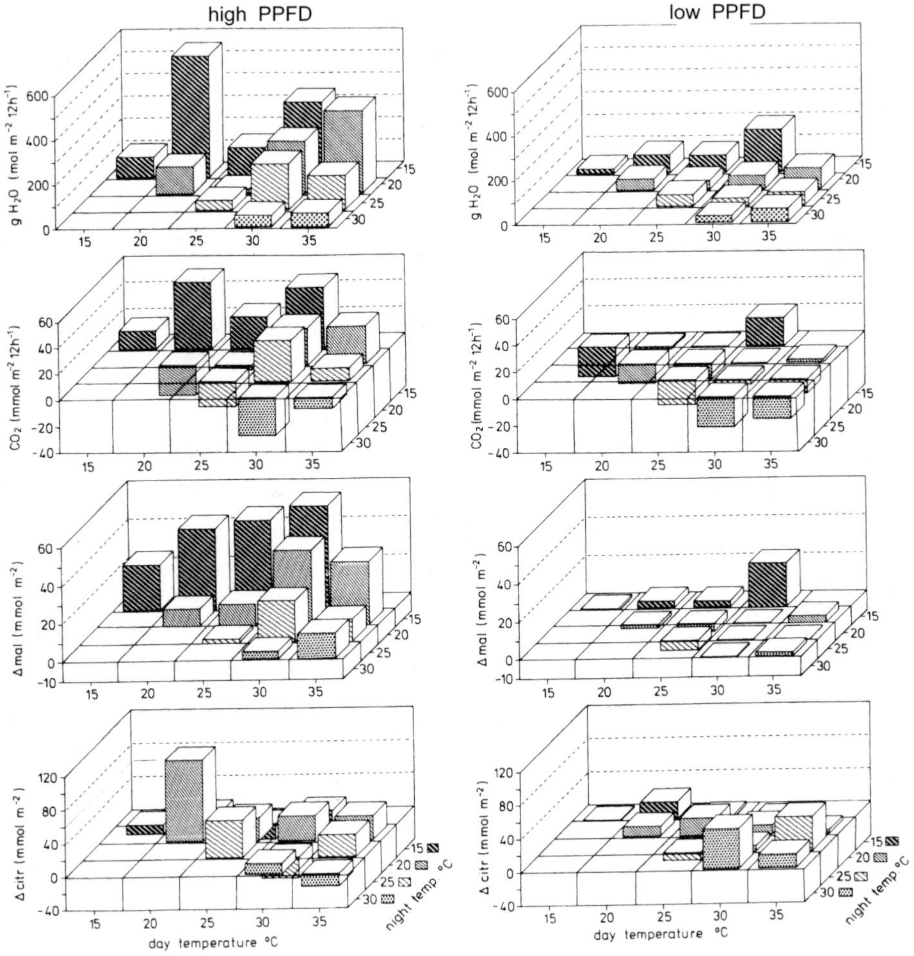

Fig. 8.29. Integrated values of night-time leaf conductance for water vapour (g_{H2O}), integrated net CO_2 exchange for the dark period (CO_2 mmol m^{-2} 12 h^{-1}) and nocturnal malate (Δmal) and citrate (Δcitr) accumulation for 13 different day/night temperature regimes in high PPFD (260–300 μmol m^{-2} s^{-1}) and low PPFD (30–50 μmol m^{-2} s^{-1}) grown plants (Fig. 4 from Haag-Kerwer et al. 1992)

advantage of C_3-photosynthesis over CAM in plants of *C. minor* when they were well watered. During the CAM to C_3 switch followed over seven days nocturnal organic acid accumulation decreased whereas CO_2-uptake during the daytime and integrated over 24 h increased. Water use efficiency was reduced. In contrast to the C_3-like photosynthesis, a pronounced reduction in the effective quantum yield of photosystem II together with a sharp increase in non-photochemical chlorophyll fluorescence quenching were observed during CAM at the beginning and the end of the light period. The results suggested that daily photon utilization increases when there is unrestricted C_3-like CO_2-uptake directly from the atmosphere. Thus, provided water is not limiting, and therefore the plant can afford increased transpiration, *C. minor* performing C_3-photosynthesis may overcome the limitations of the storage capacity of the vacuole for nocturnal organic acid accumulation, improving its daily carbon balance.

8.8.4 Light and Nitrogen

In *C. minor* grown at the lower and the higher PPFD of 60 to 70 and 220 to 320 μmol m^{-2} s^{-1}, respectively, the effect of nitrogen on the expression of the modes of photosynthesis was also investigated (Franco et al. 1991). High and low N-status, respectively, of the plants was maintained by regularly watering

Table 8.10. Effects of nitrogen nutrition (±24 mol m^{-3} NO_3^-) and photosynthetically active radiation (PPFD of 60–70 and 220–320 μmol m^{-2} s^{-1}, respectively) during growth on integrated net CO_2 exchange in the light period and the dark period and nocturnal organic acid accumulation (malate=Δmal, citrate=Δ citr) in *C. minor* (Franco et al. 1991)

	–N		+N	
	Low PPFD	High PPFD	Low PPFD	High PPFD
Integrated net CO_2 exchange				
(mmol CO_2 m^{-2} leaf area)				
Light period	62.7	59.6	65.8	209.0
Dark period	–4.0	– 0.8	–1.9	12.1
Δorganic acids				
(mmol m^{-2} leaf area)				
Δmal	–0.4	10.2	–0.2	99.9
Δcitr	24.9	45.9	41.8	32.5
Δorganic acids				
(mol m^{-3} tissue water)				
Δmal	–0.6	14.4	–0.3	150.7
Δcitr	39.1	64.6	68.9	49.0

with a solution containing 24 mol m^{-3} NO_3^- or no nitrogen. Only a weak nocturnal CO_2-uptake was observed in the high PPFD plants plus nitrogen, and in the other three treatments there was some loss of CO_2 during the night. In all treatments there was considerable nocturnal citrate accumulation which was increased by N-nutrition in the low PPFD plants and decreased in the high PPFD plants. Nocturnal malate accumulation was absent in the low PPFD plants and weak in the high PAR plants without nitrogen but was considerably increased by nitrogen supply in the high PPFD plants (Table 8.10). Here we have yet another example of differential effects of environmental parameters on nocturnal accumulation of malate and citrate (see Sects. 8.8.1 and 8.8.3), but which does not yet allow a clear ecophysiological explanation of the role of citrate.

References

Ball E, Hann J, Kluge M, Lee HSJ, Lüttge U, Orthen B, Popp M, Schmitt A, Ting IP (1991) Ecophysiological comportment of the tropical CAM-tree *Clusia* in the field. II. Modes of photosynthesis in trees and seedlings. New Phytol 117:483–491

Berg A, Orthen B, de Mattos EA, Duarte HM, Lüttge U (2004) Expression of crassulacean acid metabolism in *Clusia hilariana* Schlechtendal in different stages of development in the field. Trees 18:553–558

Bilger W, Schreiber U, Bock M (1995) Determination of quantum efficiency of photosystem II and of non-photochemical quenching of chlorophyll fluorescence in the field. Oecologia 102:425–432

Björkman O, Demmig B (1987) Photon yield of O_2 evolution and chlorophyll fluorescence characteristics at 77 K among vascular plants of diverse origins. Planta 170:489–504

Black CC, Chen J-Q, Doong RL, Angelov MN, Sung SJS (1996) Alternative carbohydrate reserves used in the daily cycle of crassulacean acid metabolism. In: Winter K, Smith JAC (eds) Crassulacean acid metabolism: biochemistry, ecophysiology and evolution. Ecological Studies, vol 114. Springer, Berlin Heidelberg New York, pp 31–45

Borland AM, Griffiths H (1996) Variations in the phases of crassulacean acid metabolism and regulation of carboxylation patterns determined by carbon-isotope-discrimination techniques. In: Winter K, Smith JAC (eds) Crassulacean acid metabolism: biochemistry, ecophysiology and evolution. Ecological Studies, vol 114. Springer, Berlin Heidelberg New York, pp 230–249

Borland AM, Griffiths H (1997) A comparative study on the regulation of C_3 and C_4 carboxylation processes in the constitutive crassulacean acid metabolism (CAM) plant *Kalanchoë daigremontiana* and the C_3-CAM intermediate *Clusia minor*. Planta 201:368–378

Borland AM, Griffiths H, Maxwell C, Broadmeadow MSJ, Griffiths NM, Barnes JD (1992) On the ecophysiology of the Clusiaceae in Trinidad: expression of CAM in *Clusia minor* L. during the transition from wet to dry season and characterization of three endemic species. New Phytol 122:349–357

Borland AM, Griffiths H, Broadmeadow MSJ, Fordham MC, Maxwell C (1993) Short-term changes in carbon-isotope discrimination in the C_3-CAM intermediate *Clusia minor* L. growing in Trinidad. Oecologia 95:444–453

Borland AM, Griffiths H, Broadmeadow MSJ, Fordham MC, Maxwell C (1994) Carbon-isotope composition of biochemical fractions and the regulation of carbon balance in leaves of the C_3-crassulacean acid metabolism intermediate *Clusia minor* L. growing in Trinidad. Plant Physiol 106:493–501

Borland AM, Griffiths H, Maxwell C, Fordham MC, Broadmeadow MSJ (1996) CAM induction in *Clusia minor* L. during the transition from wet to dry season in Trinidad: the role of organic acid speciation and decarboxylation. Plant Cell Environ 19:655–664

Borland AM, Técsi LI, Leegood R, Walker RP (1998) Inducibility of crassulaean acid metabolism (CAM) in *Clusia* species: physiological/biochemical characterisation and intercellular localization of carboxylation and decarboxylation processes in three species which exhibit different degrees of CAM. Planta 205:342–351

Brandon PC (1967) Temperature features of enzymes affecting crassulacean acid metabolism. Plant Physiol 42:977–984

Buchanan-Bollig IC, Kluge M (1981) Crassulacean acid metabolism (CAM) in *Kalanchoë daigremontiana*: temperature response of phosphoenolpyruvate (PEP) - carboxylase in relation to allosteric effectors. Planta 152:181–188

Buchanan-Bollig IC, Kluge M, Müller D (1984) Kinetic changes with temperature of phosphoenolpyruvate carboxylase from a CAM plant. Plant Cell Environ 7:63–70

Carter PJ, Wilkins MB, Nimmo HG, Fewson CA (1995a) Effects of temperature on the activity of phosphoenolpyruvate carboxylase and on the control of CO_2 fixation in *Bryophyllum fedtschenkoi*. Planta 196:375–380

Carter PJ, Wilkins MB, Nimmo HG, Fewson CA (1995b) The role of temperature in the regulation of the circadian rhythm of CO_2 fixation in *Bryophyllum fedtschenkoi*. Planta 196:381–386

Carter PJ, Fewson CA, Nimmo GA, Nimmo HG, Wilkins MB (1996) Roles of circadian rhythms, light and temperature in the regulation of phosphoenolpyruvate carboxylase in crassulacean acid metabolism. In: Winter K, Smith JAC (eds) Crassulacean acid metabolism: biochemistry, ecophysiology and evolution. Ecological studies, vol 114. Springer, Berlin Heidelberg New York, pp 46–52

Chen R (1998) Plant NADP-dependent isocitrate dehydrogenases are predominantly localized in the cytosol. Planta 207:280–285

Demmig-Adams B (1990) Carotenoids and photoprotection: a role for the xanthophyll zeaxanthin cycle. Biochim Biophys Acta 1020:1–24

Demmig-Adams B, Adams WW (1992) Photoprotection and other responses of plants to high light stress. Annu Rev Plant Physiol Plant Mol Biol 43:599–626

Duarte HM (2006) Chronobiologie von *Clusia minor*: circadianer Rhythmus in einer Pflanze mit C_3/CAM-intermediärem photosytnthetischen Verhalten. Dissertation. Darmstadt

Duarte HM, Lüttge U (2006) Gas-exchange, photorespiration and spatiotemporal dynamics of relative quantum use efficiency in leaves of the C_3-photosynthesis/crassulacean acid metabolism-intermediate species *Clusia minor* L. (Clusiaceae) in both modes of photosynthesis. Trees (in press)

Eller BM, Ruess BR (1986) Modulation of CAM and water balance of *Senecio medley-woodii* by environmental factors and age of leaf. J Plant Physiol 125:295–309

Eller BM, Ferrari S, Ruess BR (1992) Spatial and diel variations of water relations in leaves of the CAM-plant *Senecio medley-woodii*. Bot Helvetica 102:193–200

Evans JR, Sharkey TD, Berry SA, Farquhar GD (1986) Carbon-isotope discrimination measured concurrently with gas exchange to investigate CO_2 diffusion in leaves of higher plants. Aust J Plant Physiol 13:281–292

Fatland BL, Ke J, Anderson MD, Mentzen WL, Cui LW, Allred CC, Johnston JL, Nikolau BJ, Wurtele ES (2002) Molecular characterization of a heteromeric ATP-citrate lyase that generates cytosolic acetyl-coenzyme A in *Arabidopsis*. Plant Physiol 130:740–756

Fetene M, Lüttge U (1991) Environmental influences on carbon recycling in a terrestrial CAM bromeliad, *Bromelia humilis* Jacq. J Exp Bot 42:25–31
Franco AC, Ball E, Lüttge U (1990) Patterns of gas exchange and organic acid oscillations in tropical trees of the genus *Clusia*. Oecologia 85:108–114
Franco AC, Ball E, Lüttge U (1991) The influence of nitrogen, light and water stress on CO_2 exchange and organic acid accumulation in the tropical C_3-CAM tree, *Clusia minor*. J Exp Bot 42:597–603
Franco AC, Ball E, Lüttge U (1992) Differential effects of drought and light levels on accumulation of citric and malic acids during CAM in *Clusia*. Plant Cell Environ 15:821–829
Friemert V, Heininger D, Kluge M, Ziegler H (1988) Temperature effects on malic-acid efflux from the vacuoles and on the carboxylation pathways in Crassulacean acid metabolism plants. Planta 174:453–461
Genty B, Harbinson J, Baker RN (1990) Relative quantum efficiencies of the two photosystems of leaves in photorespiratory and non-photorespiratory conditions. Plant Physiol Biochem 28:1–10
Gil F (1986) Origin of CAM as an alternative photosynthetic carbon fixation pathway. Photosynthetica 20:494–507
Grams TEE, Thiel S (2002) High light-induced switch from C_3-photosynthesis to crassulacean acid metabolism is mediated by UV-A/blue light. J Exp Bot 53:1475–1483
Grams TEE, Haag-Kerwer A, Olivares E, Ball E, Arndt S, Popp M, Medina E, Lüttge U (1997) Comparative measurements of chlorophyll *a* fluorescence, acid accumulation and gas exchange in exposed and shaded plants of *Clusia minor* L. and *Clusia multiflora* H.B.K. in the field. Trees 11:240–247
Grams TEE, Herzog B, Lüttge U (1998) Are there species in the genus *Clusia* with obligate C_3-photosynthesis? J Plant Physiol 152:1–9
Haag-Kerwer A (1994) Photosynthetische Plastizität bei *Clusia* und *Oedematopus*. Ph.D.-Thesis, Darmstadt
Haag-Kerwer A, Franco AC, Lüttge U (1992) The effect of temperature and light on gas exchange and acid accumulation in the C_3-CAM plant *Clusia minor* L. J Exp Bot 43:345–352
Haag-Kerwer A, Grams TEE, Olivares E, Ball E, Arndt S, Popp M, Medina E, Lüttge U (1996) Comparative measurements of gas-exchange, acid accumulation and chlorophyll *a* fluorescence of different species of *Clusia* showing C_3 photosynthesis, or crassulacean acid metabolism, at the same field site in Venezuela. New Phytol 134:215–226
Hafke JB, Hafke Y, Smith JAC, Lüttge U, Thiel G (2003) Vacuolar malate uptake is mediated by an anion-selective inward rectifier. Plant J 35:116–128
Hager A (1980) The reversible, light-induced conversions of xanthophylls in the chloroplast. In: Czygan FC (ed) Pigments in plants. G. Fischer, Stuttgart, pp 57–79
Heber U (2002) Irrungen, Wirrungen? The Mehler reaction in relation to cyclic electron transport in C_3 plants. Photosyn Res 73:223–231
Heber U, Bukhov NG, Shuvalov VA, Kobayashi Y, Lange O (2001) Protection of the photosynthetic apparatus against damage by excessive illumination in homoiohydric leaves and poikilohydric mosses and lichens. J Exp Bot 52:1999–2006
Herzog B, Hoffmann S, Hartung W, Lüttge U (1999) Comparison of photosynthetic responses of the sympatric tropical C_3 species *Clusia multiflora* H.B.K. and the C_3-CAM intermediate species *Clusia minor* L. to irradiance and drought stress in a phytotron. Plant Biol 1:460–470
Holtum JAM, Aranda J, Virgo A, Gehrig HH, Winter K (2004) $\delta^{13}C$ values and crassulacean acid metabolism in *Clusia* species from Panama. Trees 18:658–668

Holtum JAM, Smith JAC, Neuhaus HE (2005) Intracellular transport and pathway of carbon flow in plants with crassulacean acid metabolism. Funct Plant Biol 32:429–449
Horton P, Ruban AV, Walters RG (1994) Regulation of light harvesting in green plants. Indication by nonphotochemical quenching of chlorophyll fluorescence. Plant Physiol 106:415–420
Hütt MT, Neff R (2001) Quantification of spatiotemporal phenomena by means of cellular automata techniques. Physica A 289:498
Kalt W, Osmond CB, Siedow JN (1990) Malate metabolism in the dark after carbon-13-labeled carbon dioxide fixation in the crassulacean plant *Kalanchoë tubiflora*. Plant Physiol 94:826–832
Kaplan A, Gale J, Poljakoff-Mayber A (1976) Resolution of net dark fixation of carbon dioxide into its respiration and gross fixation components in *Bryophyllum daigremontianum*. J Exp Bot 27:220–230
Kluge M (1968) Untersuchungen über den Gaswechsel von *Bryophyllum* während der Lichtperiode. II. Beziehungen zwischen dem Malatgehalt des Blattgewebes und der CO_2-Aufnahme. Planta 80:359–377
Kluge M, Ting IP (1978) Crassulacean acid metabolism. Analysis of an ecological adaptation. Ecological Studies, vol 30. Springer, Berlin Heidelberg New York
Kluge M, Lange OL, Eichmann M, Schmid R (1973) Diurnaler Säurerhythmus bei *Tillandsia usneoides*: Untersuchungen über den Weg des Kohlenstoffs sowie die Abhängigkeit des CO_2-Gaswechsels von Lichtintensität, Temperatur und Wassergehalt der Pflanze. Planta 112:357–372
Kluge M, Fischer A, Buchanan-Bollig I (1982) Metabolic control of CAM. In: Ting IP, Gibbs M (eds) Crassulacean acid metabolism. American Society of Plant Physiology, Rockville, Maryland, pp 31–50
Kusumi K, Arata H, Iwasaki I, Nishimura M (1994) Regulation of PEP-carboxylase by biological clock in a CAM plant. Plant Cell Physiol 35:233–242
Lee HSJ, Schmitt AK, Lüttge U (1989) The response of the C_3-CAM tree *Clusia rosea* to light and water stress. II. Internal CO_2 concentration and water use efficiency. J Exp Bot 40:171–179
Lüttge U (1986) Nocturnal water storage in plants having crassulacean acid metabolism. Planta 168:287–289
Lüttge U (1988) Day-night changes of citric-acid levels in crassulacean acid metabolism: phenomenon and ecophysiological significance. Plant Cell Environ 11:445–451
Lüttge U (1991) Morphogenetische, physiologische und biochemische Strategien von Baumwürgern im tropischen Wald. Naturw 78:49–58
Lüttge U (1996) *Clusia*: plasticity and diversity in a genus of C_3/CAM intermediate tropical trees. In: Winter K, Smith JAC (eds) Crassulacean acid metabolism: biochemistry, ecophysiology and evolution. Ecological Studies, vol 114. Springer, Berlin Heidelberg New York, pp 296–311
Lüttge U (1999) One morphotype, three physiotypes: sympatric species of *Clusia* with obligate C_3 photosynthesis, obligate CAM and C_3-CAM intermediate behaviour. Plant Biol 1:138–148
Lüttge U (2000) Light stress and crassulacean acid metabolism. Phyton 40:(65)–(82)
Lüttge U (2002) CO_2 concentrating: consequences in crassulacean acid metabolism. J Exp Bot 53:2131–2142
Lüttge U (2004) Ecophysiology of crassulacean acid metabolism (CAM). Ann Bot 93:629–652
Lüttge U, Smith JAC (1984) Mechanism of passive malic-acid efflux from vacuoles of the CAM plant *Kalanchoë daigremontiana*. J Membr Biol 81:149–158
Martin CE (1996) Putative causes and consequences of recycling CO_2 via crassulacean acid metabolism In: Winter K, Smith JAC (eds) Crassulacean acid metabolism: bio-

chemistry, ecophysiology and evolution. Ecological Studies, vol 114. Springer, Berlin Heidelberg New York, pp 192–203

Mattos EA de, Lüttge U (2001) Chlorophyll fluorescence and organic acid oscillations during transition from CAM to C_3-photosynthesis in *Clusia minor* L. (Clusiaceae). Ann Bot 88:457–463

Mattos EA de, Herzog B, Lüttge U (1999) Chlorophyll fluorescence during CAM-phases in *Clusia minor* L. under drought stress. J Exp Bot 50:253–261

Medina E, Delgado M, Troughton JH, Medina JD (1977) Physiological ecology of CO_2 fixation in Bromeliaceae. Flora 166:137–152

Milburn TR, Pearson DJ, Ndegwe NA (1968) Crassulacean acid metabolism under natural tropical conditions. New Phytol 67:883–897

Murphy R, Smith JAC (1998) Determination of cell water-relation parameters using the pressure probe: extended theory and practice of the pressure-clamp technique. Plant Cell Environ 21:637–657

Neales TF (1973) Effect of temperature on the assimilation of carbon dioxide by mature pineapple plants, *Ananas comosus* (L.) Merr. Aust J Biol Sci 26:539–546

Nimmo GA, Wilkins MB, Fewson CA, Nimmo HG (1987) Persistent circadian rhythms in the phosphorylation state of phosphoenolpyruvate carboxylase from *Bryophyllum fedtschenkoi* leaves and its sensitivity to inhibition by malate. Planta 170:408–415

Nobel PS (1988) Environmental biology of agaves and cacti. Cambridge University Press, Cambridge

Olivares E (1997) Prolonged leaf senescence in *Clusia multiflora* H.B.K. Trees 11:370–377

Olivares E, Faist K, Kluge M, Lüttge U (1993) $^{14}CO_2$ pulse-chase labelling in *Clusia minor* L. J Exp Bot 44:1527–1533

Osmond CB (1978) Crassulacean acid metabolism: a curiosity in context. Ann Rev Plant Phys 29:379–414

Osmond CB, Grace CE (1995) Perspectives on photoinhibition and photorespiration in the field: quintessential inefficiencies of the light and dark reactions on photosynthesis? J Exp Bot 46:1351–1362

Pfündel E, Bilger W (1994) Regulation and possible function of the violaxanthin cycle. Photosynth Res 42:89–109

Popova TN, Appenroth K-J (2002) Cytosolic and chloroplastic NADP-dependent isocitrate dehydrogeanses in *Spirodela polyrhiza*. II. Regulation of enzyme activity. J Plant Physiol 159:239–244

Popova TN, Rakhmanova TI, Appenroth K-J (2002) Cytosolic and chloroplastic NADP-dependent isocitrate dehydrogenases in *Spirodela polyrhiza*. I. Regulation of activity by metabolites in vitro. J Plant Physiol 159:231–237

Popp M, Kramer D, Lee H, Diaz M, Ziegler H, Lüttge U (1987) Crassulacean acid metabolism in tropical dicotyledonous trees of the genus *Clusia*. Trees 1:238–247

Rangasami D, Ratledge C (2000) Compartmentation of ATP:citrate lyase in plants. Plant Physiol 122:1225–1230

Roberts A, Borland AM, Griffiths H (1997) Discrimination processes and shifts in carboxylation during the phases of crassulacean acid metabolism. Plant Physiol 113:1283–1292

Roberts A, Borland AM, Maxwell K, Griffiths H (1998) Ecophysiology of the C_3-CAM intermediate *Clusia minor* L. in Trinidad: seasonal and short-term photosynthetic characteristics of sun and shade leaves. J Exp Bot 49:1563–1573

Robinson SA, Yakir D, Ribascarbo M, Giles L, Osmond CB, Siedow JN, Berry JA (1992) Measurements of the engagement of cyanide-resistant respiration in the crassulacean acid metabolism plant *Kalanchoë daigremontiana* with the use of online oxygen isotope discrimination. Plant Physiol 100:1087–1091

Ruess BR, Ferrari S, Eller BM (1988) Water economy and photosynthesis of the CAM plant *Senecio medley-woodii* during increasing drought. Plant Cell Environ 11:583–589

Rustin P, Lance C (1986) Malate metabolism in leaf mitochondria from the crassulacean acid metabolism plant *Kalanchoë blossfeldiana* Poelln. Plant Physiol 81:1039–1043

Rustin P, Queiroz-Claret C (1985) Changes in oxidative properties of *Kalanchoë blossfeldiana* leaf mitochondria during development of crassulacean acid metabolism. Planta 164:415–422

Schindler C, Lichtenthaler HK (1996) Photosynthetic CO_2-assimilation, chlorophyll fluorescence and zeaxanthin accumulation in field grown maple trees in the course of a sunny and a cloudy day. J Plant Physiol 148:399–412

Schmitt A, Lee HSJ, Lüttge U (1988) The response of the C_3-CAM tree, *Clusia rosea*, to light and water stress. J Exp Bot 39:1581–1590

Schreiber U, Bilger W (1993) Progress in chlorophyll fluorescence research: major developments during the past years in retrospect. Prog Bot 54:151–173

Sipes DL, Ting IP (1985) Crassulacean acid metabolism and crassulacean acid metabolism modifications in *Peperomia camptotricha*. Plant Physiol 77:59–63

Smith JAC, Lüttge U (1985) Day-night changes in leaf water relations associated with the rhythm of crassulacean acid metabolism in *Kalanchoë daigremontiana*. Planta 163:272–282

Spalding MH, Stumpf DK, Ku MSB, Burris RH, Edwards GE (1979) Crassulacean acid metabolism and diurnal variations of internal CO_2 and O_2-concentrations in *Sedum praealtum* DC. Aust J Plant Physiol 6:557–567

Sternberg L da SL, Ting IP, Price D, Hann J (1987) Photosynthesis in epiphytic and rooted *Clusia rosea* Jacq. Oecologia 72:457–460

Thiele A, Krause GH, Winter K (1998) In situ study of photoinhibition of photosynthesis and xanthophyll cycle activity in plants growing in natural gaps of the tropical forest. Aust J Plant Physiol 25:189–195

Ting IP, Hann J, Holbrook NM, Puth FE, Sternberg L da SL, Price D, Goldstein G (1987) Photosynthesis in hemiepiphytic species of *Clusia* and *Ficus*. Oecologia 74:339–346

Veste M, Herppich WB, Willert DJ von (2001) Variability of CAM in leaf-deciduous succulents from the Succulent Karoo (South Africa). Basic and Applied Ecology 2:283–288

Wanek W, Huber W, Arndt SK, Popp M (2002) Mode of photosynthesis during different life stages of hemiepiphytic *Clusia* species. Funct Plant Biol 29:725–732

Winter K, Holtum JAM (2002) How closely do the ^{13}C values of crassulacean acid metabolism plants reflect the proportion of CO_2 fixed during day and night? Plant Physiol 129:1843–1851

Winter K, Lesch M, Diaz M (1990) Changes in xanthophyll-cycle components and in fluorescence yield in leaves of a crassulacean-acid-metabolism plant, *Clusia rosea* Jacq., throughout a 12 hour photoperiod of constant irradiance. Planta 182:181–185

Winter K, Zotz G, Baur B, Dietz KJ (1992) Light and dark CO_2-fixation in *Clusia uvitana* and the effects of plant water status and CO_2 availabiliy. Oecologia 91:47–51

Winter K, Aranda J, Holtum JAM (2005) Carbon isotope composition and water-use efficiency in plants with crassulacean acid metabolism. Funct Plant Biol 32:381–388

Ziegler H (1994) Stable isotopes in plant physiology and ecology. Progr Bot 56:1–24

Zotz G, Winter K (1993) Short-term regulation of crassulacean acid metabolism activity in a tropical hemiepiphyte, *Clusia uvitana*. Plant Physiol 102:835–841

Zotz G, Winter K (1994a) Annual carbon balance and nitrogen-use efficiency in tropical C_3 and CAM epiphytes. New Phytol 126:481–492

Zotz G, Winter K (1994b) A one-year study on carbon, water and nutrient relationships in a tropical C_3-CAM hemi-epiphyte, *Clusia uvitana* Pittier. New Phytol 127:45–60

9 Physiological Ecology

Ulrich Lüttge

9.1 Ecological Amplitude

After having unfolded the principal potentials of photosynthesis in *Clusia* in Chap. 8 we will now look at its performance at the plant community level. The simple single morphotype of *Clusia* gains diversity by the various life forms it produces (Sect. 2.3) and by its high physiological and biochemical plasticity (Chap. 8). *Clusia* is extraordinarily successful in various tropical habitats as shown by its very large ecological amplitude. As illustrated in Fig 9.1 *Clusia*s are found:

- In coastal sand dune ridges in the so-called Atlantic restinga, where *Clusia* species may determine the physiognomy of the vegetation to such an extent, that it has even been named "*Clusia* scrub" (Ule 1901)
- At coastal rocky sites
- In savannas
- In rocky rupestrian fields
- In gallery forests of savannas or cerrados
- In dry scrub in arid interandean valleys
- In secondary scrub on white sand and laterite in lowland rainforest areas
- In different types of tropical forests, e.g.:
  - Dry forests at sea level or in karstic limestone mountains
  - Wet forests in an altitudinal gradient from:
    - Lowland rainforest
    - Lower montane rainforest
    - Upper montane rain forest, to
    - Fog or cloud forest and "elfin forest"
- On inselbergs in savannas or rain forests

Moreover, *Clusia* trees are used for afforestation of secondary savannas and wasteland and appreciated as ornamental plants (Fig. Syn. 4) where they can be encountered for example in places like the banking centre of Rio de Janeiro, along beach side hotels in Waikiki, Hawaii, and in many places in

Ecological Studies, Vol. 194
U. Lüttge (Ed.)
Clusia: A Woody Neotropical Genus
of Remarkable Plasticity and Diversity

Florida (Sternberg et al. 1987; Lüttge 2000). It is noteworthy that it is the single morphotype of *Clusia* that covers this wide range, but even for individual species a large ecological amplitude is observed and the width of the ecological amplitude is not different for species with C_3-photosynthesis and CAM, respectively (Table 9.1; see also Chap. 3).

Table 9.1. Ecological amplitude of individual species of *Clusia, C. multiflora* H.B.K. *(C. mu.), C. parviflora* Saldanha et Engl. *(C.p.), C. rosea* Jacq. *(C.r.), C. fluminensis* Planch. et Triana *(C.f.), C. minor* L. *(C.mi.), C. criuva* Camb. *(C.c.)* with their modes of photosynthesis indicated. (Information in Diaz et al. 1996 and elsewhere as cited in Sec. 9.4)

	C.mu. C_3	*C.p.* C_3	*C.r.* CAM	*C.f* CAM	*C.mi.* C_3/CAM	*C.c.* weak CAM inducible
Restinga		●		●		
Coastal rocks			●			
Savanna/cerrado	●				●	
Gallery forest – cerrado ecotone						●
Open shrub land	●					
Dry low land forest			●	●	●	
Secondary shrub forest	●				●	
Dry montane karstic limestone forest	●		●		●	
Montane rain forest	●		●		●	
Upper montane rain forest	●					
Cloud forest/fog forest/ elfin forest	●					
Inselberg		●				●

Fig. 9.1A–I. Ecological amplitude of *Clusia*: **A** vegetation islands and bushes of *Clusia* on the sand dunes of the Atlantic Restinga de Jurubatiba, Brazil (see Table 9.4); **B** *Clusia rosea* Jacq. on coastal granite rocks, Virgin Gorda, British Virgin Islands, Lesser Antilles; **C** *Clusia arrudae* Planch. et Triana, rupestrian fields at Tiradentes, Minas Gerais State, Brazil; **D** *Clusia criuva* Camb. in a gallery forest in the savanna-like cerrados near Brasilia, Brazil; **E** *C. criuva* at the gallery forest cerrado ecotone (same site as D); **F** *C. fluminensis* Planch. et Triana in a semideciduous dry forest at the Atlantic coast, Brazil; **G** *C. multiflora* H.B.K. at 630 m a.s.l. in the montane rain forest on Cerro Santa Ana, Paraguana Peninsula, Falcon State, Venezuela; **H** bushes of unidentified and presumably still undescribed species of *Clusia* in the cloud forest of the Sierra Maigualida, 2040 m a.s.l., Territorio Federal Amazonica, Venezuela; **I** dwarf forms of *C. multiflora* together with the palm *Geonema paraguanensis* on the top of Cerro Santa Ana (see G above) at 830 m a.s.l. For *Clusia*s on carstic limestone mountains and on inselbergs see Figs. 9.13/9.15 and 9.19, respectively

9.2 Expression of Modes of Photosynthesis of *Clusia* Species under Field Conditions

Chapter 8 has revealed that *Clusia* species may express four different photosynthetic physiotypes, i.e. C_3-photosynthesis, crassulacean acid metabolism (CAM), CAM cycling and C_3/CAM intermediate behaviour (Sect. 8.1). The question is to what extent different options are followed in the field if they are intrinsically possible. We have already learnt from stable carbon isotope analyses (Sect. 8.2) that in the field *Clusia*s mainly adopt the C_3-mode of photosynthesis even if an optional CAM capacity is available. The C_3-mode appears to be always preferable for optimising carbon gain provided there is little stress, especially no drought stress, which would favour CAM (de Mattos and Lüttge 2001, Sects. 8.8.2 and 8.8.3).

Although *Clusia* species do occur in seasonally very dry habitats, such as rocky rupestrian fields, restingas and lowland dry forests, they never grow in extremely arid habitats like deserts. This may explain why C_3-photosynthesis over time often is the prevailing favourable option as it is indicated by the stable carbon isotope analyses which always integrate the performance of the sample analysed over its life time (Sect. 8.2). In the initial stages of *Clusia* research 30 years ago it has therefore been assumed, that the CAM option might be particularly important for the hemi-epiphytic life form in its early epiphytic stage when there is a particular challenge with respect to water relations (see Sect. 2.3). It was thought that in *Clusia rosea* "CO_2 is fixed mostly via CAM during its epiphytic stage, when water availability is low, and via both CAM and C_3 during its rooted stage" (Sternberg et al. 1987). CAM phases II and IV with C_3-type CO_2 fixation by RubisCO in addition to C_4-type fixation by PEPC (for details see Sect. 8.1) were expressed in rooted plants of *C. rosea* in both the dry season and the wet season in Florida, where this *Clusia* species is used for horticultural purposes and escaping into the wild. Conversely, in epiphytic plants all four CAM phases were only expressed in the wet season, and phases II and IV were suppressed in the dry season. Moreover, it is also evident that the water saving features of CAM provide an advantage for the shaping of the hydraulic architecture in hemi-epiphytes (Sect. 2.4). However, extensive investigations of *Clusia*s including *C. rosea* (Ball et al. 1991 a, b), *Clusia osaensis* Hammel-ined., *Clusia peninsulae* Hammel-ined. and *Clusia valerioi* Standl. (Wanek et al. 2002b) have now shown that the expressions of modes of photosynthesis, where the different options exist in an individual species, are not determined by life forms or developmental stages but are always governed by environmental factors such as those listed in Sect. 8.8. Berg et al. (2004) have studied the expression of CAM in terrestrial rooted plants of *C. hilariana* Schlecht. at different stages of development in the Brazilian restinga. They found a pattern of physiological traits typical of CAM to be more pronounced in early growth and young plants than in mature

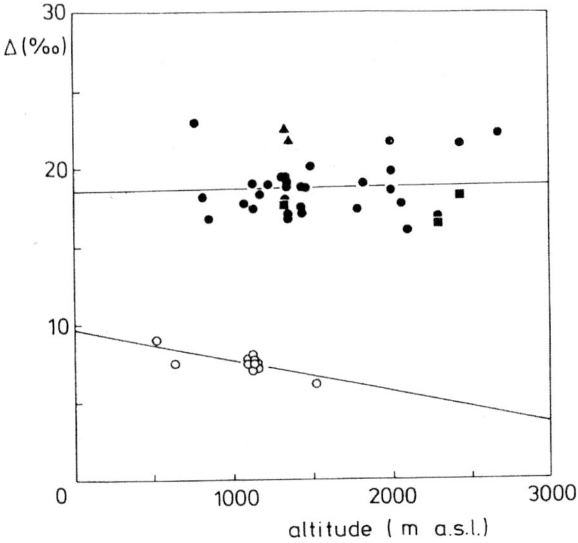

Fig. 9.2. Relation of $\Delta^{13}C$ values of leaves of Clusiaceae species to the altitude of sampling. $\Delta^{13}C$ was calculated after the Farquhar-equation (Farquhar et al. 1989)

$$\Delta^{13}C = \frac{\delta^{13}C_a - \delta^{13}C_p}{1000 + \delta^{13}C_p} \times 10^3\ (‰)$$

where $\delta^{13}C_p$ is the carbon isotope ratio of the plant sample and $\delta^{13}C_a$ the value for the CO_2 in the ambient atmosphere (–8‰). High values of $\Delta^{13}C$ indicate C_3-photosynthesis and low values CAM. *Closed symbols*: C_3-samples, *open symbols* CAM samples. *Circles*: *Clusia* species, *triangles*: *Dystovomita clusiifolia*, *squares*: *Oedematopus* species (Fig. 1 from Diaz et al. 1996)

shrubs. This may be due to a developmental regulation but perhaps is better explained by water relations. In the seasonally dry restinga the young plants with smaller root systems may suffer drought stress while the adult shrubs have access to the ground water.

With respect to the altitudinal range covered by *Clusias* (Table 9.1) there are two studies. A survey of Clusiaceae species (*Clusia, Oedematopus*, which is now also incorporated in the genus *Clusia*, and *Dystovomita*) has been performed in Venezuela (Diaz et al. 1996). In the northern cordillera, where *Clusias* may range up to an altitude of almost 2700 m, among the species tested based on stable carbon isotope analysis CAM is not expressed at altitudes above 1500 m (Fig. 9.2), which usually are the wetter habitats. A similar altitudinal limitation of CAM was observed among 25 *Clusia* species in Panamá (Holtum et al. 2004). Strong CAM species were restricted to altitudes below 680 m a.s.l. Most weak CAM species were found up to 1100 m a.s.l. and two species with weak CAM up to 1591 m a.s.l. (*C. odorata* Seem.) and 1689 m a.s.l.

(*C. croatii* D'Arcy), respectively, but not higher up. Some C_3 *Clusias* may reach altitudes of 3000–3200 m a.s.l. The factors that limit the distribution of *Clusia* species with CAM at high altitudes are not known.

9.3 Mineral Nutrition

Not much systematic work has been performed on mineral nutrition of *Clusias* and the information available is very limited. However, contents of minerals were analysed in a number of investigations. Concentrations of some mineral cations in leaves of *Clusia* species are compiled in Table 9.2. Interestingly, the levels of calcium are quite high in all species. The ratios of potassium/calcium obtained range between 0.2 and 0.8, i.e. they are mostly well below 1, and therefore as argued by Olivares and Aguiar (1999) all species can be considered to be calciotrophic. It is noteworthy that sodium contents are rather low even in the species of coastal restingas (*C. fluminensis* and *C. lanceolata*, if the values are correct; see footnote to Table 9.2). Thus, at the coastal sites the *Clusias* are probably not too much subjected to salt spray. In Sects. 8.3.3, 8.4, 8.7, 8.8.1 and 8.8.3 we have discussed the role of citrate in *Clusias*. An additional interesting idea that has been brought forward is that as an effective metal chelator citrate might also confer metal tolerance, and that *C. multiflora*, which has high non-oscillating background levels of citrate might be a manganese hyperaccumulator (Chacón et al. 1998). Leaves of *C. multiflora* contained 8 mM Mn kg^{-1} dry weight (ca. 1.5 mM Mn kg^{-1} fresh weight). However, in contrast in *C. uvitana*, Mn levels were only ca. 0.1 mM kg^{-1} fresh weight, i.e. quite low (Zotz and Winter 1994a).

Most work regarding mineral nutrition of *Clusias* has been performed with respect to nitrogen. Experiments and observations with *C. minor*, *C. osaensis* and *C. valerioi* have shown that these plants can take up inorganic nitrogen in the form of HH_4^+ and NO_3^- as well as organic nitrogen, e.g. in the form of glycine. In *C. minor* there was a strong preference for NH_4^+, however (Arndt et al. 2002; Wanek et al. 2002a). Incorporation of nitrogen in proteinogenic and non-proteinogenic compounds has also been analysed for *C. fluminensis*, *C. hilariana* and *C. parviflora* (Scarano et al. 2005).

Diaz et al. (1996) found a range of total N-levels of 550–1 215 µmol g^{-1} dry weight in leaves of various *Clusia* species in northern Venezuela. More analyses from the literature are surveyed in Table 9.3. Most of the values recorded are within this range. One notable exception is that of *C. parviflora*. These plants were growing on an inselberg (Pão de Açúcar, Rio de Janeiro). The granitic rock of the inselbergs is densely covered by crusts of cyanobacteria capable of fixing atmospheric di-nitrogen, and it has been shown that leachate from the cyanobacteria and run off from inselberg rocks significantly contributes to nitrogen input in the surrounding vegetation (Lüttge

Table 9.2. Mineral cation concentrations in leaves of *Clusia* species in mmol/L^{-1} or mmol kg^{-1} fresh weight, where litre and kilogram fresh weight as available as base units in the literature are taken to give a roughly similar assessment

Species	Na	K	Mg	Ca
C. columnaris Engl.[b]		37–59	13–31	9–105
C. fluminensis Planch. et Triana[c]	18	30	88	53
C. lanceolata Camb.[c]	35	21	29	110
C. minor L.[b]		41	26	91
C. multiflora H.B.K.[b]		11–23	14–21	45–75
C. krugiana Urb.[b]		21	33	52
C. obovata (Spruce ex Planch. et Triana) Pipoly[b]		45	17	121
C. rosea Jacq.[b]		19–24	13–57	91–145
C. rosea Jacq.[a]	2	49	91	149
C. uvitana Pittier[d]	18	60	27	77

References: [a]Ball et al. 1991a; [b]Olivares and Aguiar 1999; [c]Reinert et al. 1997; [d]Zotz and Winter 1994a

(Olivares and Aguiar (1999, Table 1) and Reinert et al. (1997), did not specify if the base unit was fresh weight or dry weight, so these data may be useless for our comparison. However, in data transformation I noticed that the ion levels and also nitrogen (see Table 9.3) would have been extraordinarily high if it had been fresh weight, but with dry weight they were well in the range of those found in other publications. Hence, I chose dry weight for data transformation, and I used the dry weight/fresh weight ratios of Table 2.1 (and where not specifically available for a species the average value of all *Clusias*) for calculation of concentrations)

1997b; Dojani et al. 2006). In fact the top soil (0–5 cm) beneath these plants of *C. parviflora* had a nitrogen content of 1.5 % of dry weight, while, e.g. the soil around the plants of *C. fluminensis* and *C. hilariana* listed in Table 9.3 had nitrogen levels of 0.3 % (Scarano et al. 2005). Compared with compatriot woody plants of similar life form in the restingas of Brazil (see Sect. 9.4.2.1) species of *Clusia* have relatively low N-levels in their leaves, e.g. in relation to dry weight there was 0.6–0.8 % N in *C. fluminensis* and *C. hilariana*, 1.0–1.1 % N in *Calophyllum brasiliense* Cambess., *Rheedia brasiliensis* (Mart.) Planch. et Triana and *Myrsine parvifolia* A. DC. and 1.3–1.9 % N in *Andira legalis* Vell. Toledo, where the latter species, however, is a nodulated dinitrogen fixing Leguminosae species (Duarte et al. 2005; Geßler et al. 2005a; Scarano et al. 2005). Lüttge et al. (1993) and Diaz et al. (1996) found some correlation of nitrogen levels of leaves in species of *Clusia* and carbon isotope discrimination. Increasingly negative values of $\delta^{13}C$ were correlated with increasing N levels which suggests enhanced nitrogen supply to the leaves when transpiration is higher (more negative $\delta^{13}C$-values). An interesting condition in respect to mineral nutrition is given when *Clusia* seedlings grow in the tanks of bromeliads. It can be observed that some epiphytes contain less nitrogen in their leaves than compatriot terrestrial plants of the same species, e.g. the two aroids shown in Fig. 9.3. This is not so, however, for tank forming bromeliads which collect a lot of litter and water in their tanks and *Clusia* growing in their tanks can benefit from that (Fig. 9.3).

Table 9.3. Nitrogen levels (μmol g^{-1} dry weight) in leaves of *Clusia* species (errors where given are SE)

Species	Irradiance during growth		
	No comment	Exposed	Shaded
C. alata Planch. et Triana	780±20		
C. columnaris Engl.*	825±19		
C. fluminensis Planch. et Triana*	565±65		
C. hilariana Schlecht.	500		
C. krugiana Urb.*	428		
C. lanceoloata Camb.*	805		
C. minor L.*	732±17		
C. minor L.		850 1 045 (+N) 490 (–N)	910 1 420 (+N) 705 (–N)
C. multiflora H.B.K.*	771±144		
C. multiflora H.B.K.	670±110	685	995
C. obovata (Spruce ex Planch. et Triana) Pipoly*	500±5		
C. parviflora Saldanha et Engl.	1 715		
C. rosea Jacq.*	742±56		
C. rosea Jacq.	770±16	730±2	1065±2
C. uvitana Pittier		642±0	605±35
C. spec.	725±25		

The ±N data are from an experiment with *C. minor*, where –N-plants did not receive supplemental nitrogen and +N-plants were watered weekly with 24 mmol L^{-1} NO_3^- and plants were grown at a high irradiance (220–320 μmol m^{-2} s^{-1}) and low irradiance (60–70 μmol m^{-2} s^{-1}), respectively (Franco et al. 1991). *C.* spec. is the same as discussed in Sect. 9.4.2.6. For the data of the plants labelled by *: see footnote of Table 9.2. References: Ball et al. 1991a; Franco et al. 1991; Lüttge et al. 1993; Zotz and Winter 1994a; Haag-Kerwer et al. 1996; Grams et al. 1997; Reinert et al. 1997; Olivares and Aguiar 1999; Scarano et al. 2005

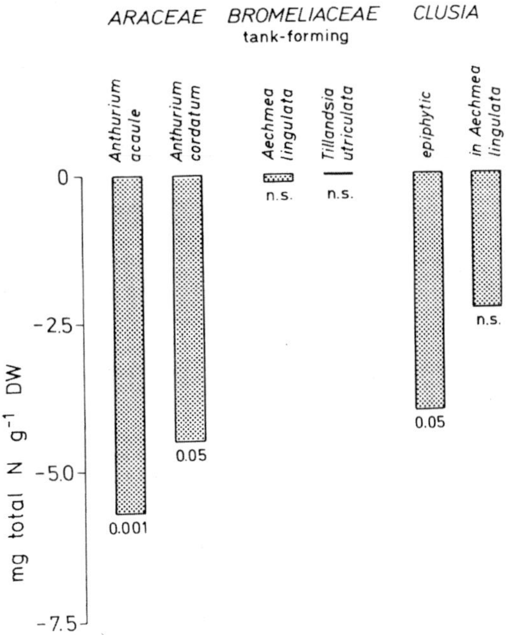

Fig. 9.3. Comparison of nitrogen levels in cohabitant epiphytic and terrestrial life-forms of two aroid and two bromeliad species and of *Clusia rosea*. Data show N content in leaves of epiphytic minus leaves of terrestrial plants of the same species growing close to each other. (Numbers are p-values of a t-test for statistically significant differences; n.s. non-significant. (Ball et al. 1991a: Lüttge 1997a)

Where a distinction has been made in Table 9.3 in all cases (except *C. uvitana*, see below) shaded plants had higher nitrogen levels in their leaves than exposed plants. This agrees well with the general observation that shade plants have larger total N-contents in their biomass than sun plants (Lüttge 1997a). It is also known that relatively higher levels of nitrogen in the leaves are correlated with higher rates of CO_2 assimilation (Field 1988; Evans 1988; Lüttge 1997a). In experiments of Franco et al. (1991) with *C. minor* this was only observed in high light grown and not in low light grown plants (±N data in Table 9.3). In the high light grown plants the rates of CO_2 uptake without N-nutrition were 59.6 and –0.8 mmol m^{-2} by day and night, respectively, and with N-nutrition (weekly watering with 24 mM NO_3^-) 209.0 and 12.1 mmol m^{-2}, respectively (Table 8.10). Thus, in the high light plants N-nutrition highly increased daytime CO_2-uptake and also induced some nocturnal CO_2 fixation in this C_3/CAM-intermediate species (Sect. 8.8.4).

The exception that in *C. uvitana* shaded plants did not contain higher nitrogen levels than exposed plants may be related to the hemi-epiphytic habitat. Possibly N was so limiting that the shaded plants could not acquire higher levels. This may be supported by the observation that the N-levels in *C. uvitana* are among the lowest recorded (Table 9.3). The P-levels in *C. uvitana*, however, were quite large (25 µmol g^{-1} dry weight; Zotz and Winter 1994a) and much larger than the values for *C. columnaris, C. fluminensis, C. krugiana, C. lanceolata, C. minor, C. multiflora, C. obovata, C. rosea* which ranged between 2 and 7 μmol g^{-1} dry weight (Reinert et al. 1997; Olivares and Aguiar 1999). S-levels in *C. uvitana* were 55 μmol g^{-1} dry weight (Zotz and Winter 1994a).

9.4 Habitat Related Performance of Clusias

9.4.1 Aims and Approaches of Assessment

Aut-ecology studies the potential and actual performance of given plants or species in the ecological context. Most physiological ecology is aut-ecology. Syn-ecology aims at the synthetic understanding of larger entities with numerous species and their synergistic and antagonistic interactions. Most syn-ecological studies are based on floristic inventories and phytosociological approaches. However, with appropriate sampling and analyses together with miniaturized and readily portable measuring equipment physiological syn-ecology has now become feasible (Lüttge and Scarano 2004). Aiming at an assessment of the habitat performance of *Clusia*s we may therefore consider the potential performance of species at the community level as well as the impact of *Clusia*s in plant communities, e.g. dominance, where *Clusia* scrubs are formed in the restinga vegetation (Sect. 9.4.2.1), or nurse effects, where *Clusia*s protect the establishment of other plants (Chap. 4). Hence, in this section for a selection of habitats, where studies have been performed, we shall compare sympatric species of *Clusia*s among each other and species of *Clusia* with other species of similar shrub and tree life-form where possible.

Laboratory studies of aut-ecology mostly cover the dynamics of short term acclimation, while ecophysiological measurements in the field due to logistic constraints in measuring campaigns often reflect 'snap shots' of the performance of plants under a momentarily given set of environmental conditions. Long term laboratory observations are as rare as recordings of gradual responses under gradually varying, e.g. seasonally, natural conditions. Repetitions in different seasons are important but also necessary over different years as the expression of seasonal conditions is not constant year after year. It is an enormous task to fulfil these requirements and comparative studies extended over years are extremely rare. An alternative is to try and assess the potential intrinsic performance of plants by integrating all the information available from both measurements in the field and in the laboratory. When a set of several ecophysiological parameters is considered, one can develop ecophysiological fingerprints of various species. I adopt here an attempt which I had used for *Clusia* before (Lüttge 1999), where ranges of values in laboratory and field studies were so similar that a separate treatment was not warranted and averaging of laboratory and field measurements was acceptable. It is important to document the potential performance or intrinsic capabilities of plants in relation to the habitats where they occur and not their average behaviour. Therefore, in many of the fingerprint diagrams below (Figs. 9.4–9.7. 9.9, 9.12, 9.14, 9.17, and 9.18) maximum values of measured parameters will be assessed when a given reference source reports a range of values for different conditions in the laboratory or due to effects of seasons, sites and habitats in the field. This allows integration of quite a large amount of published information. It is not easy and straightforward to do this though, because published studies were performed in different seasons and with different intensity (e.g., sample sizes, etc.). Moreover, units of reference varied so that appropriate conversion factors had to be worked out from all the published data. The approach chosen, however, appears to be the only one to arrive at a somewhat quantitative overview of a large amount of published work without just restricting us to verbal generalizations. It is a reasonable attempt to develop patterns for understanding the potential performance of *Clusia*s at the community level.

The parameters chosen here are those for which enough comparative information is available in the literature on *Clusia* and which are most relevant for describing the physiological ecology of *Clusia*s as listed below:

- Carbon isotope ratio, $\delta^{13}C$. This parameter gives information about the mode of photosynthesis (C_3 or CAM), and where other information indicates that the plant has a capacity for CAM, the extent of primary dark fixation of CO_2 may be assessed (Sect. 8.2).
- Day night changes in titratable acidity (ΔH^+) and levels of malate and citrate, Δmal and Δcitr, respectively. These parameters are the most reliable indicators of CAM performance (Sect. 8.1). Since it has been generally shown that the sum of $2 \times \Delta$mal$+3 \times \Delta$citr is equal to ΔH^+ due to the number of carboxyl groups in the two organic acids, I plot the acid oscillations in this way.
- Rate of CO_2 uptake in the dark period. This is also a parameter directly related to CAM.
- Rate of CO_2 uptake in the light period. This parameter provides information on photosynthetic capacity.
- Effective quantum yield of photosystem II (PS II). This parameter is obtained by measurement of chlorophyll *a* fluorescence as $\Delta F/F_m' = (F_m'-F)/F_m'$, where F is the steady-state fluorescence of a light adapted leaf and F_m' the maximal fluorescence of a light adapted leaf under a saturating light flash. $\Delta F/Fm'$ is highly dependent on photosynthetic photon flux density (PPFD), and all values from the literature were taken for the reasonably high and often saturating PPFD of 1000 $\mu mol\ m^{-2}\ s^{-1}$.
- Apparent electron transport rate of PS II, $ETR=0.5 \times \Delta F/F_m' \times PPFD$. This parameter provides information on photosynthetic capacity; the factor 0.5 accounts for equal distribution of PPFD to both photosystem II and photosystem I.

- Potential quantum yield of PS II, $F_v/F_m=(F_m-F_o)/F_m$. This parameter is also obtained by measurement of chlorophyll *a* fluorescence, where F_o is the ground fluorescence of a dark adapted leaf and F_m its maximum fluorescence under a saturating light flash. Ratios of F_v/F_m below 0.83 indicate photoinhibition (Sect. 8.6, Björkman and Demmig 1987). They are presented for measurements after darkening over night (predawn measurements) to allow the assessment of chronic photoinhibition not reversible overnight. Measurements during midday after darkening for mostly 10–20 min allow the assessment of acute photoinhibition not reversible after 10–20 min of darkening.

(Literature for chlorophyll fluorescence analysis: e.g. Genty et al. 1989; van Kooten and Snel 1990; Schreiber and Bilger 1993. In the fingerprint diagrams below averages of average values recorded from the different sources in the literature are presented with standard errors where possible.)

9.4.2 Habitats and Sites

9.4.2.1 Restingas

The restinga (Chap. 3, Fig. 9.1A) of the sandy coastal plains of Brazil consists of a mosaic of plant communities ranging from open formations to forest ecosystems (Lacerda et al. 1993). The open formation is characterized by shrub islands surrounded by white sand. There are dune ridges with dry forest in between and on fixed dunes (Cirne and Scarano 2001). Species of *Clusia* are an important floristic element of the restingas, so that some restinga physiognomies have been called *Clusia* scrub (Ule 1901), and *Clusia*s can function as nurse plants for other vegetation growing on the shrub islands which they established (Zaluar and Scarano 2000; Liebig et al. 2001; Chap. 4). The hydrology of restingas is determined by seasonal annual rainfall and the ground water table and the degree of salinity in the ground water. The rainy season lasts from the end of spring (November) to summer (January/February) and the dry season extends through the winter months (May/June to August). Taking these various factors into consideration, in the coast of the state of Rio de Janeiro, where most of the studies reviewed here have been undertaken, we can distinguish wet, dry and intermediate restingas (Scarano 2002). Extensive ecophysiological field work on *Clusia*s was performed at largely four different restinga sites listed in Table 9.4. I shall first characterise the performance of *Clusia*s at these four different sites in comparison to other woody species of a similar life form of shrubs and small trees, and then try to obtain a general overview on the ecophysiology of the *Clusia*s in the restingas.

Table 9.4. Restingas at the Atlantic coast of Brazil, where ecophysiological field studies on species of *Clusia* were performed

Location	Rainfall (mm year^{-1})	Water table (m)	Denotation
State Ecological Reserve of Jacarepiá (22° 47'–22° 57' S 42° 20'–42° 43' W)	1000	0.5–1.0	Wet
Restinga de Jurubatiba National Park (22°00'–22° 23' S 41° 15'–41° 45' W)	1165	2–3	Intermediate
Barra de Maricá (22° 53' S; 42° 52' W)	1000		Intermediate
Restinga of Massambaba (22° 56' S; 42° 13'W)	800	0.9–1.2	Dry

9.4.2.1.1 Wet Restinga of Jaracepiá

At this restinga site there is a swale between forest and the strand line that is covered by marsh vegetation dominated by Poaceae and Cyperaceae, except where a spit of sandy soil, some 200 m long by 70 m wide (maximum) extends from the edge of a dry forest almost to the beach (Sá 1992; Cirne and Scarano 2001; Cirne et al. 2003; Geßler et al. 2005a; Scarano et al. 2005). Due to the high rainfall and high ground water table it is a wet restinga (Table 9.4). Here, *C. fluminensis* occurs both as a tree up to 8 m in height at the ecotone between forest and open restinga, and as a shrub up to 2 m in height in the open restinga. As the δ^{13}C-value shows, *C. fluminensis* is a species mostly performing CAM in the field, and we can compare it with the legume shrub *Andira legalis* (Fabaceae) performing C_3-photosynthesis (Fig. 9.4; Scarano et al. 2005; Geßler et al. 2005a). In terms of the maximum values of ETR and $\Delta F/F_m'$ the CAM species *C. fluminensis* shows a stronger performance than the C_3-species *A. legalis*. Regarding acute and chronic photoinhibition, both species appear to be similarly affected, the values of F_v/F_m are clearly lower than 0.83 both at midday and before dawn.

9.4.2.1.2 Intermediate Restinga of Jurubatiba

Unlike most sandy coastal plains in the state of Rio de Janeiro that date from the Holocene, the Jurubatiba sandy plains consist of a series of parallel Pleistocenic beach ridges (Henriques et al. 1986; Araujo et al. 1998; Duarte et al. 2005). They are an intermediate restinga, because although the annual rainfall is as high or even somewhat higher as in the wet restinga (Sect. 9.4.2.1.1) the

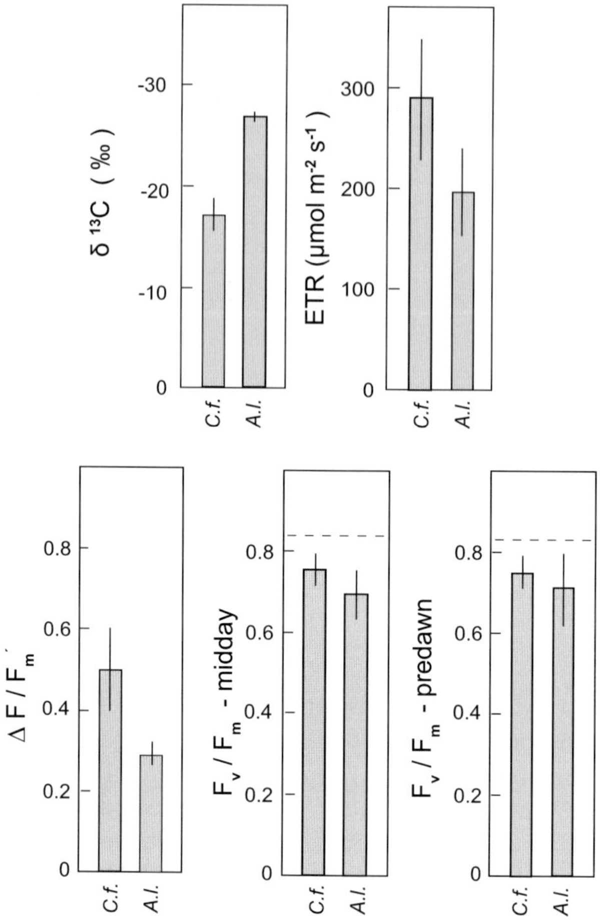

Fig. 9.4. Ecophysiological fingerprint: wet restinga of Jaracepiá. *C.f.=Clusia fluminensis, A. l.=Andira legalis.* For detailed explanation see Sect. 9.4.1 (data of Scarano et al. 2001, 2005; Geßler et al. 2005a)

ground water table is much lower (Table 9.4). Here *C. parviflora* was studied in one to two investigations but a lot of work was devoted to *C. hilariana* (Franco et al. 1996, 1999; de Mattos et al. 1997; Herzog et al. 1999 c; Liebig et al. 2001; Berg et al. 2004; Scarano et al. 2005), so that the data base is much larger for the latter than the former. δ^{13}C-values show that *C. parviflora* is performing C_3-photosynthesis and *C. hilariana* CAM (Fig. 9.5). According to $\Delta F/Fm'$ at 1000 mol m^{-2} s^{-1} and maximum ETR *C. hilariana* is performing a little better than *C. parviflora*, but this is not the same for maximum rate of CO_2 uptake in the light and acute and chronic photoinhibition (F_v/F_m-values at midday and before dawn, respectively). In Fig. 9.5 the two *Clusia* species are compared with three other woody species of similar life form, *A. legalis*, *Rheedia brasiliensis* (Clusiaceae) and *Myrsine parvifolia* (Myrsinaceae), all performing C_3-photosynthesis. The comparison is interesting in that the CAM species *C. hilariana* in respect to ETR and $\Delta F/F_m'$ shows superior performance to all the C_3 species, but the C_3 *Clusia*, *C. parvifolia*, seems to perform better than

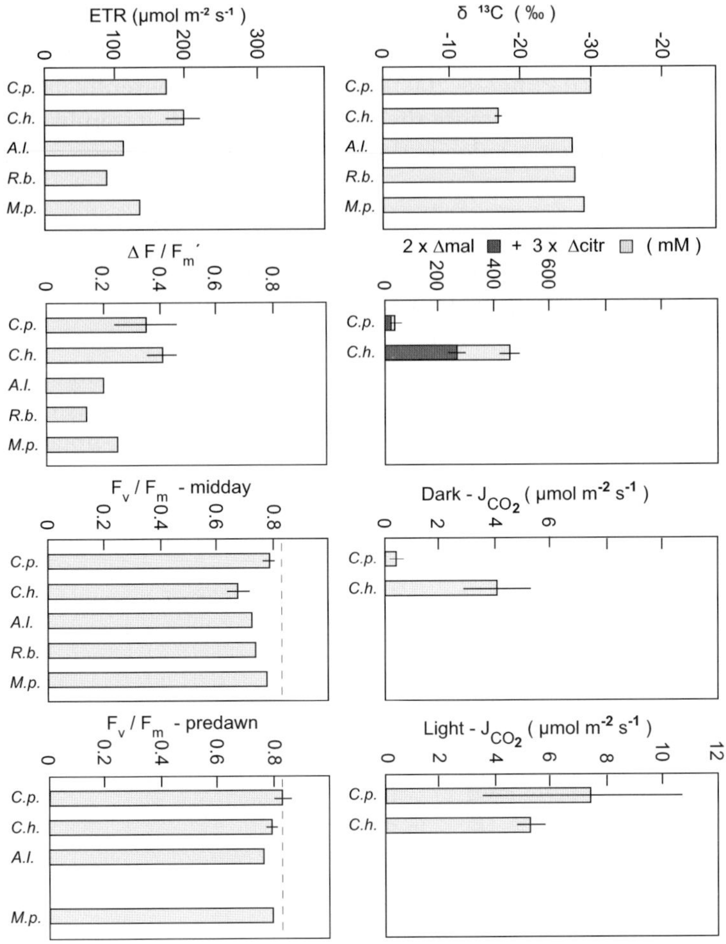

Fig. 9.5. Eco-physiological fingerprint: intermediate restinga of Jurubatiba. *C.p.=Clusia parviflora*, *C.h.=Clusia hilariana*, *A.l.=Andira legalis*,, *R.b.=Rheedia brasiliensis*, *M.p.=Myrsine parvifolia*. For detailed explanation see Sect. 9.4.1 and for references Sect. 9.4.2.1.2

the other C_3 shrubs. Chronic photoinhibition is absent or low in all species, but most noteworthy, an advantage of CAM in avoiding acute photoinhibition is not seen.

The data of Fig. 9.5 are integrated over the whole literature. A more subtle comparison is given by just considering one study where *C. hilariana* was measured along a transect from the sea inland on a first and a second beach ridge and a dry forest in between (Scarano et al. 2005). Instantaneous measurements of $\Delta F/F_m'$ and ETR suggested that the plants closer to the sea in the first beach ridge ($\Delta F/F_m'$ and ETR at 400 µmol m^{-2} s^{-1} irradiance are 0.38 and 72 µmol m^{-2} s^{-1}, respectively) had an inferior performance to those in the dry forest and the second beach ridge ($\Delta F/F_m'$ and ETR at 400 µmol m^{-2} s^{-1} irradiance are 0.55 and 97 µmol m^{-2} s^{-1}, respectively). They also had lower carbohydrate concentrations in the leaves. This may be due to salt spray closer to the sea and a higher ground water table farther inland (Scarano et al. 2005). However, *C. hilariana* also is the dominant tree species on the first beach ridge (Zaluar 1997; Pimentel 2002) and despite its apparently inferior performance as compared to the more sheltered inland plants it is unlikely that it really has reduced ecological vigour closer to the sea.

9.4.2.1.3 Intermediate Restinga of Barra de Maricá

Figure 9.6 shows the fingerprints of the two CAM performing species *Clusia lanceolata* and *Clusia fluminensis* at the intermediate restinga of Barra de Maricá (Roberts et al. 1996; Reinert et al. 1997). *C. lanceolata* is more in the range of δ^{13}C-values indicating prevailing C_3-photosynthesis and *C. fluminensis* is shifted more towards CAM. *C. lanceolata* presents the highest levels of nocturnal acid accumulation and the lowest net CO_2-uptake in the dark. This discrepancy may be due to internal recycling of respiratory CO_2. However,

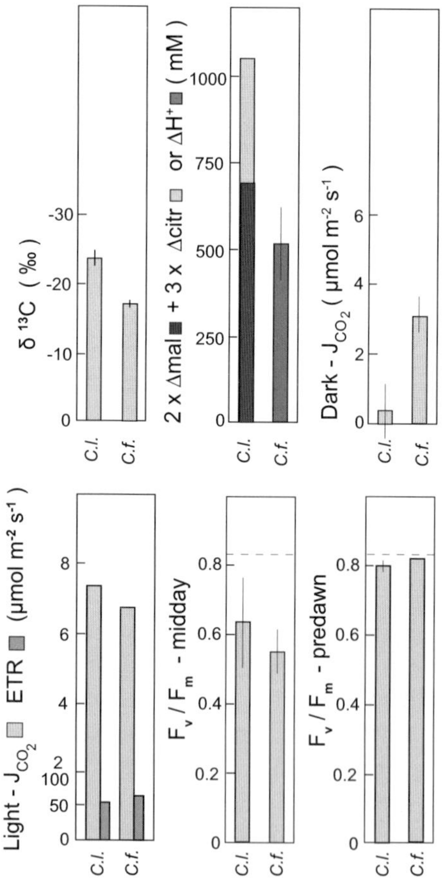

Fig. 9.6. Ecophysiological fingerprint: intermediate restinga of Barra de Maricá. *C.l.*=*Clusia lanceolata*, *C.f.*=*Clusia fluminensis*. For detailed explanation see Sect. 9.4.1 (data of Roberts et al. 1996; Reinert et al. 1997)

data available are limited. Day-time behaviour (CO_2-uptake in the light, ETR) is very similar in both species. Both experience acute photoinhibition where they grow, with F_v/F_m-midday well below 0.8, but no chronic photoinhibition (F_v/F_m-predawn close to 0.8). Thus, the overview of their potential performance in Fig. 9.6 does not separate the two species very much, although Roberts et al. (1996) also noted that *C. lanceolata* behaved more C_3-like and *C. fluminensis* more CAM like at the Barra de Maricá field site.

9.4.2.1.4 Dry Restinga of Massambaba

This restinga is located in the Cabo Frio region, where the climate is drier due to a cold oceanic upwelling (Araujo 1997). On the east side, where the plants of *C. fluminensis* of Fig. 9.7 were studied (Scarano et al. 2005), the Pleistocenic innermost beach ridge (Martin and Suguio 1989) was covered by a dune system ca. 2000 years BP (FEEMA 1988). Here the distance between the ocean and the

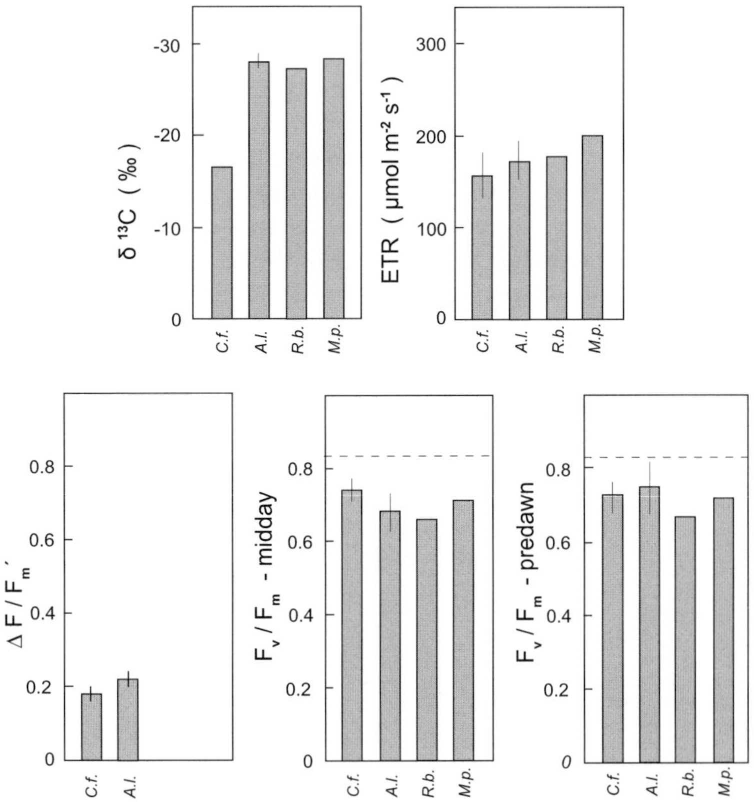

Fig. 9.7. Ecophysiological fingerprint: dry restinga of Massambaba. *C.f.=Clusia fluminensis, A.l.=Andira legalis, R.b.=Rheedia brasiliensis, M.p.=Myrsine parvifolia.* For detailed explanation see Sect. 9.4.1 (data of Scarano et al. 2001, 2005; Duarte et al. 2005; Geßler et al. 2005a)

Aruama lagoon, the largest hypersaline lake in the world, is ca. 1.5 km. Although the water table is high, the result is a soil water deficit throughout the year with a partially saline ground water, so that together with the low precipitation this makes it a dry restinga. *C. fluminensis*, which does not rank among the dominant plants in this vegetation, is a small shrub up to 2 m in height. It is mainly performing CAM and in Fig. 9.7 its performance is compared with three similar C_3-shrubs, *A. legalis, R. braslliensis* and *M. parvifolia.* In terms of photosynthetic capacity and also photoinhibition all four species were rather similar, so that no immediate advantage of CAM was visible here.

9.4.2.1.5 Some Inter- and Intra-specific Comparisons among the *Clusias* of the Restingas

Using expression of maximum ETR as an ecophysiological trait some more general comparisons are presented in Table 9.5. In a more subtle comparative study restricted to one year, September to October 1999, Scarano et al. (2005) noted that the performance of *C. fluminensis* was influenced by a moisture gradient, which is also seen in the data in Table 9.5 where ETR is 290 µmol m^{-2} s^{-1} in the wet restinga and 160 µmol m^{-2} s^{-1} in the dry restinga and with the measurement in the dry forest (Sect. 9.4.2.5) in the same period of 1999 of an ETR of 100 µmol m^{-2} s^{-1} this tendency is perfectly confirmed. The same authors also noted from the measurements in 1999 that the performance of the three species *C. parviflora, C. fluminensis* and *C. hilariana* was increasing in this order. For the comparison of the two CAM species *C. fluminensis* and *C. hilariana* the better performance of the latter in the intermediate restinga was discussed to be in harmony with the phytosociological patterns as *C.*

Table 9.5. Intra- and inter-specific comparison of maximum ETR observed for various species of *Clusia* at various restinga sites (Sect. 9.4.2.1), an inselberg (Sect. 9.4.2.11) and a dry forest (Sect. 9.4.2.5)

Species	Mode of Photosynthesis	Site	ETR (µmol m^{-2} s^{-1})
C. parviflora	C3	Inselberg	255
		Intermediate restinga	130
C. fluminensis	CAM	Wet restinga	290
		Intermediate restinga	64
		Dry restinga	160
		Dry forest	100
C. hilariana	CAM	Intermediate restinga	200
C. lanceolata	CAM	Intermediate restinga	60

hilariana is dominant throughout the open restinga vegetation and *C. fluminensis* is less conspicuous. Taking together all figures obtained over the different sites and years from the different studies as available from the literature surveyed in this Sect. 9.4.2.1 on the restingas (Table 9.5) it turns out to be impossible, however, to arrive at a gross distinction of the capacity of the various species. Niche differences of the various *Clusia* species must be based on rather more subtle expressions of ecophysiological traits.

9.4.2.2 Northern Coastal Range of Venezuela Near Caracas: Secondary Savanna, Secondary Forest, Cloud Forest

The pair of *C. multiflora* and *C. minor* with obligate C_3-photosynthesis und C_3/CAM-intermediate behaviour, respectively, has been subject to detailed comparative ecophysiological studies in a phytotron demonstrating different potentials of adaptation to irradiance in the two species (Herzog et al. 1999a, Sect. 8.6). The territory of the Instituto Venezolano de Investigaciones Científicas near Caracas, Venzuela (10° 24' N, 66° 58' W) offered the great opportunity of a field based comparison (Franco et al. 1994; Grams et al. 1997). Exposed plants of both species were found to occur together jointly forming a *Clusia* shrubbery in a sun exposed open secondary savanna at 1500 m a.s.l. (Fig. 9.8), shaded plants of *C. minor* were occurring nearby in a secondary for-

Fig. 9.8. *C. minor* (small leaves, right side and background) and *C. multiflora* (large leaves foreground left) sympatrically in a shrubbery of a secondary savanna, area of the Instituto de Investigaciones Científicas, Caracas, Venezuela

est at 1540 m a.s.l. and shaded plants of *C. multiflora* were found at a short distance but higher elevation at 1740 m a.s.l. in a cloud forest.

The fingerprint diagram of the potential maximum performance of both species (Fig. 9.9) reveals only one difference, which is the CAM capacity of *C. minor* with nocturnal CO_2 uptake and organic acid accumulation which lacks in *C. multiflora*. The $\delta^{13}C$ signature shows, however, that in the field *C. minor* is mainly using the C_3-option, its $\delta^{13}C$ values are not significantly less negative than those of *C. multiflora*. For all the other parameters both species are also very similar, i.e. maximum day time CO_2 uptake rates, effective quantum use efficiency and acute photoinhibition and chronic photoinhibition as indicated by the F_v/F_m values below 0.83 at midday and before dawn, respectively.

It is interesting in this case, however, to compare the actual behaviour of sun exposed and shaded plants of both species. CO_2-exchange patterns are shown in Fig. 9.10 from two different studies. In the measurements of Grams

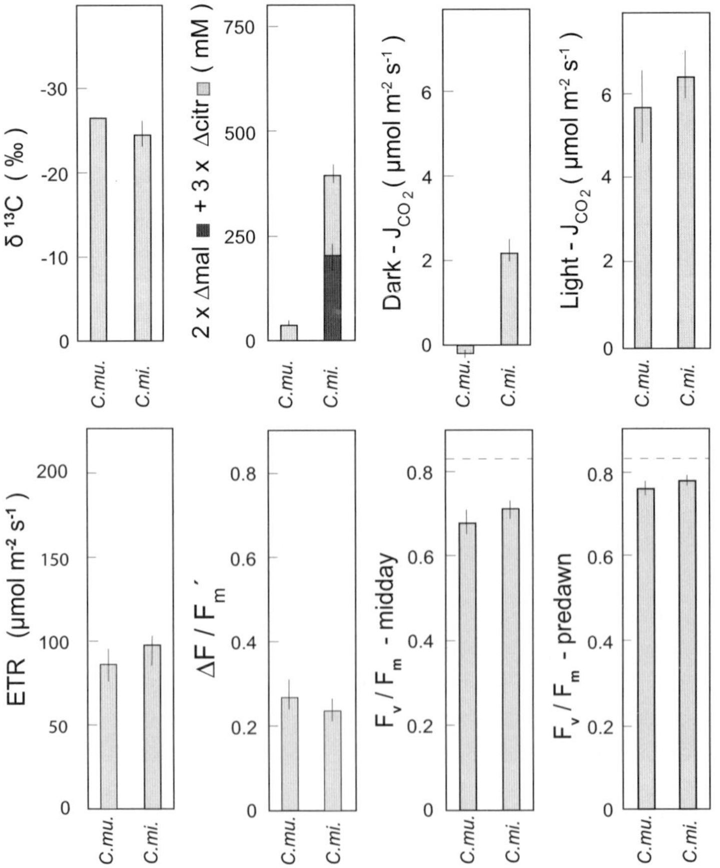

Fig. 9.9. Eco-physiological fingerprint: northern coastal range of Venezuela. *C.mu.=Clusia multiflora, C.mi.=Clusia minor.* For detailed explanation see Sect. 9.4.1 (data of Franco et al. 1994; Grams et al. 1997)

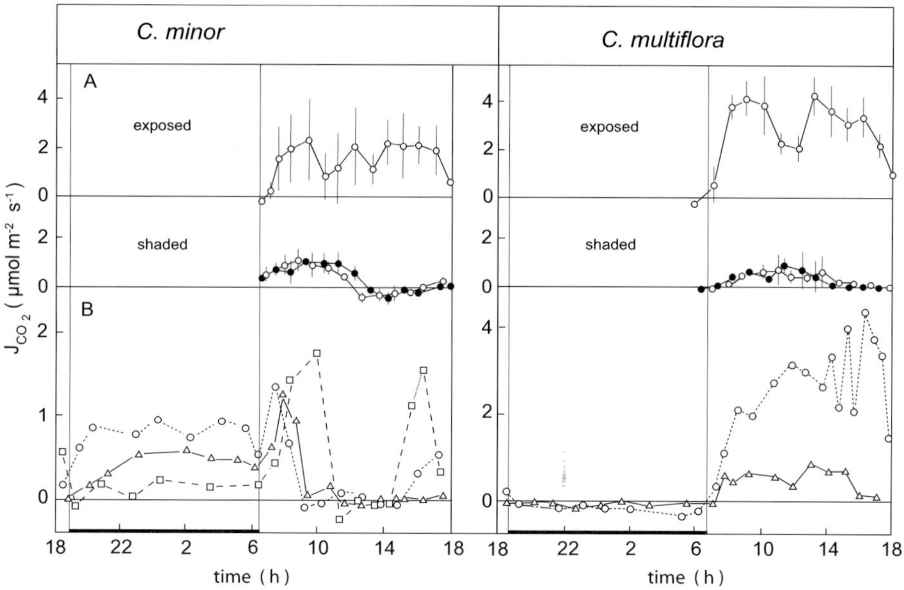

Fig. 9.10A,B. CO_2-exchange (J_{CO2}) of *C. minor* and *C. multiflora* in the field in the northern coastal range of Venezuela observed in two different studies: **A** in the study of Grams et al. (1997) exposed and shaded plants of *C. minor* received a daily irradiance of 34.3 and 1.5 mol m^{-2}, respectively, and *C. multiflora* 2.0 and 28.6 mol m^{-2}, respectively; **B** in the study of Franco et al. (1994) the daily irradiances received by *C. minor* and *C. multiflora*, respectively, were 2.2 and 0.8 (*triangles*), 18.8 and 34.1 (*circles*), and 34.1 (*squares*). *Dark bars* on the abscissa indicate dark periods

et al. (1997) both species followed C_3-photosynthesis. The exposed plants of *C. multiflora* showed somewhat higher rates than those of *C. minor* but were subject to a pronounced midday depression. CO_2 exchange by the shaded plants was much lower in both species. In the measurements of Franco et al. (1994) the pattern and the rates of CO_2 uptake for the exposed and shaded plants of *C. multiflora* were similar. However, in their measurements *C. minor* was seen to perform CAM. In both studies the total daily irradiance received by both species was 1–2 and circa 30 mol m^{-2} for the shaded and exposed plants, respectively, while in the study of Franco et al. (1994) an additional irradiance of circa 20 mol m^{-2} was included in the measurements of *C. minor*. It is seen that nocturnal CO_2 uptake in the CAM-performing *C. minor* increased from the low to the medium irradiance and then decreased again at the highest irradiance. This is also reflected in the total integrated CO_2 uptake for the night periods and night/day oscillations of organic acids (Fig. 9.11). Integrated day time CO_2 uptake was much higher in the exposed than in the shaded plants of *C. multiflora*. In *C. minor* at the highest irradiance day time CO_2-uptake was also much higher and night time CO_2 uptake was reduced as

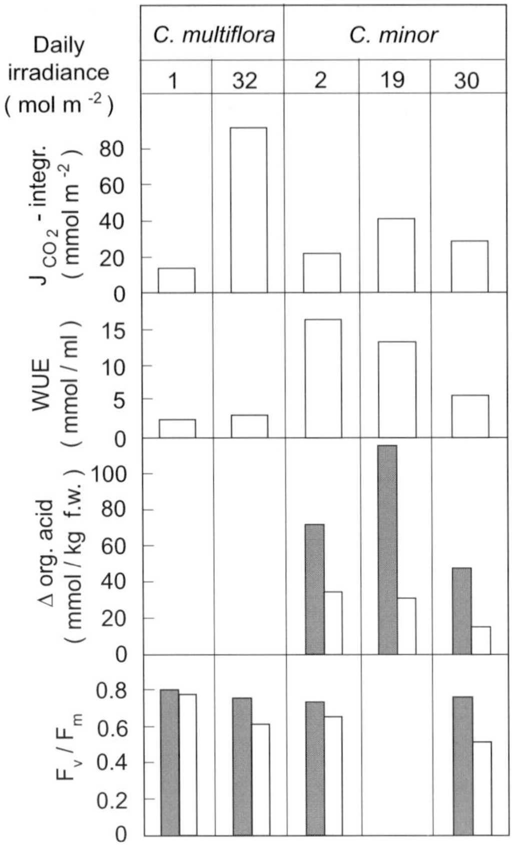

Fig. 9.11. Photosynthetic performance of *C. multiflora* and *C. minor* in the field with different daily irradiances as indicated. J_{CO2}-integr., net CO_2 uptake integrated over 24 h; WUE, water use efficiency; Δorg. acid, night/day changes of malate (*dark bars*) and citrate (*open bars*); F_v/F_m, potential quantum yield of chlorophyll *a* of photosystem II, predawn (*dark bars*) and at midday (*open bars*) (after data of Franco et al. 1994; Grams et al. 1997)

compared with the shaded plants at the lowest irradiance. These observations agree very nicely with those of the phytotron studies (Sect. 8.6) and underline the flexibility of *C. minor* using both the C_3 and the CAM option and expressing CAM to different degrees depending on irradiance. Both species control chronic photoinhibition in a similar way, while acute photoinhibition at midday is a little more pronounced in *C. minor*. On the other hand *C. minor* always has a better water use efficiency than *C. multiflora* (Fig. 9.11).

Now it is very interesting to compare this behaviour in the phytotron experiments and in the field measurement with the observation that *C. minor* seemed to perform better in the partially shaded than in the exposed habitats but that it can also invade the exposed habitats of *C. multiflora* (Fig. 9.8). When *C. multiflora* was transferred from low growth irradiance to high irradiance its leaves became necrotic (Sect. 8.6). The plants could grow new leaves from dormant buds which were then adapted to the high irradiance. Thus, whether the plant is fit for low and high irradiance, respectively, depends on its growth history and the development of the photosynthetic apparatus of light harvesting and electron transport at the irradiance prevailing during growth. Conversely *C. minor* did not suffer such damage most likely due to the flexibility of using rapid switches of its metabolic machinery to CAM in addition to thermal energy dissipation under light stress (Sect. 8.6). Thus, its flexibility gives *C. minor* a larger niche width in the secondary savanna/secondary forest complex than *C. multiflora*. The CAM option is not per se an advantage for growth in exposed sites. If that had been the case, one would have expected *C. minor* to be superior under the exposed rather than the shaded situation. It is rather the C_3/CAM plasticity which appears to be the decisive advantage.

9.4.2.3 Rupestrian Fields

We have unpublished ecophysiological data still to be processed on *Clusia arrudae* Planch. et Triana studied in rupestrian fields at Tiradentes (Fig. 9.1C) and in the Cerro do Cipó (Minas Gerais State, Brazil) and an unidentified C_3-*Clusia* at Tiradentes. Gas exchange measurements suggested that *C. arrudae* was performing some kind of CAM-cycling also supported by rather negative $\delta^{13}C$-values (see Sect. 8.1).

9.4.2.4 Gallery Forest to Cerrado Transect

Clusia criuva was studied in a gallery forest to cerrado transect near Brasilia, Brazil (15° 55' S, 47° 52' W) at 1050 m a.s.l. (Herzog et al. 1999b; Fig. 9.1D,E). The fingerprint data are compiled in Table 9.6. The transect goes from highly shaded and near the ground water level inside the gallery forest to semi-shaded and exposed and somewhat above the groundwater level at the gallery forest/cerrado ecotone and a few meters into the cerrado, respectively. *C. criuva* was not observed in the vegetation of the cerrado itself. We can see that *C. criuva* is C_3/CAM intermediate and performs weak CAM at the exposed site with some nocturnal organic acid accumulation and less negative $\delta^{13}C$ values. The latter still indicate that over time it largely performs C_3-photosynthesis but a certain CAM-component is certainly given. There was no photoinhibition at any of the sites.

Thus, *C. criuva* may be similar to *C. minor* with a flexibility allowing it to occupy both more shaded and more exposed sites. The question arises, however, why it then does not penetrate more deeply into the cerrado. This could be due to a water problem when it grows higher above the ground water table. However, we know that *C. criuva* also occurs on the highly sun exposed and temporarily very dry granitic rock outcrops of inselbergs in Brazil (Martii 1889, and personal observations). Unfortunately there are no ecophysiological measurements for *C. criuva* on inselbergs and we do not know if under the stronger light and drought stress given there it may express the CAM-option

Table 9.6. Performance of *Clusia criuva* in a gallery forest to cerrado transect

	Gallery forest shaded	→	Gallery forest cerrado ecotone semi-shaded	→	Cerrado exposed
Ground-water level (m)	0		0.4		1.0
Daily PPFD (mol m^{-2} day^{-1})	2.3		9.0		32.4
$\delta^{13}C$ (‰)	−30.78		−29.05		−26.61
Δmal (mM)	0		−2		45
Δcitr (mM)	0		−2		38
Succulence (l H_2O m^{-2})	0.42		0.43		0.41
max. ETR (μmol m^{-2} s^{-1})	140		335		390
$\Delta F/F_m'$ at 1000 μmol m^{-2} s^{-1} PPFD	0.25		0.55		0.50
F_v/F_m midday	0.80		0.83		0.78
F_v/F_m predawn	0.83		0.83		0.83

more strongly like *C. minor*. Its absence in the cerrado then might have other reasons, such as the frequent fires in the cerrado, when *C. criuva* may not be fire resistant.

9.4.2.5 Semi-deciduous Dry Forest at the Atlantic Coast in the State of Rio de Janeiro, Brazil

Clusia fluminensis has a broad range of distribution in different coastal habitats in the state of Rio de Janeiro including the last remnant of a typical semi-deciduous forest (22° 49' S, 41° 59' W, Fig. 9.1F). This site is quite arid with a mean annual rainfall of only 800 mm, a rainy season from the end of November to January and a dry season with monthly rainfall around 40 mm from June to August. In the forest *C. fluminensis* occurs as a free standing tree and as a hemi-epiphytic strangler. *C. fluminensis* is a CAM species and the $\delta^{13}C$-value of its leaves shows that it mostly makes use of primary CO_2-fixation via PEPC at this site (Fig. 9.12). To obtain an idea about its relative performance in the ecosystem we may compare it with other trees, the C_3-legume trees *Caesalpinia echinata* Lam., *Caesalpinia ferrea* Mart. ex. Tul. and *Machaerium obovatum* Kuhlm. et Hoehne, and the euphorb shrub *Croton compressus* Lam.

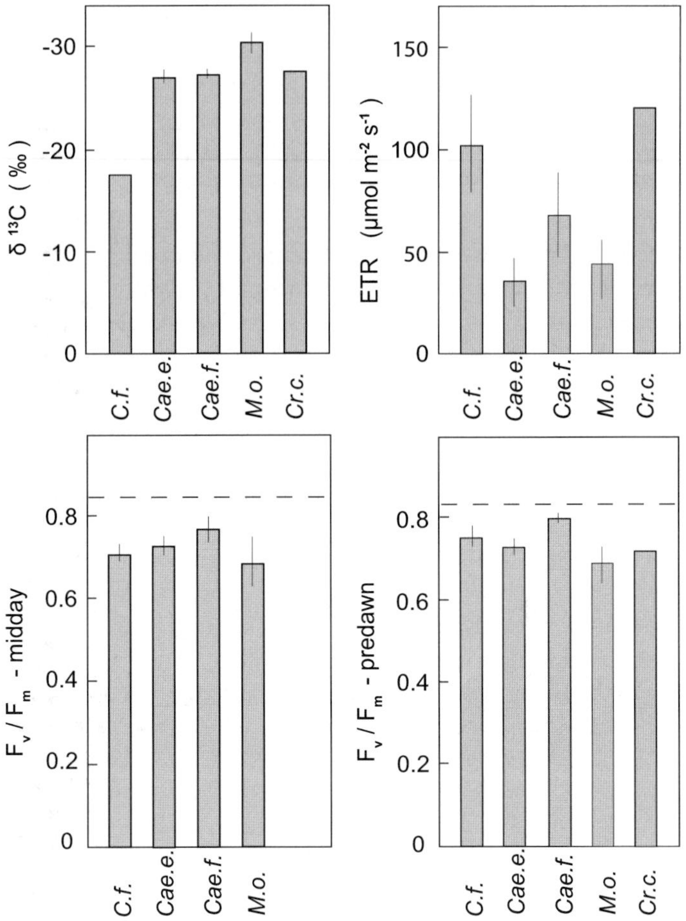

Fig. 9.12. Ecophysiological fingerprint: semi-deciduous dry forest Atlantic coast of Rio de Janeiro. *C.f.=Clusia fluminensis, Cae.e.=Caesalpinia echinata, Cae.f.=Caesalpinia ferrea, M.o.=Machaerium obovatum, Cr.c.=Croton compressus.* For detailed explanation see Sect. 9.4.1 (data of Scarano et al. 2001, 2005; Duarte et al. 2005; Geßler et al. 2005b)

(Fig. 9.12). The data available (Duarte et al. 2005; Scarano et al. 2005; Geßler et al. 2005b) show that *C. fluminensis* achieves somewhat lower maximum rates of apparent photosynthetic electron transport (ETR) than *C. compressus* and higher rates than the other species. In terms of acute and chronic photoinhibition as given by the values of F_v/F_m at midday and predawn, respectively, it is not better off than the two *Caesalpinia* species but somewhat superior predawn to *M. obovatum* and *C. compressus*. *C. fluminensis* is not very abundant in this forest, and a comparison with other sites in the area where it also occurs shows that it performs better in a wet restinga. Presumably in the very dry forest it is limited by the moisture regime, and most interestingly CAM does not provide it much advantage in comparison to similar life forms of woody species.

9.4.2.6 Karstic Limestone Mountains, Sierra de San Luis, Venezuela

The Sierra de San Luis, south of Coro in the state of Falcón, Venezuela (11° 18' N, 69° 45' W) is a dry karstic limestone ridge. The vegetation is predominantly a drought deciduous forest, in which evergreen species of *Clusia* are scattered. Ecophysiology of *Clusias* was mainly studied by Popp et al. (1987), Franco et al. (1994) and Haag-Kerwer et al. (1996). At an elevation of 1200 m a.s.l. four species of *Clusia* were found growing in an open sun exposed patch very close to each other (Fig. 9.13), *C. multiflora, C. rosea, C. alata* and *C.* sp. The latter species remained unidentified. It has all the characteristics of *C. minor*, white latex, short petiolate leaves and small pinkish-white flowers. It is not one of the *Clusia* species reported in Steyermark and Huber (1978) and might be possibly a new non-described species (Franco et al. 1994).

The carbon isotope ratios of leaf material collected in the field agree with the C_3-type nature of *C. multiflora* (highly negative $\delta^{13}C$ values) and show that the other three species are CAM species and under the conditions in the Sierra de San Luis actually make extensive use of the CAM option (significantly less negative $\delta^{13}C$ values). The ecophysiological fingerprints (Fig. 9.14) unanimously demonstrate the CAM nature of these three species, with nocturnal acid accumulation and CO_2 uptake. The expression of CAM is very similar in the three species. Photosynthetic capacity as given by maximal rates of

Fig. 9.13. Four sympatric species of *Clusia* (*C. multiflora, C. rosea, C. alata, C.* sp.) in the karstic limestone mountains, Sierra de San Luis, Falcon State, Venezuela

CO_2-uptake in the light period is rather similar in all four species. Maximum rates of photosynthetic electron transport are higher in the three CAM species than in the C_3-species *C. multiflora*, which reflects the high rates obtained at high internal CO_2-concentrations in phase III of CAM. This is also seen in the higher maximal effective quantum yields of the CAM-species. With respect to photoinhibition, the performance of all species is rather similar. There is slight chronic photoinhibition (reduced predawn values of F_v/F_m) independent of the mode of photosynthesis as well as pronounced acute photoinhibition at midday. The comparisons do not give clear hints for differential advantages of either C_3-photosynthesis or CAM among the *Clusia* plants sympatrically co-occurring at this site. In different ways, the plants seem to be similarly fit or similarly affected by stress given at the same site in these dry karstic mountains.

One would expect that different sympatric species growing close by each other at a given site have different niche requirements. Therefore, it may be

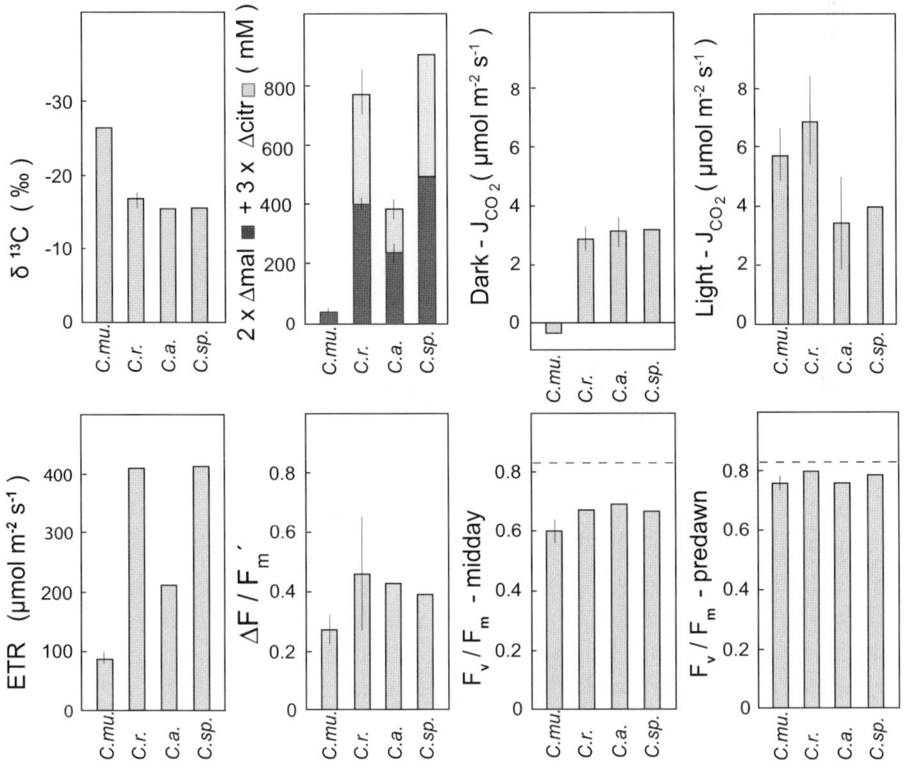

Fig. 9.14. Ecophysiological fingerprint: Karstic limestone mountains, Venezuela. *C.mu.=Clusia multiflora, C.r.=Clusia rosea, C.a.=Clusia alata, C.* sp.=unidentified *Clusia* species. For detailed explanation see Sect. 9.4.1 (data of Popp et al. 1987; Franco et al. 1994; Haag-Kerwer et al. 1996)

Fig. 9.15. Adventitious roots of *Clusia* growing down cracks in the karstic limestone mountains of Serra de San Luis at the site shown in Fig. 9.13. A small opening of the cave is seen on the top. On the bottom the roots reach rotting litter and humus

somewhat astonishing to find three CAM species of *Clusia* so close by each other with similar morphotype, similar life form as free standing trees with adventitious roots establishing contact with humus of rotting leaves and soil in the cracks and hollows of the karstic limestone (Fig. 9.15) and very similar performance of the CAM mode of photosynthesis. Other traits, which have not been studied so far may be involved.

9.4.2.7 Semi-evergreen Moist Tropical Forest, Barro Colorado Island, Panama

For the hemi-epiphyte *Clusia uvitana* in the semi-evergreen moist tropical lowland forest on Barro Colorado Island, Panama (9° 10' N, 79° 51' W) we have the most unique data set, because so far this appears to be the only case where a large number of different parameters were measured continuously during a whole year. A wealth of information was obtained for the performance of this single species over the seasons by Zotz and Winter (1994a, b). *C. uvitana* seems to be an obligate hemi-epiphyte, it was never seen to occur as a free standing tree.

From the work of these authors Fig. 9.16 extracts some essential features of the behaviour of this C_3/CAM-intermediate species. CO_2-gas exchange curves are shown together with night/day oscillations of titratable acidity and two important environmental parameters, i.e. daily irradiance and weekly rainfall. Seasonality is given for the hemi-epiphyte *C. uvitana* not only by the general climatic conditions with a pronounced dry season from late December to April but also by the responses of the host tree (*Ceiba pentandra* (L.) Gaertn., Bombacaceae). The host tree was shedding its leaves in the dry season, so that there was no shading of the hemi-epiphyte from October to mid-January as it is clearly seen in the irradiance data in Fig. 9.16. Irradiance proved to be one of the major constraints for the performance of *C. uvitana* on the host tree. In the wet season from late April/May to September *C. uvitana* showed a clear C_3 character as seen by the typical C_3-type gas exchange, often with a pronounced midday depression but no nocturnal CO_2-uptake, and the low diurnal acid oscillations (from end of April to August). In the other months *C. uvitana* showed clear diurnal acid oscillations and nocturnal CO_2-fixation although daytime CO_2-uptake still made important contributions to total carbon gain over 24 h (phases II and IV of CAM). The season of dominating C_3-photosynthesis was correlated with the season of production of new leaves which for the first three months were restricted to performance of C_3-photosynthesis. Thus, in addition to the responses to seasonal changes there is a developmental component in the C_3/CAM-intermediate behaviour of *C. uvitana*.

The work of Zotz and Winter (1994a, b) also allows a comparison of the performance of the C_3/CAM-intermediate *C. uvitana* with other species of a similar hemi-epiphytic life form, namely the C_3-orchid *Catasetum viridiflavum* Hook. and the C_3-fern *Polypodium crassifolium* L. (Table 9.7). The possible advantage of the CAM option of *C. uvitana* in structuring hydraulic architecture was already noted in Sect. 2.4. We can now see that the performance of *C. uvitana* was always superior in the dry season both with respect to 24 h CO_2 uptake and maximum rates of CO_2-uptake, while in the wet season performance of *C. uvitana* was close to the fern and the orchid showed superior rates of carbon acquisition. Water use efficiency was always much higher

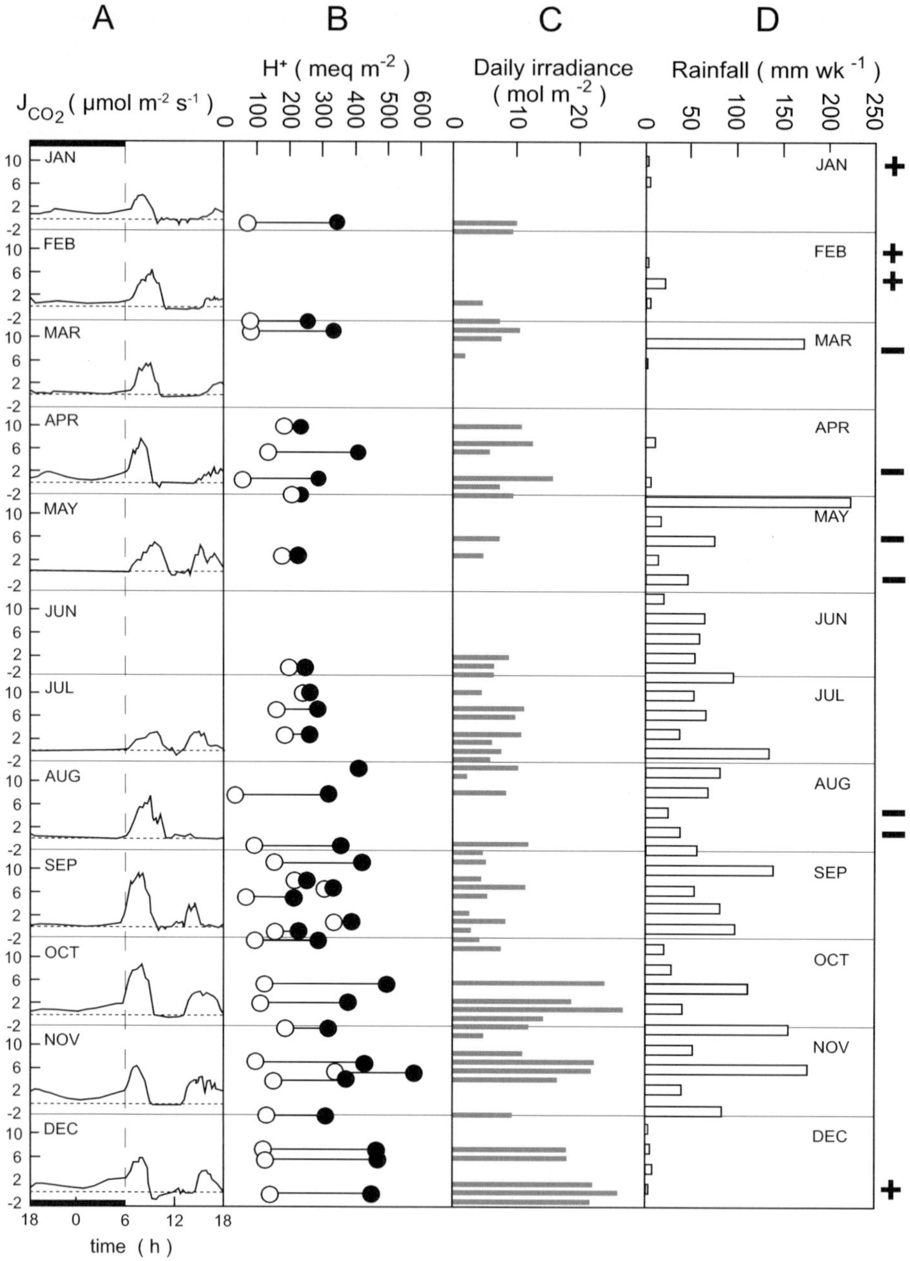

Fig. 9.16A–D. Seasonal changes of CO_2-gas exchange pattern (J_{CO2}) of mature sun leaves of the hemi-epiphyte *Clusia uvitana* as documented by: **A** representative diurnal curves for the 12 months of the year 1991 in Panamá, where *dark bars* on the abscissa indicate dark periods; **B** night/day oscillations of titratable acidity (*closed circles* dawn, *open circles* dusk); **C** daily irradiance; **D** weekly rainfall. At the *right hand margin* – indicates events of production of new leaves and + senescence and abscission of leaves of the host tree (after Zotz and Winter 1994a, b)

Table 9.7. Comparison of the ecophysiological performance of the C_3/CAM-intermediate *Clusia uvitana* with two C_3-hemi-epiphytes, the orchid *Catasetum viridiflavum* and the fern *Polypodium crasssifolium* in the dry and wet season in Panamá. (Data from Zotz and Winter 1994a)

	C. viridiflavum		*P. crassifolium*		*C. uvitana*	
	Dry S.	Wet S.	Dry S.	Wet S.	Dry S.	Wet S.
Net exchange of CO_2 for 24 h (mmol m^{-2} day^{-1})	29	108	35	83	50	80
Maximum rate of CO_2 uptake (µmol m^{-2} s^{-1})	3.1	5.5	1.9	3.9	3.6	4.5
Water use efficiency (WUE) (mmol CO_2/mol H_2O)	4	5	6	5	13	10
Nocturnal carbon gain (% of total)					50	18
Night/day acid oscillation, ΔH^+ (mmol m^{-2})					280	95

in *C. uvitana* and this was related to CAM-performance with nocturnal carbon gain and night/day oscillations of acidity, which was more pronounced in the dry season than in the wet season. In comparison with its host tree the annual primary productivity of *C. uvitana* was only 30 % lower, productivity was 1.78 kg CO_2 m^{-2} $year^{-1}$ for *C. uvitana* and 2.64 kg m^{-2} $year^{-1}$ for C. *pentandra.*

9.4.2.8 Secondary Tropical Forest on a Caribbean island, St. John, US-Virgin Island, Lesser Antilles

The Island of St. John was largely reinvaded by secondary tropical forest after colonial sugar cane plantations were abandoned at the turn of the nineteenth to the twentieth century. The northern coast and the central mountains, with a peak elevation of 387 m a.s.l., are moist due to trade winds, while moisture declines on the southern slope down to the south coast. The mountains are covered by montane rain forest while the dry south-coast has semi-deciduous thornbush vegetation. The distribution of the only species of *Clusia* on the island, *C. rosea*, follows the moisture gradient from north to south. It is abundant on the northern shore up to the mountain ridge and down the southern slope to an elevation of 165 m, but it is much less common at the lower eleva-

Table 9.8. Carbon isotope ratios, $\delta^{13}C$, and net CO_2-uptake in the CAM phases I, II and IV by *Clusia rosea* on the central mountain ridge and the southern slope of the island of St. John, Lesser Antilles. (Rounded values of measurements by Ball et al. 1991b; Lüttge et al. 1993)

Altitude (m a.s.l.)	Annual rainfall (mm)	$\delta^{13}C$ (‰)	Net CO_2 uptake in the CAM-phases (mmol m^{-2})		
			I	II	IV
340–385	1270–1400	–20.0	50	35	5
230		–15.5			
180		–14.9			
165		–18.2			
0	1015–1140		35	15	0

tions near the south shore. Thus, *C. rosea* on the island appears to be favoured by moisture. Some field measurements (Ball et al. 1991 a, b; Lüttge et al. 1993) which are available, however, do not bear out a clearly superior performance at the moister sites (Table 9.8). $\delta^{13}C$-values show that over time *C. rosea* makes intensive use of primary CO_2 acquisition by PEPC and relatively somewhat more so on the mountain ridge than further down the south slopes. Total CO_2 uptake also seemed to be somewhat larger on the mountains than at the south shore directly by the sea. However, sample size is rather small and these differences are not statistically significant.

9.4.2.9 Tropical Forest on the Island of Trinidad: *Clusia minor* L. and Three Endemic Species

For *C. minor* in Trinidad we have data sets nearly as complete over the seasons as for *C. uvitana* in Panama (see Sect. 9.4.2.7). Studies were performed for many weeks during the transition from wet to dry season at a site in the Arima Valley which runs north to south subtending the Northern Mountain Range (10° 41' N, 61° 17' W, 250 m a.s.l.). The climate is characterized by an annual dry season from February to mid March and an annual rainfall of 2500 mm. The transition from the wet to the dry season in the semi-deciduous forest is associated by a 60 % increase in light intensity with increased daily photon flux density from 24 to 40 mmol m^{-2} (Roberts et al. 1998). Exposed and shaded plants were compared during the seasons; the exposed plants were growing on a rocky limestone outcrop in the deciduous seasonal forest. Fingerprint diagrams compiled for the exposed and shaded plants in the wet and the dry season from the data of Borland et al. (1992, 1993, 1994, 1996) and Roberts et al. (1998) are presented in Fig. 9.17.

The rather negative $\delta^{13}C$-values show that in the field site the C_3/CAM-intermediate species *C. minor* over time predominantly made use of the C_3-option, perhaps with a little more primary CO_2-fixation by PEPC in the exposed than in the shaded plants. The night/day changes of organic acid levels show that in the short dry season the performance of CAM by both the exposed and the shaded plants was much stronger than in the wet season. CAM was more strongly expressed in the exposed than in the shaded plants. The large standard deviations in the organic acid data of the shaded plants in the wet season are due to the more sporadic occurrence of organic acid oscillations. In this case the field observations agree with the basic expectation that performance of CAM is a useful behaviour under stress of drought (seasons) and irradiance (exposure). In conformity with the requirements of CAM succulence is higher in the exposed than in the shaded leaves. The ratios

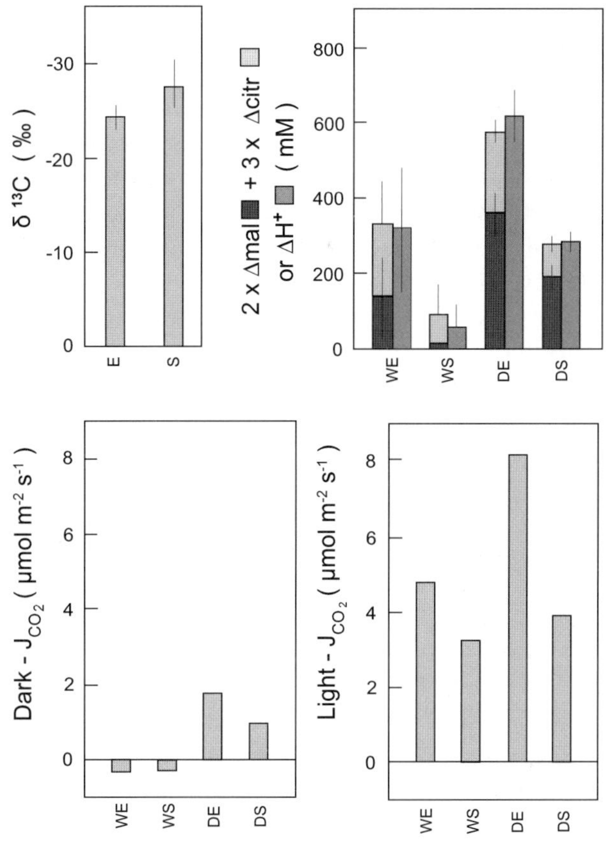

Fig. 9.17. Ecophysiological fingerprint: *Clusia minor* in Trinidad. E=exposed, S=shaded, W=wet season, D=dry season. For detailed explanation see Sect. 9.4.1 and for references Sect. 9.4.2.9

of succulence of exposed/shaded leaves were ca. 1.2 and ca. 1.1 in the wet and dry season, respectively, while the ratios of nocturnal organic acid accumulation were ca. 3. Rates of net CO_2-exchange presented in the fingerprints are maximum rates found in the four publications; they are not necessarily directly comparable with the organic acid data, which are averages of data given in these papers. The maximum rates found in the dark are clear indicators of CAM in the dry period. Maximum rates in the light period are higher in the exposed than in the shaded plants, showing that higher irradiance can be utilized by the exposed plants for higher CO_2 uptake, the high value given for the exposed plants in the dry season is due to a highly expressed phase II of CAM in this particular case.

Regarding the organic acids it is noteworthy that increased performance of CAM in the dry season is exclusively due to increased malate oscillations. Thus, the behaviour of *C. minor* in Trinidad does not support a hypothesis which postulates a special role of citrate oscillations under drought stress (see Sect. 8.8.1).

In addition to *C. minor* there are three endemic species of *Clusia* in Trinidad, namely *C. tocuchensis* Britt. and *C. aripoensis* Britt. growing hemi-epiphytically in transitional lower montane/upper montane rain forest and *C. intertexta* Britt., which is exclusively found on the summit of the highest mountain in Trinidad, Cerro del Aripo, 941 m a.s.l. Their carbon isotope ratios and night/day acid oscillations at the end of the dry season are given together with values for *C. minor* in Table 9.9 (Borland et al. 1992), although the comparison is not strictly adequate because the annual precipitation at the sites of the three endemic species higher up in the mountains is higher and may be well over 4000 mm on the summit. The low C_3-like values of $\delta^{13}C$ indicate that over time in all species CAM does not significantly contribute to carbon gain. However, like *C. minor* the three endemic species also show the capacity for CAM induction although they are restricted to moist habitats in Trinidad. This underlines the plasticity and the flexibility in possible responses to fluctuations of habitat conditions which is so frequent in the genus *Clusia*.

Table 9.9. $\delta^{13}C$-values and night/day oscillations in titratable acidity (ΔH^+) of leaves of intermediate age of three endemic species of *Clusia* in comparison to *C. minor* in Trinidad (Borland et al. 1992)

Species	$\delta^{13}C$ (‰)	ΔH^+ (mM)
C. minor (exposed)	−23.2	147
C. minor (shaded)	−25.3	180
C. aripoensis	−23.0	60
C. tocuchensis	−24.6	13
C. intertexta	−28.1	103

9.4.2.10 Atlantic Rain Forest, Brazil

In a reserve of natural Atlantic Rain Forest of Brazil (Estação Biologica de Santa Luis, Tiradentes, Espirito Santo State) we obtained ecophysiolgical data yet unpublished and still to be processed for four species of *Clusia*. *C. intermedia* Mariz and *C. spiritu sanctensis* Mariz et Weinberg have a wide ecological amplitude in the rain forest complex, the former occurring in moist riverine forest, hill forest and on exposed rock outcrops, the latter occurring both in the riverine forest and on the rock outcrops. Both are CAM species. A third CAM species *C. aemygdioi* Silva et Weinberg was studied only in the riverine and hill forests. The fourth species *C. marizii* Silva et Weinberg was a C_3 species of the hill forest.

9.4.2.11 Inselbergs

Inselbergs are rock-outcrops providing very conspicuous isolated habitats amidst different vegetation such as savannas, cerrados or tropical forests (Lüttge 1997a; Porembski and Barthlott 2000). Species of *Clusia* often form

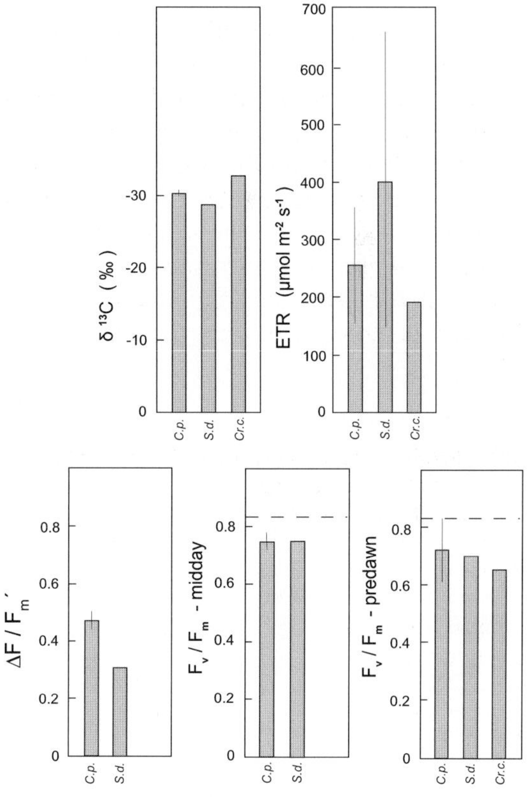

Fig. 9.18. Ecophysiological fingerprint: inselberg, Rio de Janeiro. *C.p.=Clusia parviflora, S.d.=Styllingia dichotoma, Cr. c.=Croton compressus.* For detailed explanation see Sect. 9.4.1 (data of de Mattos et al. 1997; Duarte et al. 2005; Scarano et al. 2005)

Fig. 9.19.A The inselberg Pão de Açúcar, Rio de Janeiro, Brazil. **B** With bushes of *Clusia criuva*. **C** With bushes of *Croton compressus* (background left) and *Styllingia dichotoma* (background right and foreground)

dense scrubs on inselbergs, e.g. *C. minor* and *C. nemorosa* G. Mey. on an inselberg in a rainforest in French Guiana, *Oedematopus obovatus* Spruce ex Planch. et Triana (syn. *Clusia obovata*, Spruce ex Planch. et Triana) Pipoly on an inselberg at the upper Orinoco, *C. criuva* and *C. parviflora* on inselbergs in the cerrado region of Brazil. The Pão de Açúcar in the municipality of Rio de Janeiro (22° 57' S, 43° 59' W) is a typical inselberg with the C_3-species *C. parviflora*. In Fig. 9.18 its performance is shown in comparison with two Euphorbiaceae, *Styllingia dichotoma* Muell. Arg. and *Croton compressus*, which form very similar shrubs on the inselberg (Fig. 9.19). The electron transport rates obtained for *S. dichotoma* were very variable. Overall the performance of *C. parviflora* was quite similar to the other two shrub species.

9.5 Plasticity and Diversity of *Clusia*s

The genotype comprises the complete information of the plant. It represents a given constitution. In ontogeny it generates a certain phenotype based on a genetic developmental programme. Differences in phenotypes can be seen in the formation of both different morphotypes and different physiotypes. The complete set of phenotypical traits generated by a genotype in the morphological domain is the morphotype and in the physiological domain is the physiotype (Kinzel 1972, 1982). The morphotypes are structural life forms as delineated by comparative morphology and anatomy. The physiotypes are physiological life forms as delineated by comparative physiology, biochemistry and molecular biology.

Thus, morphotypes and physiotypes are important units in ecophysiology. In ecology and physiological ecology the phenotype is always the direct receiver of environmental input. Via the phenotypes there may then be feedback of environmental input on the genotype. By gene regulation in response to environmental cues such feedback modulates phenotype expression, because phenotypes are generated by the genotype under the pressure of environmental parameters (Fig. 9.20). This implies that various genotypes may be capable of generating phenotypic plasticity to a smaller or larger extent. Phenotypic plasticity must be considered in relation to a co-occurrence of different genotypes within a population which are adapted to a slightly different environment. Genetic variation is reflected in phenotypic plasticity (Booy et al. 2000).

In the genus *Clusia* with its one leaf morphotype (Sect. 2.1) mainly the photosynthetic physiotypes reveal plasticity. Having reviewed the various life forms (Sect. 2.3), the large ecological amplitude (Sect. 9.1) and the habitat specific performance (Sect. 9.4) of *Clusia* we realize the great richness of physiotypical appearance based in *Clusia* genomes. While it is the phenotype, where environmental input and signals first elicit ecophysiological responses, reac-

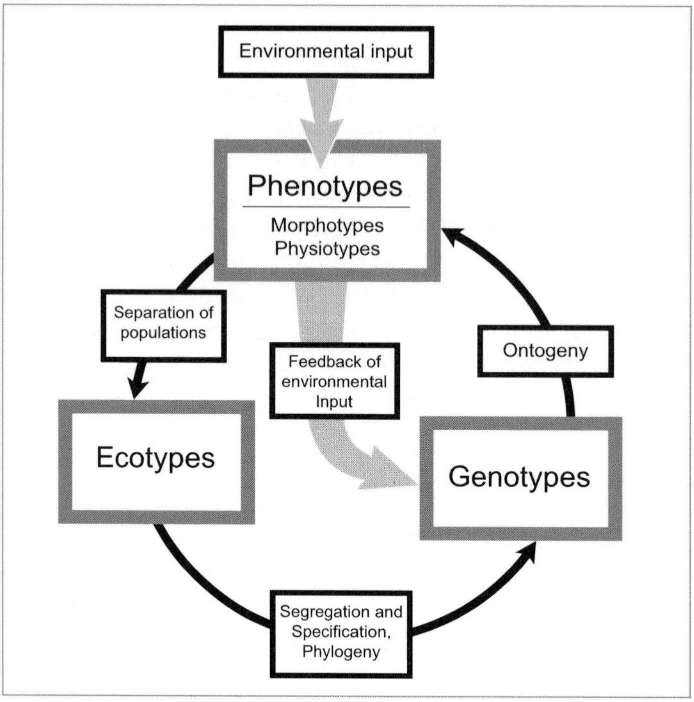

Fig. 9.20. Network-like relationships between genotypes, ecotypes and phenotypes (Lüttge 2005)

tions of phenotypes affect expressions of genotypically anchored traits. This is phenotype/genotype feedback in environmental responses. Ecophysiologically performing phenotypes may also separate into genetically stable populations and ecotypes (Turesson 1922; Kinzel 1982). Such separated phenotypes by segregation may evolve new genotypes. Evidently phenotypes, ecotypes and genotypes are integrated in a network with feed back loops (Fig. 9.20, Lüttge 2005). On this basis I have argued previously that the high plasticity and physiotypical diversity of the genus *Clusia* may be a basis for generation of a large species diversity (Lüttge 1995, 2000, 2005). This is controversial. The relations between structures and sizes of populations and speciation rates are very complex (Rieseberg et al. 2003). On the one hand, it is evident that large ecological amplitudes supported by plasticity can separate populations with reduced sets of genotypes and genetically stable new populations and ecotypes, and thus, plasticity can support speciation and the development of diversity and as Schuster (1998) writes:

> *... the development of phenotypes from genotypes is the real origin of complexity in biology ...*

(see also West-Eberhard 1986, 1989, 2003; Solbrig 1994; Gehrig et al. 2001). A strong potential of radiation in the genus *Clusia* is born out for example by the establishment of species of *Clusia* originating from the Brazilian Atlantic rain forest in plant communities marginal to the rainforest (Chap. 3, Lüttge 2005). *C. parviflora* in the state of Rio de Janeiro in Brazil is occurring at two rather different types of sites, i.e. inselbergs and restingas (see Table 9.1). Molecular population analysis shows genetic differences between the inselberg and the restinga populations (Vaasen 2005, Chap. 7). Separation is possibly due to the very different conditions of the two sites, but it is noteworthy that human activities of rural development along the coastline also contribute to habitat fragmentation preventing genetic exchange between the populations (Vaasen 2005). It is also evident that the species rich genus *Clusia* in fact does have a high speciation rate and there is an extraordinary inherent evolutionary plasticity in the genus (Chap. 6). On the other hand, high plasticity also can protect genotypes from selection under variable fluctuations of environmental cues, and thus, plasticity can hinder development of diversity. Both possibilities are supported by the scheme of Fig. 9.20.

A basic question which first needs to be asked in this context is if plasticity is adaptive or not. Plasticity of a trait in itself does not necessarily imply that it is adaptive because plasticity may be associated with high costs. A key question to be asked regards fitness, i.e. the relations between gene or protein expression, plasticity in such expression, phenotypic plasticity and fitness (van Kleunen and Fischer 2005). Thus, with respect to the plasticity of *Clusi*a to express different photosynthetic physiotypes we may evaluate the different costs, e.g. of C_3-photosynthesis as compared to CAM. In discussion of the evolution of CAM from ancestors with C_3-photosynthesis it was argued repeatedly that metabolically this did not require any new selection of particular enzyme reactions and metabolic pathways. CAM just requires a different management of existing metabolic house keeping reactions (Lüttge 2004, 2005). This may also explain the polyphyletic emergence of CAM in many different branches of the vascular plants (Lüttge 2005) including polyphyletic evolution of CAM within the genus *Clusia* (Vaasen et al. 2002, 2006; Gehrig et al. 2003; Vaasen 2005; Chap. 6) as well as the ready switches between C_3-photosynthesis and CAM in *Clusia*s (Sect. 8.1). There are, however, CAM specific isoenzymes of some CAM-functions which are different from the respective house-keeping enzymes (Lüttge 2005). Particular research efforts in this respect were directed towards the key enzyme of CAM, the carboxylating enzyme PEPC (Vaasen 2005; Vaasen et al. 2006). The obligate CAM-species *C. hilariana* produces three isoforms of PEPC, one of which is root-specific and probably is a house-keeping isoform in this organ, while the C_3-species *C. multiflora* and the C_3/CAM-intermediate species *C. minor* only have one isoform. Transcript levels (mRNA) of PEPC in *C. minor* are increased in the CAM-state (Taybi et al. 2004; Vaasen 2005; Vaasen et al. 2006). There is an important similarity to the lack of more PEPC isoforms in the relations of PEPC-kinase (PEPCK) which activates/phosphorylates PEPC at night and is thus a key regulator of CAM expression (see Sect. 8.3.2). Taybi et al. (2004) showed that there was only one isoform of PEPCK in CAM- and C_3-*Clusia*s.

For *C. minor* this is somewhat surprising but as far as we can see the one isoform of both PEPC and PEPCK can serve house-keeping functions in the C_3-state as well as the extra requirements given in the CAM-state, and this may facilitate the flexible shifts between C_3-photosynthesis and CAM that are typical of this species.

Figure 6.1 (Chap. 6) shows a phylogenetic tree of *Clusia* where both a polyphyletic gain and loss of CAM during evolution is seen. The obligate C_3 species *C. multiflora* is on a branch where CAM never developed. The C_3/CAM intermediate species *C. minor* is on a branch where CAM originated at the basis but rather close neighbours of *C. minor* on that major branch are on secondary branches where CAM was lost. This implies the possibility that the evolution as well as the loss of CAM may have had rather different molecular bases in different taxons. However, even the more general question of to be or not to be CAM in Fig. 6.1 is still speculative (see Chap. 6). In the case of both PEPC and PEPCK, the same isoforms appear sufficient to manage both CAM performance and house keeping functions. As to whether this phenomenon is based on the lack of an evolution of more isoforms or on the loss of redundant isoforms during evolution remains entirely unsolved.

In any case, overall it does not appear that the photosynthetic physiotype plasticity may involve particularly high costs in producing the required enzyme complements. On the other hand, evidently CAM is energetically more costly than C_3-photosynthesis (Table 8.4). The question is, however, if this is decisive because the major problem in ecophysiological performance often is control of over-energization of the photosynthetic apparatus and dissipation of excess energy rather than limited excitation (Sect. 8.6). Hence, it may even be an advantage of the higher energy demand of CAM that this could save costs of building up mechanisms of harmless thermal energy dissipation.

A different problem is that it has been questioned that photosynthetic capacity is related to ecophysiological fitness:

> *Direct correlations between photosynthetic rates and fitness are rarely observed in natural populations; .... it is impossible to say that variation in photosynthetic rate, per se, contributed directly to individual fitness* (Ackerly et al. 2000).

On the other hand:

> *ecophysiological traits ...... are likely to influence fitness and undergo adaptive evolution* (also Ackerly et al. 2000).

Basically this depends on how we define and especially how we quantify ecophysiological fitness, which is a very difficult problem and much more complex than measuring sexual reproduction and seed production (Lüttge and Scarano 2006). This is, for example, underlined by a study on salinity tolerance in *Sorghum*, where improvement of salinity tolerance during early vegetative development caused a strong decrease in reproductive fertility (Amzallag 2005). It leads to the phenophase concept, and Amzallag (2005) writes:

> *Adaptation becomes a relative notion, inherent to each phase of development. A change occurring during a critical period may be adaptive for the emerging phenophase, but distorting for the later developmental events. Consequently, measurements of adaptation through the number of viable seeds may be useful in ecology, but misleading in physiology.*

Thus, considering the network-like interactions of phenotypes, ecotypes and genotypes with phenotypes being the receivers of environmental input and with phenotype-genotype feedback (Fig. 9.20), we remain interested in photosynthetic capacity as the most outstanding function of primary producers in relation to ecophysiological fitness (Lüttge and Scarano 2006).

A modular concept of phenotypic plasticity in plants interestingly suggests the importance of plasticity in the expression of modules, i.e. plasticity at the sub individual level (de Kroon et al. 2005). This brings into discussion the question of whether the unit for evolutionary selection is the individual organism (Ch. Darwin) or the gene (R. Dawkins) or even the species (St. J. Gould). (See discussion in Gould 2002, who strongly criticises the position of R. Dawkins.) de Kroon et al. (2005) propose that

> *plasticity of whole plants is a by-product of modular responses, shaped by hierarchical selection.*

In relation to *Clusia* this is an interesting question because we noted the biochemical plasticity in CAM expression with the involvement of different solutes, i.e. organic acid and carbohydrate species (Sect. 8.3). Taybi et al. (2004) investigated the expression of phospho*enol*pyruvate carboxylase (PEPC) and PEPC-kinase genes in four species of *Clusia*, i.e. *C. rosea* a CAM species, *C. minor* a C_3/CAM-intermediate species, *C. aripoensis* a weak CAM-inducible species, and *C. multiflora* a C_3-species, in order to assess the genotypic capacity and phenotypic plasticity in the expression of CAM. We recall that PEPC is the key enzyme of primary carbon fixation in CAM (Sect. 8.1). It needs to be phosphorylated to be active. Taybi et al. (2004) found that transcriptional control of PEPC abundance is a key factor in determining genotypic capacity of CAM and the control of the PEPC-kinase gene (*Ppck*) is a consequence of metabolite transport across the tonoplast because cytosolic malate levels exert a feedback regulation on *Ppck* expression. There are no C_3- or CAM-specific isoforms of *Ppck* genes, and the abundance of PEPC-kinase transcripts are comparable in both modes of photosynthesis in the *Clusia*s. The day/night regulation of PEPC-kinase is a consequence of CAM in response to metabolic cycling of malate and also sugars rather than representing a major controlling factor. Hence, in this particular example there is no evidence for modular plasticity being involved in the diversity of photosynthetic physiotypes in *Clusia*. The modular concept may culminate in the caricature of

> *a tree is not a tightly integrated organism but a by-product of its parts* (Hankioja 1991, quoted by de Kroon et al. 2005).

Returning to the question raised above of whether plasticity supports or hinders evolution of diversity we may note that it is posed in a somewhat different way by de Jong (2005) who asks to distinguish between phenotypic plasticity (i) as a trait subject to selection, as any other trait, and (ii) a developmental mechanism as important as selection in evolution. de Jong (2005) himself supports (i) but, in contrast to West-Eberhard (2003), he strongly opposes (ii). Is it not possible to consider a combination of both? Phenotypic plasticity is a trait subject to selection. At the same time it offers material for selection. This does not necessarily mean that it itself is an evolutionary mechanism. This is partially admitted by de Jong (2005):

Niche width is potentially higher for the individual phenotypically plastic genotypes. Selection therefore favours plasticity, if it is physiologically possible. However, plasticity is itself not an evolutionary mechanism and does not promote different evolutionary solutions. Phenotypic plasticity does not contribute a major alternative view of evolutionary biology, but takes a legitimate place in the neo-Darwinian modern synthesis.

A good illustration of niche width of phenotypically plastic genotypes is the comparison of the C_3-species *C. multiflora* and the very flexible C_3/CAM-intermediate species *C. minor* discussed in Sect. 9.4.2.2. Determination of ecological niche widths relates aut-ecological plasticity to ecological amplitude. Thus, ecophysiological evaluations are increasingly important for the assessment of specific adaptive differences in relation to speciation (Gottlieb 2003), where ecological speciation is based on ecological attributes including habitat and resource utilization (Levin 2003). Nevertheless, de Jong (2005), evaluating various mathematical models, definitely denies that such plasticity may be involved in speciation:

> *Phenotypic plasticity as a major mechanism for evolution – such as invading new niches, speciation or macroevolution – has, at present, neither empirical nor model-support.*

Perhaps it is more than a semantic difference calling it a "major mechanism" or considering it an element in a feedback network such as that presented in Fig. 9.20. We may then conclude this final speculative section with the hope and outlook that the wonderfully flexible, plastic and diverse genus of *Clusia* will present further insights and help us to sharpen questions relating to the mechanisms underpinning evolution and diversity.

References

Ackerly DD, Dudley SA, Sultan SE, Schmitt J, Coleman JS, Linder CR, Sandquist DR, Geber MA, Evans AS, Dawson TE, Lechowicz MJ (2000) The evolution of plant ecophysiological traits: recent advances and future directions. BioScience 50:979–995

Amzallag GN (2005) Perturbed reproductive development in salt-treated *Sorghum bicolour*: a consequence of modifications in regulation networks? J Exp Bot 56:2821–2829

Araujo DSD (1997) Cabo Frio region, south-eastern Brazil. In: Davis SD, Heywood VH, Herrera-Macbryde O, Villa-Lobos J, Hamilton AC (eds) Centres of plant diversity: a guide and strategy for their conservation. 3. The Americas. WWF/IUCN, Washington, pp 373–375

Araujo DSD, Scarano FR, Sá CFC, Kurtz BC, Zaluar HLT, Montezuma RCM, Oliveira RC (1998) Comunidades vegetais do Parque Nacional da Restinga de Jurubatiba. In: Esteves FA (ed) Ecologia das Lagoas Costeiras do Parque Nacional da Restinga de Jurubatiba e do Município de Macaé (RJ). Nupem-UFRJ, Rio de Janeiro, pp 39–62

Arndt SK, Wanek W, Hoch G, Richter A, Popp M (2002) Flexibility of nitrogen metabolism in the tropical C_3-crassulacean acid metabolism tree species, *Clusia minor*. Funct Plant Biol 29:741–747

Ball E, Hann J, Kluge M, Lee HSJ, Lüttge U, Orthen B, Popp M, Schmitt A, Ting IP (1991a) Ecophysiological comportment of the tropical CAM-tree *Clusia* in the field. I. Growth of *Clusia rosea* Jacq. on St. John, US Virgin Islands, Lesser Antilles. New Phytol 117:473–481

Ball E, Hann J, Kluge M, Lee HSJ, Lüttge U, Orthen B, Popp M, Schmitt A, Ting IP (1991b) Ecophysiological comportment of the tropical CAM-tree *Clusia* in the field. II. Modes of photosynthesis in trees and seedlings. New Phytol 117:483–491

Berg A, Orthen B, Mattos EA de, Duarte HM, Lüttge U (2004) Expression of crassulacean acid metabolism in *Clusia hilariana* Schlechtendal in different stages of development in the field Trees 18:553–558

Björkman O, Demmig B (1987) Photon yield of O_2 evolution and chlorophyll fluorescence characteristics at 77 K among vascular plants of diverse origins. Planta 170:489–504

Booy G, Hendriks RJJ, Smulders MJM, van Groenendael JM, Vosman B (2000) Genetic diversity and the survival of populations. Plant Biol 2:379–395

Borland AM, Griffiths H, Maxwell C, Broadmeadow MSJ, Griffiths MN, Barnes JD (1992) On the ecophysiology of the Clusiaceae in Trinidad: expression of CAM in *Clusia minor* L. during the transition from wet to dry season and characterization of three endemic species. New Phytol 122:349–357

Borland AM, Griffiths H, Broadmeadow MSJ, Fordham MC, Maxwell C (1993) Short-term changes in carbon-isotope discrimination in the C_3-CAM intermediate *Clusia minor* L. growing in Trinidad. Oecologia 95:444–453

Borland AM, Griffiths H, Broadmeadow MSJ, Fordham MC, Maxwell C (1994) Carbon-isotope composition of biochemical fractions and the regulation of carbon balance in leaves of the C_3-crassulacean acid metabolism intermediate *Clusia minor* L. growing in Trinidad. Plant Physiol 106:493–501

Borland AM, Griffiths H, Maxwell C, Fordham MC, Broadmeadow MSJ (1996) CAM induction in *Clusia minor* L. during the transition from wet to dry season in Trinidad: the role of organic acid speciation and decarboxylation. Plant Cell Environ 19:655–664

Chacón H, Aguiar G, Olivares E (1998) Manganese tolerance in *Clusia multiflora* (H.B.K.) J Plant Physiol 153:167–173

Cirne P, Scarano FR (2001) Resprouting and growth dynamics after fire of the clonal shrub *Andira legalis* (Leguminosae) in a sandy coastal plain in southeastern Brazil. J Ecol 89:351–357

Cirne P, Zaluar HLT, Scarano FR (2003) Plant diversity, interspecific associations and postfire resprouting on a sandy spit in a Brazilian coastal plain. Ecotropica 9:33–38

Diaz M, Haag-Kerwer A, Wingfield R, Ball E, Olivares E, Grams TEE, Ziegler H, Lüttge U (1996) Relationships between carbon and hydrogen isotope ratios and nitrogen levels in leaves of *Clusia* species and two other Clusiaceae genera at various sites and different altitudes in Venezuela. Trees 10:351–358

Dojani S, Lakatos M, Rascher U, Wanek W, Lüttge U, Büdel B (2006) Nitrogen input by cyanobacterial crusts of an inselberg into a tropical rainforest in French Guiana. Flora (in press)

Duarte HM, Geßler A, Scarano FR, Franco AC, de Mattos EA, Nahm M, Rennenberg H, Rodrigues PJFP, Zaluar HTL, Lüttge U (2005) Ecophysiology of six selected shrub species in different plant communities at the periphery of the Atlantic Forest of SE-Brazil. Flora 200:456–476

Evans JR (1988) Acclimation by the thylakoid membranes to growth irradiance and the partitioning of nitrogen between soluble and thylakoid proteins. Aus J Plant Physiol 15:93–106

Farquhar GD, Ehleringer JR, Hubick KT (1989) Carbon isotope discrimination and photosynthesis. Annu Rev Plant Phys Plant Mol Biol 40:503–537

FEEMA – Fundação Estadual do Meio Ambiente (1988) Importância da biota da região de Cabo Frio. Rio de Janeiro

Field CB (1988) On the role of photosynthetic responses in constraining the habitat distribution of rainforest plants. Aust J Plant Physiol 15:343–358

Franco AC, Ball E, Lüttge U (1991) The influence of nitrogen, light and water stress on CO_2 exchange and organic acid accumulation in the tropical C_3-CAM tree, *Clusia minor*. J Exp Bot 42:597–603

Franco AC, Haag-Kerwer A, Herzog B, Grams TEE, Ball E, de Mattos EA, Scarano FR, Barreto S, Garcia MA, Mantovani A, Lüttge U (1996) The effect of light levels on daily patterns of chlorophyll fluorescence and organic acid accumulation in the tropical CAM tree *Clusia hilariana*. Trees 10:359–365

Franco AC, Herzog B, Hübner C, de Mattos EA, Scarano FR, Ball E, Lüttge U (1999) Diurnal changes in chlorophyll *a* fluorescence, CO_2-exchange and organic acid decarboxylation in the tropical CAM tree *Clusia hilariana*. Tree Physiol 19:635–644

Franco AC, Olivares E, Ball E, Lüttge U, Haag-Kerwer A (1994) *In situ* studies of crassulacean acid metabolism in several sympatric species of tropical trees of the genus *Clusia*. New Phytol 126:203–211

Genty B, Briantais J-M, Baker NR (1989) The relationship between the quantum yield of photosynthetic electron transport and quenching of chlorophyll fluorescence. Biochim Biophys Acta 990:87–92

Gehrig H, Gaussmann O, Marx H, Schwarzott D, Kluge M (2001) Molecular phylogeny of the genus *Kalanchoë* (Crassulaceae) inferred from nucleotide sequences of the IST-1 and IST-2 regions. Plant Sci 160:827–835

Gehrig HH, Aranda J, Cushman MA, Virgo A, Cushman JC, Hammel BE, Winter K (2003) Cladogram of Panamanian *Clusia* based on nuclear DNA: implications for the origins of crassulacean acid metabolism. Plant Biol 5:59–70

Geßler A, Duarte HM, Franco AC, Lüttge U, de Mattos EA, Nahm M, Scarano FR, Zaluar HLT, Rennenberg H (2005a) Ecophysiology of selected tree species in different plant communities at the periphery of the Atlantic Forest of SE-Brazil II. Spatial and ontogenetic dynamics in *Andira legalis*, a deciduous legume tree. Trees 19:510–522

Geßler A, Duarte HM, Franco AC, Lüttge U, de Mattos EA, Nahm M, Rodrigues PJFP, Scarano FR, Rennenberg H (2005b) Ecophysiology of selected tree species in different plant communities at the periphery of the Atlantic Forest of SE-Brazil III. Three legume trees in a semi-deciduous dry forest. Trees 19:523–530

Gould SJ (2002) The structure of evolutionary theory. Harvard Univ Press, Cambridge, Mass

Gottlieb LD (2003) Rethinking classic examples of recent speciation in plants. New Phytol 161:71–82

Grams TEE, Haag-Kerwer A, Olivares E, Ball E, Arndt S, Popp M, Medina E, Lüttge U (1997) Comparative measurements of chlorophyll *a* fluorescence, acid accumulation and gas exchange in exposed and shaded plants of *Clusia minor* L. and *Clusia multiflora* H.B.K. in the field. Trees 11:240–247

Haag-Kerwer A, Grams TEE, Olivares E, Ball E, Arndt S, Popp M, Medina E, Lüttge U (1996) Comparative measurements of gas-exchange, acid accumulation and chlorophyll *a* fluorescence of different species of *Clusia* showing C_3 photosynthesis, or crassulacean acid metabolism, at the same field site in Venezuela. New Phytol 134:215–226

Hankioja E (1991) The influence of grazing on the evolution, morphology and physiology of plants as modular organisms. Phil Trans R Soc London, Ser B Biol Sci 333:241–247

Henriques RBP, Araujo DSD, Hay JD (1986) Descrição e classificação dos tipos de vegetação da restinga de Carapebus, Rio de Janeiro. Rev Brasil Bot 9:173–189

Herzog B, Hoffmann S, Hartung W, Lüttge U (1999a) Comparison of photosynthetic responses of the sympatric tropical C_3 species *Clusia multiflora* H.B.K. and the C_3-CAM intermediate species *Clusia minor* L. to irradiance and drought stress in a phytotron. Plant Biol 1:460–470

Herzog B, Grams TEE, Haag-Kerwer A, Ball E, Franco AC, Lüttge U (1999b) Expression of modes of photosynthesis (C_3, CAM) in *Clusia criuva* Camb. in a cerrado/gallery forest transect. Plant Biol 1:357–364

Herzog B, Hübner C, Ball E, Bastos R do N, Franco AC, Scarano FR, Lüttge U (1999 c) Comparative study of the C_3/CAM intermediate species *Clusia parviflora* Saldanha et Engl. and the obligate CAM species *Clusia hilariana* Schlecht. growing sympatrically exposed and shaded in the coastal restinga of Brazil. Plant Biol 1:453–459

Holtum JAM, Aranda J, Virgo A, Gehrig HH, Winter K (2004) $\delta^{13}C$ values and crassulacean acid metabolism in *Clusia* species from Panama. Trees 18:658–668

Jong G de (2005) Evolution of phenotypic plasticity: patterns of plasticity and the emergence of ecotypes. New Phytol 166:101–118

Kinzel H (1972) Biochemische Pflanzenökologie. Schriften des Vereins zur Verleihung naturwissenschaftlicher Kenntnisse in Wien 112:77–98

Kinzel H (1982) Pflanzenökologie und Mineralstoffwechsel. Eugen Ulmer, Stuttgart

Kleunen M van, Fischer M (2005) Constraints on the evolution of adaptive phenotypic plasticity in plants. New Phytol 166:49–60

Kooten van O, Snel JFH (1990) The use of chlorophyll fluorescence nomenclature in plant stress physiology. Photos Res 25:147–150

Kroon H de, Huber H, Stuefer JF, Groenendal JM van (2005) A modular concept of phenotypic plasticity in plants. New Phytol 166:73–82

Lacerda LD, Araujo DSD, Maciel NC (1993) Dry coastal ecosystems of the tropical Brazilian coast. In: Van der Maarel E (ed) Dry coastal ecosystems: Africa, America, Asia and Oceania. Elsevier, Amsterdam, pp 477–493

Levin DA (2003) The ecological transition in speciation. New Phytol 161:91–96

Liebig M, Scarano FR, de Mattos EA, Zaluar HLT, Lüttge U (2001) Ecophysiological and floristic implications of sex expression in the dioecious neotropical CAM tree *Clusia hilariana* Schltdl. Trees 15:278–288

Lüttge U (1995) Ecophysiological basis of the diversity of tropical plants: the example of the genus *Clusia*. In: Heinen HD, San José JJ, Caballero-Arias H (eds): Nature and human ecology in the neotropics. Scientia Guaianae 5:23–36

Lüttge U (1997a) Physiological ecology of tropical plants. Springer, Berlin Heidelberg New York

Lüttge U (1997b) Cyanobacterial Tintenstrich communities and their ecology. Naturw 84:526–534

Lüttge U (1999) One morphotype, three physiotypes: Sympatric species of *Clusia* with obligate C_3 photosynthsis, obligate CAM and C_3-CAM intermediate behaviour. Plant Biol 1:138–148

Lüttge U (2000) Photosynthese-Physiotypen unter gleichen Morphotypen, Species und bei Klonen: Kann ökophysiologische Plastizität zur Entstehung von Diversität beitragen? Ber Reinh Tüxen Ges 12:319–334

Lüttge U (2004) Ecophysiology of crassulacean acid metabolism (CAM). Ann Bot 93:629–652

Lüttge U (2005) Genotypes – phenotypes – ecotypes: relations to crassulacean acid metabolism. Nova Acta Leopoldina NF 92(342):177–193

Lüttge U, Scarano FR (2004) Ecophysiology. Revta Brasil Bot 27:1–10

Lüttge U, Scarano FR (2006) Synecological comparisons sustained by ecophysiological fingerprinting of intrinsic photosynthetic capacity of plants as assessed by measurements of light response curves. Brazilian J Bot (in press)

Lüttge U, Ziegler H, Ting IP (1993) Ecophysiological comportment of the tropical C_3/CAM-intermediate tree *Clusia rosea* in the field as assessed by analyses of stable carbon- and hydrogen-isotope ratios. J Plant Physiol 142:497–501

Martii CFP (1889) CII Guttiferae. In: Flora Brasiliensis, vol. XII Pars I. Lipsiae apud. Frid. Fleischer in Comm. Oldenbourg

Martin L, Suguio K (1989) Excursion route along the Brazilian coast between Santos (State of São Paulo) and Campos (northern state of Rio de Janeiro). In: International symposium on global changes in South America during the Quaternary: past, present and future. Special publication no.2, São Paulo

Mattos EA de, Grams TEE, Ball E, Franco AC, Haag-Kerwer A, Herzog B, Scarano FR, Lüttge U (1997) Diurnal patterns of chlorophyll *a* fluorescence and stomatal conductance in species of two types of coastal tree vegetation in southeastern Brazil. Trees 11:363–369

Mattos EA de, Lüttge U (2001) Chlorophyll fluorescence and organic acid oscillations during transition from CAM to C_3-photosynthesis in *Clusia minor* L. (Clusiaceae). Ann Bot 88:457–463

Olivares E, Aguiar G (1999) Total and water-soluble calcium in six species of Clusiaceae. Flora 194:179–188

Pimentel M (2002) Variação espacial na estrutura de comunidades vegetais da formação aberta de *Clusia* no Parque Nacional da Restinga de Jurubatiba, RJ. DSc Thesis, Universidade Federal do Rio de Janeiro, Rio de Janeiro

Popp M, Kramer D, Lee H, Diaz M, Ziegler H, Lüttge U (1987) Crassulacean acid metabolism in tropical dicotyledonous trees of the genus *Clusia*. Trees 1:238–247

Porembski S, Barthlott W (eds) (2000) Inselbergs. Ecological studies, vol 146. Springer, Berlin Heidelberg New York

Reinert F, Roberts A, Wilson MJ, de Ribas L, Cardinot G, Griffiths H (1997) Gradation in nutrient composition and photosynthetic pathways across the restinga vegetation of Brazil. Bot Acta 110:135–142

Rieseberg LH, Church SA, Morjan CL (2003) Integration of populations and differentiation of species. New Phytol 161:59–69

Roberts A, Borland AM, Maxwell K, Griffiths H (1998) Ecophysiology of the C_3-CAM intermediate *Clusia minor* L. in Trinidad: seasonal and short-term photosynthetic characteristics of sun and shade leaves. J Exp Bot 49:1563–1573

Roberts A, Griffiths H, Borland AM, Reinert F (1996) Is crassulacean acid metabolism activity in sympatric species of hemi-epiphytic stranglers such as *Clusia* related to carbon cycling as a photoprotective process? Oecologia 106:28–38

Sá CFC (1992) A vegetação da restinga de Ipitangas, Reserva Ecológia Estadual de Jacarepía, Saquarema (RJ); fisionomia e listagem das angiospermas. Arq J Bot Rio de Janeiro 31:87–102

Scarano FR (2002) Structure, function and floristic relationships of plant communities in stressful habitats marginal to the Atlantic rain forest. Ann Bot 90:517–524

Scarano FR, Duarte HM, Ribeiro KT, Rodrigues PJFP, Barcellos EMB, Franco AC, Brulfert J, Deléens E, Lüttge U (2001) Four sites with contrasting environmental stress in southeastern Brazil: relations of species, life form diversity, and geographic distribution to ecophysiological parameters, Bot J Linn Soc 136:345–364

Scarano FR, Duarte HM, Franco AC, Geßler A, de Mattos EA, Nahm M, Rennenberg H, Zaluar HLT, Lüttge U (2005). Ecophysiology of selected tree species in different plant communities at the periphery of the Atlantic Forest of SE-Brazil I. Performance of three different species of *Clusia* in an array of plant communities. Trees 19:497–509

Schreiber U, Bilger W (1993) Progress in chlorophyll fluorescence research: major developments during the past years in retrospect. Progr Bot 54:151–172

Schuster P (1998) Evolution in molekularer Auflösung. Berichte und Abhandlungen. Berlin-Brandenburgische Akademie der Wissenschaften. Akademie Verlag 6:187–215

Solbrig OT (1994) Plant traits and adaptive strategies: their role in ecosystem function. In: Schulze E-D, Mooney HA (eds) Biodiversity and ecosystem function. Ecological Studies, vol 99. Springer, Berlin Heidelberg New York, pp 97–116

Sternberg L da SL, Ting IP, Price D, Hann J (1987) Photosynthesis in epiphytic and rooted *Clusia rosea* Jacq. Oecologia 72:457–460

Steyermark JA, Huber O (1978) Flora del Avila. Sociedad Venezolana de Ciencias Naturales y Ministerio del Ambiente y de los Recursos Naturales Renovables, Caracas

Taybi T, Nimmo HG, Borland AM (2004) Expression of phosph*enol*-pyruvate carboxylase and phospho*enol*-pyruvate carboxylase kinase genes. Implications for genotypic capacity and phenotypic plasticity in the expression of crassulacean acid metabolism. Plant Physiol 135:587–598

Turesson G (1922) The genotypic response of the plant species to the habitat. Hereditas 3:211–350

Ule E (1901) Die Vegetation von Cabo Frio an der Küste von Brasilien. Botanische Jahrbücher für Systematik, Pflanzengeschichte und Pflanzengeographie 28:511–528

Vaasen A (2005) Verwandtschaftsbeziehungen, Photosynthesemechanismen und Phospho*enol*pyruvat-Carboxylase bei der südamerikanischen Pflanzengattung *Clusia*. Ph.D. thesis, Tübingen

Vaasen A, Begerow D, Hampp R (2006) Phosphoenolpyruvate carboxylase genes in C_3, crassulacean acid metabolism (CAM) and C_3/CAM intermediate species of the genus *Clusia*: rapid reversible C_3/CAM switches are based on the C_3 housekeeping gene. Plant Cell Environ 29:2113–2123

Vaasen A, Begerow D, Lüttge U, Hampp R (2002) The genus *Clusia* L.: Molecular evidence for independent evolution of photosynthetic flexibility. Plant Biol 4:86–93

Wanek W, Arndt SK, Huber W, Popp M (2002a) Nitrogen nutrition during ontogeny of hemiepiphytic *Clusia* species. Funct Plant Biol 29:733–740

Wanek W, Huber W, Arndt SK, Popp M (2002b) Mode of photosynthesis during different life stages of hemiepiphytic *Clusia* species. Funct Plant Biol 29:725–732

West-Eberhard MJ (1986) Alterntive adaptations, speciation, and phylogeny (a review). Proc Nat Acad Sci USA 83:1388–1392

West-Eberhard MJ (1989) Phenotypic plasticity and the origins of diversity. Annu Rev Ecol Syst 20:249–278

West-Eberhard MJ (2003) Developmental plasticity and evolution. Oxford University Press, Oxford

Zaluar HLT (1997) Espécies focais e a formação de moitas na restinga aberta de *Clusia*, Carapebus, RJ. MSc Dissertation, Universidade Federal do Rio de Janeiro, Rio de Janeiro

Zaluar HLT, Scarano FR (2000) Facilitação em restingas de moitas: um século de buscas por espécies focais. In: Esteves FA, Lacerda LD (eds) Ecologia de Restingas e Lagoas Costeiras. NUPEM-UFRJ, Rio de Janeiro, pp 3–23

Zotz G, Winter K (1994a) Annual carbon balance and nitrogen-use efficiency in tropical C_3 and CAM epiphytes. New Phytol 126:481–492

Zotz G, Winter K (1994b) A one year study on carbon, water and nutrient relationships in a tropical C_3-CAM hemi-epiphyte, *Clusia uvitana* Pittier. New Phytol 127:45–60

10 Mycorrhiza of *Clusia* Species: Types, Abundance, Responses to Environmental Conditions

MEIKE KREUZER, ANJA VAASEN, FABIO RUBIO SCARANO, and RÜDIGER HAMPP

10.1 Introduction

Mycorrhiza is formed by more than 90 % of terrestrial plants between their roots and soil fungi (Molina et al. 1992). It is commonly accepted that mycorrhizal associations are beneficial for both partners in that the fungus depends on plant derived carbohydrates while the plant in return receives nutrients, refined products such as amino acids and also water (Hampp et al. 1995; Smith and Read 1997; Chalot and Brun 1998). As plants can manage without fungal partners for example in well-fertilised fields or under artificial culture conditions, mycorrhizas are not a strictly obligate symbiosis. They have, however, a crucial role for development and stability of plant communities and are regarded as "ecologically obligate" (Read 1993). Two major types of interaction exist, the arbuscular mycorrhiza (AM) and the ectomycorrhiza (ECM). These differ in morphological features and in the type of fungi. AM are typical for most herbaceous plants, including crop plants, and also for the majority of tropical tree species (Smith and Read 1997).

Koske and Gemma (1997) reviewed a number of papers conducted in both temperate and tropical sand dune vegetations and showed that at least since 1959 the possible importance of AM in succession has been recognized. They argued that during primary succession in dune vegetations, AM arrive early, often with the first plant colonizers. Despite the obvious potential relevance of this information for both conservation and restoration of sandy plain vegetation in Brazil, which includes restingas and dunes, there have been a very limited number of publications on mycorrhiza occurrence and function in such sandy areas (e.g., Cordoba et al. 2001; Stürmer and Bellei 1994; Tavares 1998; Trufem et al. 1989).

Ecological Studies, Vol. 194
U. Lüttge (Ed.)
Clusia: A Woody Neotropical Genus
of Remarkable Plasticity and Diversity

The genus *Clusia* plays an important role in the colonization of sandy areas, such as restingas, but also extends into the coastal rain forest. There exist, however, only a few studies about *Clusia*-mycorrhiza associations, which examined the effects of fungal inocula on the development of members of the genus *Clusia*, such as *C. minor* L. (Cáceres and Cuenca 1996), *C. multiflora* H.B.K. (Cuenca et al. 2001) and *C. pusilla* Maguire (Cuenca et al. 2003). They showed that *Clusia* species generally developed much better when mycorrhized. *C. multiflora* grows naturally on acid soils, which are high in soluble Al. Inoculation with AM fungi increased resistance to acidity and Al (Cuenca et al. 2001). The increased tolerance toward Al was in part due to its accumulation in fungal cell walls. *C. pusilla* is used for the recovery of deforested areas and, in nutrient poor forest soil, mycorrhization was shown to substitute fertilizer effects (Cuenca et al. 2003). In sandy shrubland soils, addition of fungal inoculum was less effective (probably due to little retention of inocula), and required a P source (Cuenca et al. 2003).

In this chapter, we provide the first report on *Clusia*-mycorrhiza association along coastal habitats in Rio de Janeiro, Brazil, particularly in the vegetation type known as restingas (see detailed descriptions of this vegetation type in Chaps. 3 and 4). In such habitats, *Clusia* species have been reported to have an important nurse plant effect, decisively affecting vegetation structure and function, as reviewed in Chap. 4. Mycorrhizal colonisation can promote greater coexistence of plant species, and therefore diversity, apparently as a consequence of reducing competition intensity and/or promoting facilitation (Read et al. 1985; Grime et al. 1987; Bethenfalvy et al. 1991; Van der Heijden et al. 1998, Wardle 2002). Therefore, it might help explain some of the interspecific associations found between *Clusia* and other plants in the restingas. The data presented here are the result of an initial effort to determine the ecological relevance of *Clusia*-mycorrhiza associations in the Brazilian restingas.

10.2 Types of *Clusia* mycorrhizae

Development of the AM symbiosis has been described clearly at the morphological level (Bonfante-Fasolo 1984). Following entry into the root, the process of development of an AM symbiosis varies slightly with the plant and fungal species involved, and two morphologies referred to as the *Paris*-type and the *Arum*-type are recognized (Smith and Smith 1997). In the *Paris*-type, fungal hyphae grow within the host cells, while in the *Arum*-type, the fungus grows through the intercellular spaces of the root to the inner cortical cell layers and then differentiates within the cortical cells to form dichotomously branched hyphae called arbuscules.

Preliminary data from a collection of roots from different *Clusia* species showed that all species investigated exhibited endomycorrhizal structures.

This is in accordance with observations reported by Cáceres and Cuenca (1996) and Cuenca et al. (2001). According to the organization of the fungal hyphae, mycorrhiza formed by *C. parviflora* Engl. belonged to the *Arum*-type, while *C. hilariana* Schlecht. exhibited both intra- and intercellularly growing hyphae (mixed type).

Similar observations were made with mycorrhizas of Brazil pine (*Araucaria angustifolia* Bert. O. Ktze.) (Breuninger et al. 2000). Analysis of the root systems from forest and grassland (campo) sites revealed mycorrhizal structures (appressoria, penetration and coiled hyphae, vesicles, arbuscules, spores), which are characteristic for the arbuscular (endo)mycorrhiza (AM) type. However, from these structures alone, the mycorrhiza-forming fungal species cannot be identified.

In the Brazil pine study we therefore tried to quantify and identify (according to specific morphological characteristics and staining properties) spores of AM fungi at both sites – forest and campo. The biodiversity at the forest site was much higher, with 13 species, whereas only 6 different species could be identified at the campo site. *Glomus* and *Acaulospora* were the only genera present at the campo. The forest, however, contained also spores of *Entrophospora* and *Scutellospora.* In addition to the greater biodiversity, the spore

Table 10.1. General geographic and climatic information related to the study sites in the state of Rio de Janeiro, Brazil. Climate is according to Köppen. There is very little information available on the Archipelago of Santana, but since it is only at a distance of 22.2 km away from the coast of Macaé, where the Restinga de Jurubatiba National Park is partially located, we believe that the climate should be similar

Sites	Environmental Protection Area of Arquipélago de Santana	Restinga de Jurubatiba National Park	Environmental Protection Area of Massambaba
Latitude	22°24'–22°25' S	22°00'–22°23' S	22°56'–23°05' S
Longitude	41°42'–41°45' W	41°15'–41°45' W	42°10'–42°13' W
Climate	–	Hot and humid (Aw)	Transition hot and humid / semi-arid (Aw/Bsh)
Mean annual rainfall	–	1200 mm	800 mm
Water deficit	–	Two months (July-August)	Five months (May-September)
References	–	Henriques et al. (1986), Liebig et al. (2001)	Barbiére (1984), Scarano et al. (2001)

number in soil, as well as the percent mycorrhizal colonization in roots, were significantly higher at the forest site than at the campo site. Because of the low frequency of hyphal coils and the dominating intercellular growth of hyphae, these mycorrhizae could be classified as an *Arum*-type, which was the first report of this kind in gymnosperms (Breuninger et al. 2000).

For a similar approach with *Clusia*, soil samples were collected in the restingas of Jurubatiba and of Massambaba (Environmental Protection Area), both Rio de Janeiro state, Brazil (general information on locations and climatic conditions is given in Table 10.1).

Due to the sandy soils with particle sizes close to those of spores, sequential sieving appeared not appropriate to enrich spores. Instead, iso-osmotic (sorbitol, sucrose) density gradient centrifugation had to be employed. Owing to morphological characteristics and staining properties, our preliminary studies revealed at least six different types of spores with highly varying abundance with regard to site and plant species (Fig. 10.1, Table 10.2). They all belonged to the genus *Glomus*. The limited spore diversity was most probably due to the very low number of vital spores recovered from the sandy soil samples (less than 100 per 50 g of soil).

Identification and counting of spores gives an idea about the potential for mycorrhiza formation (spore potential) within a soil sample. Identification of fungi present within host roots is, however, the only way to determine the symbiotic fungus, because not all fungi produce extraradical spores (compare Sanders et al. 1996). This can be done by sequence analysis of fungal DNA contained in root samples or single spores (Breuninger 2001; Kreutzer 2004). Meanwhile, a series of family/genus specific nucleotide sequences are available for this purpose. They allow for amplification, cloning and sequencing of specific fungal DNA sequences. Using such primers for DNA extracted from *Clusia* fine roots, we were able to amplify DNA fragments from fungal ribosomal genes (18S rDNA-ITS1-5.8S rDNA-ITS-25S rDNA). In

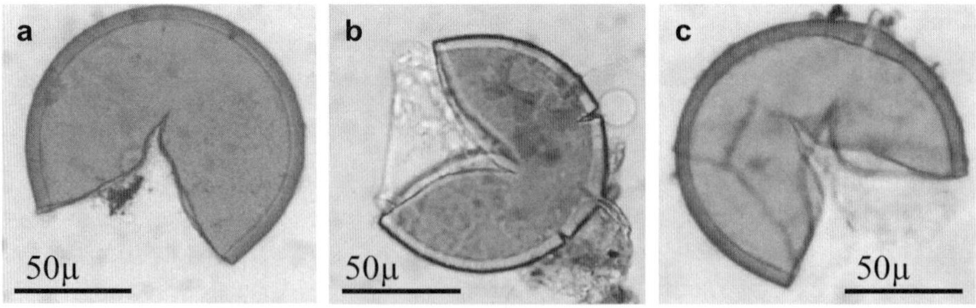

Fig. 10.1a–c. Spores collected from soil samples of the rhizospheres of: **a** *Clusia fluminensis*; **b** *Clusia hilariana*; **c** *Clusia parviflora*. Owing to size and wall characteristics all spores appear to belong to the genus *Glomus*

Table 10.2. Identification of mycorrhizal fungi from *Clusia* rhizosphere samples according to spore morphology (compare Breuninger et al. 2000)

Clusia species	Location	Fungal genus
Clusia hilariana	Restinga de Jurubatiba National Parc (JNR)	*Glomus* (possibly *Glomus macrocarpum* Tulasne et Tulasne)
Clusia hilariana	JNR	*Glomus*
Clusia hilariana	JNR	No spores recovered
Clusia hilariana	JNR	*Glomus* (possibly *Glomus macrocarpum*)
Clusia hilariana	JNR	*Glomus* (possibly *Glomus etunicatum* Becker et Gerdemann)
Clusia parviflora	JNR	*Glomus*
Clusia parviflora	JNR	*Glomus*
Clusia parviflora	JNR	*Glomus* (possibly *Glomus macrocarpum*)
Clusia fluminensis	Inselberg "Arquipelago de Santana" (Dry Forest)	*Glomus*
Clusia fluminensis	Environmental Protection Area of Massambaba	*Glomus*

order to classify better the respective fungal material, the family-specific primers GIGA5.8R and VAGIGA (Gigasporaceae), GLOM1310 and VAGLO (Glomaceae), and VAACAU (Acaulosporaceae) were combined with the primers NS5 (Redecker 2000; universal) and VANS1 (Simon et al. 1992). PCR fragments resulting from these combinations identified members of the Gigasporaceae, Acaulosporaceae and Glomaceae within roots of the three *Clusia* species (Table 10.3).

Although the data are very preliminary, they indicate large differences in fungal diversity within the rhizospheres of the respective *Clusia* plants. Roots of *C. hilariana* Schlecht. and *C. parviflora* were associated with a wider range of fungi: members of the genera *Glomus*, *Acaulospora*, *Entrophospora*, and *Gigaspora* co-existed within the same root. *C. fluminensis*, in contrast exhibited only one *Glomus* species. This could be due to the respective *Clusia* species, to other plant roots colonizing the soil around *Clusia*, or to nutrient availability within the soil (*C. hilariana* and *C. parviflora* grew in restingas in close vicinity to the coast, while *C. fluminensis* Pl. et Tr. inhabited a rather dry area). Our studies with Brazil pine showed that such parameters largely influ-

Table 10.3. Identification of fungal sequences from total *Clusia* fine root according to the NCBI data base (<http://www.ncbi.nlm.nih.gov/>). Primers used are from White et al. 1990; Simon et al. 1992, 1993; Redecker 2000

Clusia species/ location Primers	*Clusia hilariana*/Restinga de Jurubatiba National Parc	*Clusia parviflora*/Restinga de Jurubatiba National Parc	*Clusia fluminensis*/Inselberg "Archipelago de Santana", Macaé
VANS1-VAGLO	*Glomus intraradices* Schenck et Smith; *Glomus caledonium* (Nicol. et Gerd.) Trappe et Gerd	*Glomus intraradices; Glomus vesiculiferum* (Thaxter) Gerd. et Trappe	*Glomus intraradices*
VANS1-VAACAU	*Acaulospora laevis* Gerd. et Trappe; *Entrophospora* sp.	*Acaulospora* sp.	
VANS1-VAGIGA		*Gigaspora rosea* Nicol. et Schenck	
GLOM1310-GLOM5.8R	*Glomus clarum* Nicol. et Schenck; *Glomus manihotis* Howeler, Sieverding et Schenck; *Glomus intraradices*		

ence species richness and abundance of endomycorrhiza-forming fungi (Breuninger et al. 2000). *Glomus intraradices* was obviously the least sensitive fungus as it was the only one that could be identified at all sites.

10.3 Conclusions

The Brazilian Atlantic rain forest is surrounded by plant communities, which are subjected to more adverse environmental conditions (e.g., flooding, drought, oceanicity, and/or cold winter temperatures). The genus *Clusia* comprises such species, which exhibit an extraordinary physiological plasticity. In addition, *Clusia* has an important function as nurse plant/facilitator for increasing species diversity in such ecosystems. As the respective soils are generally nutrient-poor, *Clusia* depends on root symbiotic fungi for nutrient acquisition. These are of the endomycorrhiza-type, and a variety of Glomalean fungi are involved.

Acknowledgements. A.V. is indebted to the Landesgraduierten-Förderung Baden-Württemberg and the Reinhold-und-Maria-Teufel-Stiftung for financial support.

References

Barbiére EB (1984) Cabo Frio e Iguaba Grande, dois microclimas distintos a um curto intervalo espacial. In: Lacerda LD, Araujo DSD, Cerqueira R, Turcq B (eds) Restingas: origem, estrutura, processos, Niterói: CEUFF, pp 413–423

Bethenfalvy GJ, Reyes-Solis MG, Camel SB, Ferrera-Cerrato R (1991) Nutrient transfer between the root zones of soybean and maize plants connected by a common mycorrhizal mycelium. Physiol Plant 82:423–432

Bonfante-Fasolo P (1984) Anatomy and morphology of vesicular abuscular mycorrhizae. In: Powell CL, Bagyaraj DJ (eds) VA Mycorrhizae. CRC Press Boca Raton, FL, USA, pp 5–33

Breuninger M (2001) Mykorrhizierung von *Araucaria angustifolia*. PhD Thesis, Faculty of Biology, University of Tübingen, Germany

Breuninger M, Einig W, Magel E, Cardoso E, Hampp R (2000) Mycorrhiza of Brazil pine (*Araucaria angustifolia* (Bert. O. Ktze.)). Plant Biol 2:4–10

Cáceres A, Cuenca G (1996) Mycorrhizal dependence, phosphorus response and gas exchange of *Clusia* minor L., a tropical woody pioneer plant. In: Azcón-Aguilar C, Barea JM (eds) Mycorrhizas in integrated systems. From genes to plant development. Eur Commun Italy, pp 89–92

Chalot M, Brun A (1998) Physiology of organic nitrogen acquisition by ectomycorrhizal fungi and ectomycorrhizas. FEMS Microbiol Rev 22:21–44

Cordoba AS, Mendonça MM, Stürmer SL, Rygiewicz PT (2001) Diversity of arbuscular mycorrhizal fungi along a sand dune stabilization gradient: a case study at Praia de Joaquina, Ilha de Santa Catarina, South Brazil. Mycoscience 42:379–387

Cuenca G, Andrade Z, Meneses E (2001) The presence of aluminum in arbuscular mykorrhizas of *Clusia multiflora* exposed to increased acidity. Plant Soil 231: 233–241

Cuenca G, Andrade Z, Lovera M, Fajardo L, Meneses E (2003) Mycorrhizal response of *Clusia pusilla* growing in two different soils in the field. Trees 17:200–206

Grime JP, Mackey ML, Hillier SH, Read DJ (1987) Floristic diversity in a model system using experimental microcosms. Nature 328:420–422

Hampp R, Schaeffer C, Wallenda T, Stülten C, Johann P, Einig W (1995) Changes in carbon partitioning or allocation due to ectomycorrhiza formation: biochemical evidence. Can J Bot 73(Suppl. 1):S548–S556

Henriques RPB, Araujo DSD, Hay JD (1986) Descrição e classificação dos tipos de vegetação da restinga de Carapebus, Rio de Janeiro. Rev Bras Bot 9:173–189

Koske RE, Gemma JN (1997) Mycorrhizae and succession in plantings of beachgrass in sand dunes. Am J Bot 84:118–130

Kreutzer M (2004) Mykorrhizierung von Wurzeln der Gattung *Clusia*. State exam thesis, Faculty for Biology, University of Tuebingen, Germany

Liebig M, Scarano FR, Mattos EA de, Zaluar HLT, Lüttge U (2001) Ecophysiological and floristic implications of sex expression in the dioecious neotropical CAM tree *Clusia hilariana* Schltdl. Trees Struct Funct 15:278–288

Molina R, Massicotte H, Trappe JM (1992) Specificity phenomena in mycorrhizal symbioses: Community-ecological consequences and practical implications. In: Allen MF (ed) Mycorrhizal functioning, an integrative plant-fungal process. Chapman and Hall, New York, pp 357–423

Read DJ (1993) Mycorrhiza in plant communities. Adv Plant Pathol 9:1–31

Read DJ, Francis R, Finlay RD (1985) Mycorrhizal mycelia and nutrient cycling in plant communities. In: Fitter AH, Read DJ, Usher MB (eds) Ecological interactions in soil. Blackwell, Oxford, pp 193–217

Redecker D (2000) Specific PCR primers to identify arbuscular mycorrhizal fungi within colonized roots. Mycorrhiza 10:73–80

Sanders IR, Clapp JP, Wiemken A (1996) The genetic diversity of arbuscular mycorrhizal fungi in natural ecosystems – a key to understanding the ecology and functioning of the mycorrhizal symbiosis. New Phytol 133:123–134

Scarano FR, Duarte HM, Ribeiro KT, Rodrigues PJFP, Barcellos EMB, Franco AC, Brulfert J, Deléens E, Lüttge U (2001) Four sites with contrasting environmental stress in southeastern Brazil: relations of species, life form diversity, and geographical distribution to ecophysiological parameters. Bot J Linnean Soc 136:345–364

Simon L, Lalonde M, Bruns TD (1992) Specific amplification of 18S ribosomal genes from VAM - fungi colonizing roots. Appl Environ Microbiol 58:291–295

Simon L, Bousquet J, Levesque RC, Lalonde M (1993) Origin and diversification of endomycorrhizal fungi and coincidence with vascular plants. Nature 363: 67–69

Smith SE, Read DJ (1997) *Mycorrhizal Symbiosis*. 2nd edn. Academic Press, London, UK

Smith SE, Smith FA (1997) Structural diversity in (vesicular) – arbuscular mycorrhizal symbiosis. New Phytol 137:373–388

Stürmer SL, Bellei MM (1994) Composition and seasonal variation of spore populations of arbuscular mycorrhizal fungi in dune soils on the Island of Santa Catarina, Brasil. Can J Bot 72:359–363

Tavares SRL (1998) Uso de leguminosas arbóreas submetidas a inoculações com rizóbios e micorrizas arbusculares para revegetação de solo de restinga degradado. MSc Dissertation. Universidade Federal Rural do Rio de Janeiro. Seropédica, Rio de Janeiro, Brazil

Trufem SB, Otomo HS, Malatinszky SMM (1989) Fungos micorrízicos vesículo-arbusculares em rizosfera de plantas em dunas do Parque Estadual da Ilha do Cardoso, São Paulo, Brasil. (1) Taxonomia. Acta Bot Bras 3:141–152

Van der Heijden MGA, Klironomos JN, Ursic M, Moutoglis P, Streitwolf-Engel R, Boller T, Wiemken A, Sanders IR (1998) Mycorrhizal fungal diversity determines plant biodiversity, ecosystem variability and productivity. Nature 396:69–72

Wardle DA (2002) Communities and ecosystems: linking the aboveground and belowground components. Princeton University Press, Princeton

White TJ, Bruns T, Lee S, Taylor J (1990) Amplification and direct sequencing of fungal ribosomal RNA genes for phylogenetics. In: Innis MA, Gelfand DH, Sninsky JJ (eds) PCR Protocols: a guide to methods and applications. Academic Press, San Diego, pp 315–322

Section IV *Clusia*'s Clock

11 Ciracadian Rhythmicity

Heitor M. Duarte and Ulrich Lüttge

11.1 *Clusia*'s Clock: The Background of Endogenous Rhythmicity of C_3- and C_4-Photosynthesis and Crassulacean Acid Metabolism (CAM)

Does *Clusia* have an endogenous clock? Does *Clusia* need a clock? In the prokaryotic cyanobacterium *Synechococcus* almost all genes and in the vascular higher plant *Arabidopsis thaliana* (L.)Heynh. very many genes are oscillating in their transcription with an endogenous circadian period close to 24 h (from the Latin words *circa* = approximately, *dies* = day), i.e. they are clock controlled (Liu et al. 1995; Michael and McClung 2003). Circadian rhythmicity is a basic property of living organisms. Thus, we must assume that also *Clusia* has circadian rhythmicity.

In the plant kingdom most of the overt circadian oscillations are related to light and directly or indirectly involve photosynthesis (for review see Lüttge 2002a). Rhythms of C_3-photosynthesis in leaves have been analysed at the levels of gas exchange (CO_2 and water vapour), leaf conductance and stomatal opening, respiration, and metabolism of CO_2 assimilation and for example starch accumulation (Hennessey and Field 1991; Freeden et al. 1991; Li et al. 1992; Hennessey et al. 1993; Geiger et al. 1995; Lu et al. 2005; Lu and Sharkey 2006). Also in the C_4-plant *Sorghum bicolor* net photosynthesis, soluble sugar and starch accumulation showed circadian oscillations (Britz et al. 1987). At the molecular level studies are most advanced in the C_3 vascular plant *A. thaliana*. (For reviews of this rapidly and broadly developing field see, e.g., Millar 1999; Somers 1999; Staiger and Heintzen 1999; McClung 2000; Staiger 2002; Michael and McClung 2003.)

Rhythmicity of CAM has been mainly studied in obligate CAM species of the genus *Kalanchoë*. (For reviews see, e.g., Wilkins 1992; Borland et al. 1999; Lüttge 2000, 2002a, b; Nimmo 2000). The major overt output oscillations assessed generally were CO_2 and water vapour gas exchange and leaf conductance and night/day organic acid oscillations (reviews as above), but also car-

Ecological Studies, Vol. 194
U. Lüttge (Ed.)
Clusia: A Woody Neotropical Genus
of Remarkable Plasticity and Diversity

bon isotope discrimination, $\delta^{13}C$, and activities of the carboxylating enzymes phospho*enol*pyruvate carboxylase (PEPC) and ribulose-bis-phosphate carboxylase/oxygenase (RubisCO) (Grams et al. 1996) and quantum yield of photosystem II (Rascher et al. 2001; Rascher and Lüttge 2002). At the molecular level regulation of PEPC-activity via phosphorylation by PEPC-kinase (PEPCK) has been investigated, and it was seen that there was a feedback loop of regulation of PEPC by biochemical and biophysical processes involved in the metabolism and vacuolar compartmentation of malic acid during the CAM cycle (Carter et al. 1991; Hartwell et al. 1996, 1999, 2002; Borland et al. 1999; Borland and Taybi 2004; for details of the biochemistry of CAM see Sect. 8.3, Figs. 8.9, 8.12 and 8.14). However, at the molecular level insights are now more rapidly advanced using the C_3/CAM-intermediate annual species *Mesembryanthemum crystallinum* L. (Boxall et al. 2005). It has been recognized that its major upstream and central clock genes are the same as those of *A. thaliana*, namely *CCA1 (circadian clock associated), LHY (late elongated hypocotyl), TOC1 (timing of CAB expression;* "chlorophyll a, b binding protein"), *ELF 3* and *ELF 4 (early flowering 3* and *4), ZTL (zeitlupe), FKF1 (flavin-binding kelch repeat F-box 1).* Most interesting for a comparative evaluation of diurnal and circadian C_3- and CAM-oscillations are the genes *CCA1/LHY, TOC1* and *ELF 4.* TOC1 is mainly expressed in the late light and early dark period and its breakdown is regulated by CCA1 and LHY oscillating with a phase shifted by ca. 180° in relation to TOC1 (Kikis et al. 2005). *TOC1* is also thought to be involved in circadian oscillations of stomatal guard cell movements (Somers et al. 1998) which are a very important aspect of circadian oscillations of photosynthesis. Comparing C_3- and CAM-oscillations in *M. crystallinum* it is most noteworthy that some of the genes were oscillating only in either the C_3- or the CAM-mode, but *CCA1/LHY, TOC1* and *ELF 4* were oscillating in both modes of photosynthesis. The phases of *TOC1* oscillations in the C_3- and CAM-mode were strongly offset against each other and also the phases of *CAA1* and *LHY* but to a lesser extent.

While the investigations with the C_3/CAM-intermediate annual species *M. crystallinum* provide important comparative insights in C_3- and CAM-rhythmicity, it still appears essential to also study the C_3/CAM-intermediate species *Clusia minor* L. in this respect. Both species represent two highly different life forms. In contrast to the therophyte *M. crystallinum* where due to the short life time of the plants reversibility of C_3 to CAM switches is rather limited (Ratajczak et al. 1994) longevity of the leaves of the tropical trees of *Clusia* for two or more years (Olivares 1997) allows repeated shifts between the two modes of photosynthesis and *C. minor* is particularly flexible (Chaps. 8 and 9). However, considering endogenous circadian rhythmicity in *Clusia* at this stage we have to dwell much on the background given in this section, because so far the only work available on *Clusia*'s rhythmicity described in the following section (Sect. 11.2) is the dissertation of Duarte (2006).

11.2 *Clusia minor*'s Clock

11.2.1 Endogenous Oscillations of Gas Exchange and Effective Quantum Yield of Photosystem II in the C_3- and CAM-Modes of Photosynthesis

The C_3/CAM intermediate *C. minor* was adapted to perform C_3-photosynthesis and CAM, respectively, by growing plants at constant temperature day and night and at an irradiance of 120 μmol m^{-2} s^{-1} during the light period and keeping them well watered and withholding water for up to four days, respectively (see Sects. 8.5 and 8.8.1). Endogenous rhythmicity of net CO_2 exchange, J_{CO2}, and leaf conductivity for water vapour, g_{H2O}, was observed under continuous irradiance (LL) of 120 μmol m^{-2} s^{-1} at three different temperatures (Fig. 11.1). It was most pronounced at a medium temperature of 25 °C in the C_3-mode and at 25 °C but also at 30 °C in the CAM mode. This temperature dependence shows similarities with the behaviour of the obligate CAM plant *Kalanchoë daigremontiana* Hamet et Perrier where rhythmicity is only obtained within a temperature window above a lower and below an upper temperature threshold (Lüttge and Beck 1992; Grams et al. 1996). Due to homeoviscous adaptation of molecular tonoplast membrane structure governing malate compartmentation (Kluge et al. 1991; Kliemchen et al. 1993) the absolute temperatures marking the thresholds depend on growth temperature (Grams et al. 1995). As the *C. minor* plants were grown at temperatures between 20 °C (night) and 28 °C (day) rhythmicity in the CAM-mode still pertaining at 30 °C is consistent with the observations in *K. daigremontiana*.

It is noteworthy that both in the C_3- and in the CAM-mode rhythmicity is highly dampened and lost after only a few endogenous periods, which is in strong contrast to endogenous circadian rhythmicity in both obligate C_3- and obligate CAM-species (Sect. 11.1). In the C_3-mode loss of rhythmicity was more rapid than in the CAM-mode. A more detailed analysis is given for the temperature of 25 °C, where in both modes of photosynthesis rhythmicity was expressed particularly well, and where circadian oscillations of effective quantum yield of photosystem II (PS II), $\Delta F/F_m'$, measured every 20 min during the recording of gas exchange by pulse amplitude modulated chlorophyll fluorometry are also shown (Fig. 11.2). The dampening of oscillations is seen again. The oscillations of $\Delta F/F_m'$ in LL had a larger amplitude in the CAM-mode than in the C_3-mode and persisted longer. In both modes the oscillations of $\Delta F/F_m'$ dampened out more rapidly than those of J_{CO2} and g_{H2O} and the dampening out of endogenous rhythmicity was associated with an overall decline of $\Delta F/F_m'$. This was correlated with a general reduction of g_{H2O} in the arrhythmic stage attained after the loss of rhythmicity which must have been the reason for a similarly reduced J_{CO2} explaining the reduced $\Delta F/F_m'$ by sub-

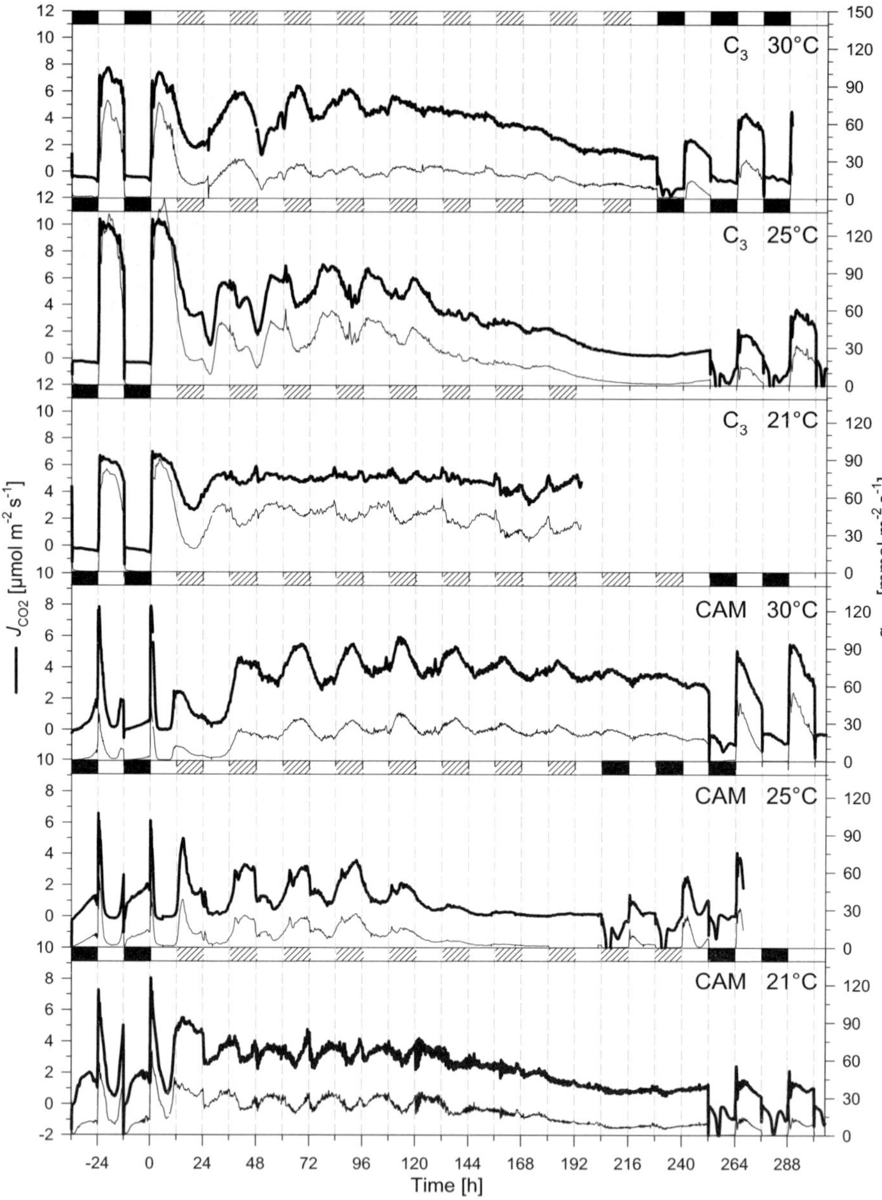

Fig. 11.1. Net exchange of CO_2, J_{CO2}, and leaf conductance for water vapour, g_{H2O}, of the leaves of *C. minor* plants adapted to perform C_3- and CAM-photosynthesis at three different temperatures as indicated. *Black bars* indicate darkness (D), *white bars* light (L) and *hatched bars* the subjective dark periods under constant illumination (LL)

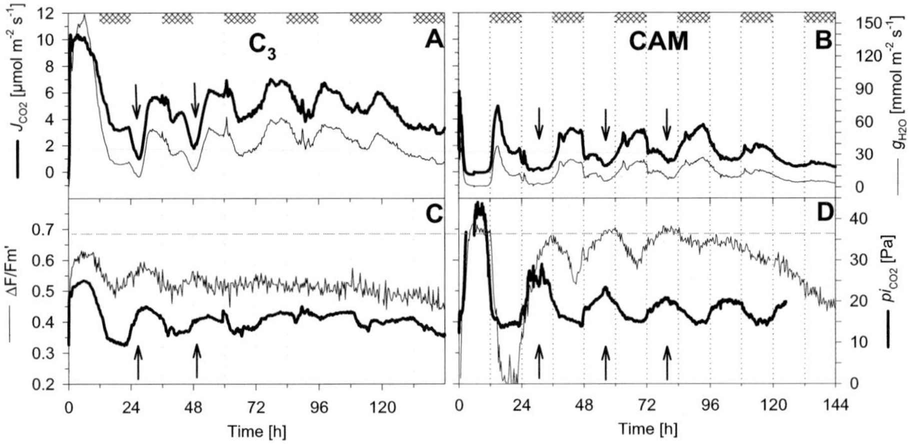

Fig. 11.2.A Gas exchange, J_{CO2} and g_{H2O}. **B** p^i_{CO2}, and effective quantum yield of PS II, $\Delta F/F_m'$, for leaves of *C. minor* plants adapted to the C_3-mode (*left*) and to the CAM-mode (*right*) at 25 °C. *Black bars* indicate darkness (D), *white bars* light (L) and *hatched bars* the subjective dark periods under continuous illumination (LL). **C** The first 60 h (C_3-mode) and **D** the first 72 h (CAM-mode), respectively, under LL are amplified. The *arrows* indicate high p^i_{CO2} and $\Delta F/F_m'$ at simultaneously low J_{CO2} and g_{H2O}. The *horizontal lines* in C and **D** mark the ambient CO_2 partial pressure, p^a_{CO2}.

strate limitation of the carboxylase activity of RubisCO, and hence energy use by PSII as indicated by $\Delta F/F_m'$. The amplification of the first periods of the oscillations (Fig. 11.2C) reveals that in the C_3-adapted plants changes of $\Delta F/F_m'$ and internal CO_2 partial pressure, p^i_{CO2}, followed each other and were inversely related to J_{CO2} and g_{H2O}. This confirms that energy use is related to the availability of internal CO_2. It is remarkable that two times in sequence in the C_3-adapted plants the increase in $\Delta F/F_m'$ and p^i_{CO2} was observed together with a reduction in J_{CO2} and g_{H2O} (arrows in Fig. 11.2C). This shows that the increase of p^i_{CO2} was not due to CO_2 uptake but rather to internal CO_2 sources. It suggests that there were internal sources of CO_2, possibly due to some organic acid mobilisation and decarboxylation, i.e. that the C_3-adapted plants had kept a residual CAM capacity possibly due to activity of the major leaf vein chlorenchyma (see Sect. 8.8.1). In the CAM-adapted leaves in the first 35 h under LL $\Delta F/F_m'$ was correlated with p^i_{CO2} and inversely correlated with J_{CO2} and g_{H2O}, corresponding to what one expects in CAM, i.e. that during organic acid mobilisation and decarboxylation p^i_{CO2} increases and J_{CO2} and g_{H2O} decrease. However, subsequently the coupling of $\Delta F/F_m'$ and p^i_{CO2} was not so tight anymore. The peaks of p^i_{CO2} showed a clear tendency to become smaller, which suggests a weakening of the CO_2 concentrating mechanism of CAM as time under LL moves on. This conclusion is very interesting in relation to the following observation. The C_3-adapted plants performing C_3-photosynthesis before application of LL as expected still performed C_3-photosynthesis when an external dark/light rhythm (DL) was given again after rhythmicity had dampened out in LL (Fig. 11.1). However, the CAM-adapted plants well expressing the four phases of gas exchange of CAM (Chap. 8.1) before LL showed C_3-type gas exchange under DL following LL (Fig. 11.1). This means that the plants had changed the mode of photosynthesis and shifted from CAM to C_3-photosynthesis during LL.

Although *K. daigremontiana* is an obligate CAM-species it shows a most noteworthy analogy to this behaviour of *C. minor*. In the circadian rhythm of *K. daigremontiana* only for the first endogenous periods night/day oscillations of malic acid levels are observed, which are highly dampened. This loss of organic acid oscillations is not reflected at all in the overt rhythmicity of gas exchange which – as noted above (Sect. 11.1) – continues for very many endogenous periods (Wyka and Lüttge 2003; Wyka et al. 2004; see also Borland and Taybi 2004). Thus, control of rhythmicity in *K. daigremontiana* is handed over from a CAM-type oscillator with a hysteresis switch based on malate metabolism and vacuolar compartmentation (Lüttge 2000) to a C_3-type oscillator. The nature of the latter is not known but could be RubisCO activation and activity or stomatal guard cell movements, which are also known to show circadian rhythmicity in C_3-plants (see Sect. 11.1).

11.2.2 Endogenous Oscillations of Oxygenase activity of RubisCO in the C_3- and CAM-Modes of Photosynthesis

Photorespiration was followed during endogenous oscillations by applying a gas mixture with only 1 % O_2 for 20 min at intervals during registration of gas exchange (J_{CO2}, g_{H2O}) and effective quantum yield of photosystem II, relΦ_{PSII}, measured via chlorophyll fluorescence imaging. This causes non-photorespiratory conditions and the difference between maximum possible CO_2 uptake obtained under 1 % O_2 (J_{CO2}max) and CO_2 uptake obtained under 21 % O_2 equals the oxygenation activity of RubisCO, J_{O2} (Figs. 11.3 and 11.4). The dampening of the endogenous oscillations again was seen to be faster in the C_3-mode than in the CAM-mode. In both modes J_{CO2}max and J_{O2} followed the curves of J_{CO2} and g_{H2O}. In the C_3-mode J_{O2} in per cent of total RubisCO-activity, %J_{O2}, was largely in phase with the other parameters. However, in the CAM-mode the pattern of %J_{O2} was much more complex. It was shifted in phase as it increased ahead of J_{CO2} and g_{H2O} when these were still low, and the highest values of %J_{O2} were reached when J_{CO2} and g_{H2O} changed from lower to higher values. %J_{O2} was also higher in the CAM-mode than in the C_3-mode, the maximum, values in the peaks were 50 and 35 %, respectively. When rhythmicity had dampened out the values of %J_{O2} of the CAM-adapted plants came close to those of the C_3-adapted plants which corresponds to the observed CAM to C_3 shift under LL.

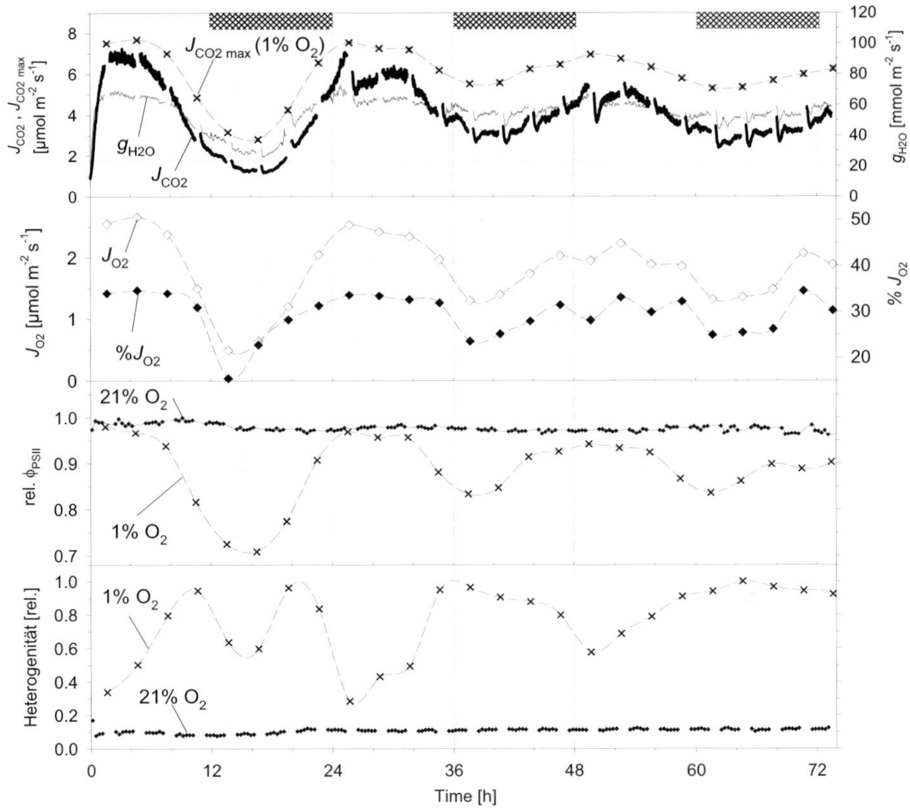

Fig. 11.3. Endogenous rhythm of photorespiratory activity of a leaf of a C_3-adapted *C. minor* plant at 25 °C with maximum carboxylation activity of RubisCO, J_{CO2}max, in the presence of 1 % CO_2, net CO_2-exchange, J_{CO2}, and g_{H2O} in the presence of 21 % O_2 (*uppermost panel*), J_{O2}, i.e. the difference between J_{CO2}max and J_{CO2} and %J_{O2} (J_{O2} in per cent of J_{CO2}max) (*second panel*), relΦ_{PSII} at 21 % and 1 % O_2 (*third panel*) and heterogeneity at 21 % and 1 % O_2 (*bottom panel*). Time 0 is the beginning of LL, *white bars* indicate light and *hatched bars* the subjective dark periods under LL

RelΦ_{PSII} under 21 % O_2 did not oscillate under LL in both modes of photosynthesis and there was no heterogeneity of relΦ_{PSII} over the leaves calculated by a nearest neighbour matrix algorithm from the images of relΦ_{PSII}. By contrast, under 1 % O_2 in both modes clear oscillations of relΦ_{PSII} were borne out. They were in phase with oscillations of J_{CO2} and g_{H2O} and also dampened out more rapidly in the C_3-mode than in the CAM-mode. RelΦ_{PSII} is a measure of the use of irradiance energy and excitation. Thus, by reducing external O_2 to 1 % it was seen that in the CAM-adapted plants the highest energy demand during an endogenous period was reached in the peaks of J_{CO2} and g_{H2O}, which was so strong that in the first peaks during LL there was almost no difference between relΦ_{PSII} at 21 % and 1 % O_2. This

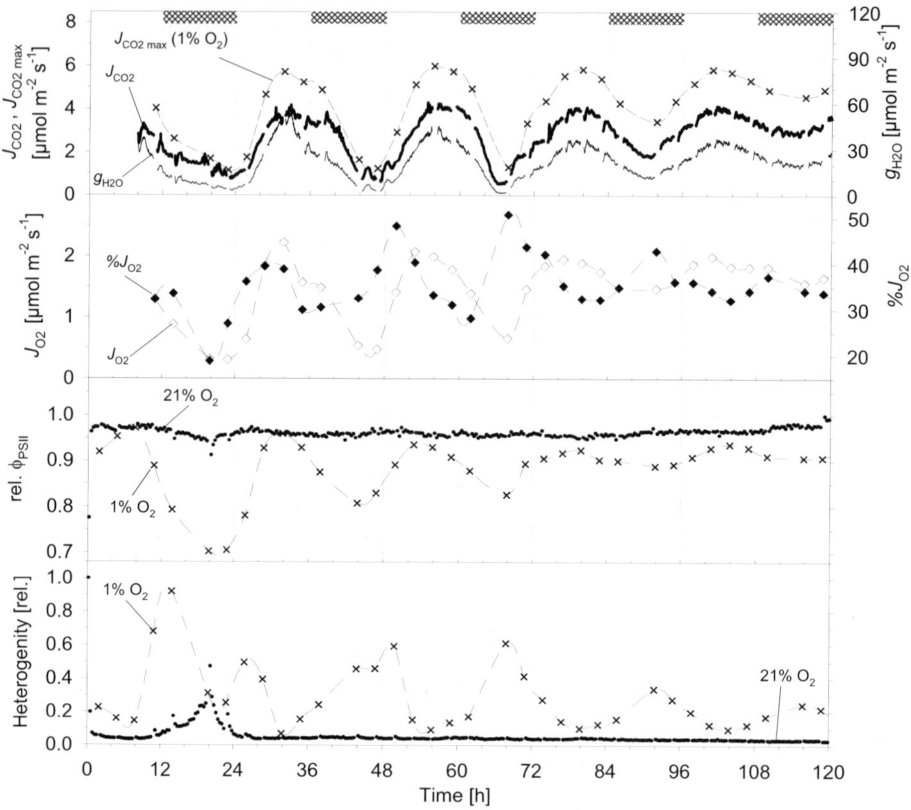

Fig. 11.4. Endogenous rhythm of photorespiratory activity of a leaf of a CAM-adapted plant at 25 °C. Further details as for Fig. 4

implies that when photorespiration was suppressed the energy consumed due to the oxygenation activity of RubisCO was deviated to other energy consuming processes. In the normal external dark/light rhythm of CAM organic acid synthesis and transport into the vacuoles mainly occur in the dark and use respiratory energy. In LL the energy for these processes can also be supplied by the light reactions of photosynthesis but now in direct competition with the Calvin cycle and photorespiration. This competition was largest in the peaks and lowest in the valleys of the rhythm. It shows that under varying energy demand photorespiration has a compensating effect on relΦ_{PSII} and confirms conclusions that photorespiration stabilizes and synchronises energy use in the whole leaf (Sect. 8.5).

Under 1 % O_2 also heterogeneity of relΦ_{PSII} was observed over the leaves in both modes and the spatial structure showed oscillations between homogenous and heterogeneous states, which were more pronounced and more regular in the CAM-mode than in the C_3-mode. Maximum values of heterogeneity were reached while J_{CO2}, g_{H2O}, and relΦ_{PSII} were in the phase of increasing from low to high values. The minimal values of heterogeneity in both modes were found at the peaks of J_{CO2}. In the CAM-mode maximum heterogeneity was attained at maximum $\%J_{O2}$.

11.3 Oscillator Elements and their Cryptic Network

Evidently the studies on *K. daigremontiana*, *M. crystallinum* and *C. minor* reveal oscillator elements at different hierarchical levels, such as the central oscillator genes, the metabolism related genes and the functional pacemakers of C_3 and CAM rhythms. In CAM we know of a feedback loop from the functional pacemaker malate accumulation/remobilisation to a metabolism gene, viz. *PEPCK* (Borland et al. 1999). However, largely it remains unknown how the various oscillator elements are connected in an operating network. *K. daigremontiana* and much more pronouncedly *C. minor* pose us the question of how such a network might function in mediating the CAM to C_3 shift during ongoing endogenous rhythmicity of gas exchange. Where are the feedback connections? Is the system feeding back to a central oscillator element like *TOC1* which is so strongly shifted in phase in oscillations in the CAM-mode as compared to the C_3-mode of photosynthesis in *M. crystallinum* (Boxall et al. 2005)? Unfortunately the so well advanced molecular studies in *M. crystallinum* are not accompanied by biochemical and physiological information to allow full assessment of the meaning of these observations for a functional network. We do not know if possibly *M. crystallinum* like *K. daigremontiana* and very conspicuously *C. minor* may also perform a CAM to C_3 shift during ongoing overt rhythmicity. The molecular study of Boxall et al. (2005) does not provide any information on concomitant malic acid oscillations and it only covers the first periods, when CAM type malate oscillations are still observed in *K. daigremontiana*. We do not know what happens later on and much more work is clearly needed. Conversely, molecular analyses like those of *M. crystallinum* are lacking for *K. daigremontiana* and *C. minor*. In particular the intriguing questions *C. minor* is asking us here would need to be addressed at the molecular level.

Evidently the answer to the first initial question (Sect. 11.1) "does *Clusia* have an endogenous clock?" is yes, albeit its oscillation is so strongly dampened. However, "does *Clusia* need a clock?" then, if it is so strongly dampened? It is widely assumed in the literature that endogenous circadian rhythmicity is important for fitness by anticipation of regularly changing environmental parameters in external night/day cycles. Hard evidence for this is not abundant, however. A competition experiment with *Synechococcus elongatus* strains of different endogenous period lengths reveals a selective advantage of circadian rhythmicity (Ouayang et al. 1998; Johnson and Golden 1999). Synchronization of the clock to external cycles promotes photosynthetic activity, biomass increases, survival and competitive capacities (Dodd et al. 2005). The well known requirement of circadian time keeping for photoperiod sensing is a necessity in environmental adaptations. On the other hand, circadian rhythmicity may be just an inescapable side-product of evolution of life under the continuous entrainment by the natural environmental rhythm of days and nights (Lüttge 2002a). If circadian rhythmicity really provides preparedness for conditions to be regularly anticipated, it could be a hindrance for plasticity. Thus, it would make sense if strongly dampened rhythmicity as observed here were an intrinsic property of a plant as versatile and flexible ecophysiologically as *C. minor* (Chaps. 8 and 9).

References

Borland AM, Taybi T (2004) Synchronization of metabolic processes in plants with Crassulacean acid metabolism. J Exp Bot 55:1255–1265

Borland AM, Hartwell J, Jenkins GI, Wilkins MB, Nimmo HG (1999) Metabolic control overrides circadian regulation of phospho*enol*pyruvate carboxylase kinase and CO_2 fixation in Crassulacean acid metabolism. Plant Physiol 121:889–896

Boxall SF, Foster JM, Bohnert HJ, Cushman JC, Nimmo HG, Hartwell J (2005) Conservation and divergence of circadian clock operation in a stress-inducible crassulacean acid metabolism species reveals clock compensation against stress. Plant Physiol 137:969–982

Britz SJ, Hungerford WE, Lee DR (1987) Rhythms during extended dark periods determine rates of net photosynthesis and accumulation of starch and soluble sugars in subsequent light periods in leaves of *Sorghum.* Planta 171:339–345

Carter PJ, Nimmo HG, Fewson CA, Wilkins MB (1991) Circadian rhythms in the activity of a plant protein kinase. EMBO J 10:2063–2068

Dodd AN, Salathia N, Hall A, Kevei E, Toth R, Nagy F, Hibberd JM, Millar AJ, Webb AAR (2005) Plant circadian clocks increase photosynthesis, growth, survival, and competitive advantage. Science 309:630–633

Duarte HM (2006) Chronobiologie von *Clusia minor*: circadianer Rhythmus in einer Pflanze mit C_3/CAM-intermediärem photosynthetischen Verhalten. Dr. rer. nat.-Dissertation, Darmstadt

Freeden AL, Hennessey TL, Field CB (1991) Biochemical correlates of the circadian rhythm in photosynthesis in *Phaseolus vulgaris.* Plant Physiol 97:415–419

Geiger DR, Shei W-J, Yu X-M (1995) Photosynthetic carbon metabolism and translocation in wild-type and starch-deficient mutant *Nicotiana sylvestris* L. Plant Physiol 107:507–514

Grams TEE, Kluge M, Lüttge U (1995) High temperature-adapted plants of *Kalanchoë daigremontiana* show changes in temperature dependence of the endogenous CAM rhythm. J Exp Bot 46:1927–1929

Grams TEE, Beck F, Lüttge U (1996) Generation of rhythmic and arrhythmic behaviour of crassulacean acid metabnolism in *Kalanchoë daigremontiana* under continuous light by varying the irradiance or temperature: measurements in vivo and model simulations. Planta 198:110–117

Hartwell J, Smith LH, Wilkins MB, Jenkins GI, Nimmo HG (1996) Higher plant phosphoenolpyruvate carboxylase kinase is regulated at the level of translatable mRNA in response to light or a circadian rhythm. Plant J 10:1071–1078
Hartwell J, Gill A, Nimmo GA, Wilkins MB, Jenkins GI, Nimmo HG (1999) Phosphoenolpyruvate carboxylase kinase is a novel protein kinase regulated at the level of expression. Plant J 20:333–342
Hartwell J, Nimmo GA, Wilkins MB, Jenkins GI, Nimmo HG (2002) Probing the circadian control of phosphoenolpyruvate carboxylase kinase expression in *Kalanchoë fedtschenkoi*. Funct Plant Biol 29:663–668
Hennessey TL, Field C (1991) Circadian rhythms in photosynthesis. Oscillations in carbon assimilation and stomatal conductance under constant conditions. Plant Physiol 96:831–836
Hennessey TL, Freeden AL, Field CB (1993) Environmental effects on circadian rhythms in photosynthesis and stomatal opening. Planta 189:369–376
Johnson CH, Golden SS (1999) Circadian programs in cyanobacteria: adaptiveness and mechanism. Annu Rev Microbiol 53:389–409
Kikis EA, Khanna R, Quail PH (2005) *ELF4* is a phytochrome-regulated component of a negative-feedback loop involving the central oscillator components *CAA1* and *LHY*. Plant J 44:300–313
Kliemchen A, Schomburg M, Galla H-J, Lüttge U, Kluge M (1993) Phenotypic changes in the fluidity of the tonoplast membrane of crassulacean acid metabolism plants in response to temperauture and salinity stress. Planta 189:403–409
Kluge M, Kliemchen A, Galla H-J (1991) Temperature effects on crassulacean acid metabolism: EPR spectroscopic studies on the thermotropic phase behaviour of the tonoplast membranes of *Kalanchoë daigremontiana*. Bot Acta 104:355–360
Li B, Geiger DR, Shieh W-J (1992) Evidence for circadian regulation of starch and sucrose synthesis in sugar beet leaves. Plant Physiol 99:1393–1399
Liu Y, Tsinoremans NF, Johnson CH, Lebedeca NV, Golden SS, Ishiura M, Kondo TI (1995) Circadian orchestration of gene expression in cyanobacteria. Genes Dev 9:1469–1478
Lu Y, Gehan JP, Sharkey TD (2005) Daylength and circadian effects on starch degradation and maltose metabolism. Plant Physiol 138:2280–2291
Lu Y, Sharkey TD (2006) The importance of maltose in transitory starch breakdown. Plant Cell Environ 29:353–366
Lüttge U (2000) The tonoplast functioning as the master switch for circadian regulation of crassulacean acid metabolism. Planta 211:761–769
Lüttge U (2002a) Circadian rhythmicity: is the "biological clock" hardware or software? Progr Bot 64:279–319
Lüttge U (2002b) CO_2 concentrating consequences in crassulacean acid metabolism. J Exp Bot 53:2131–2142
Lüttge U, Beck F (1992) Endogenous rhythms and chaos in crassulacean acid metabolism. Planta 188:28–38
McClung CR (2000) Circadian rhythms in plants: a millennium view. Physiol Plant 109:359–371
Michael TP, McClung CR (2003) Enhancer trapping reveals widespread circadian clock transcriptional control in *Arabidopsis*. Plant Physiol 132:629–639
Millar AJ (1999) Biological clocks in *Arabidopsis thaliana*. New Phytol 141:175–197
Nimmo HG (2000) The regulation of phosphoenolpyruvate carboxylase in CAM plants. Trends Plant Sci 5:75–80
Olivares E. 1997. Prolonged leaf senescence in *Clusia multiflora* H.B.K. Trees 11:370–377
Ouyang Y, Andersson CR, Kondo T, Golden SS, Johnson CH (1998) Resonating circadian clocks enhance fitness in cyanobacteria. Proc Natl Acad Sci USA 95:8660–8664

Rascher U, Lüttge U (2002) High-resolution chlorophyll fluorescence imaging serves as a non-invasive indicator to monitor the spatio-temporal variations of metabolism during the day-night cycle and during the endogenous rhythm in continuous light in the CAM plant *Kalanchoë daigremontiana*. Plant Biol 4:671–681

Rascher U, Hütt MT, Siebke K, Osmond B, Beck F, Lüttge U (2001) Spatiotemporal variation of metabolism in a plant circadian rhythm: the biological clock as an assembly of coupled individual oscillators. Proc Natl Acad Sci USA 98:11801–11805

Ratajczak R, Richter J, Lüttge U (1994) Adaptation of the tonoplast V-type H^+-ATPase of *Mesembryanthemum crystallinum* to salt stress, C_3-CAM transition and plant age. Plant Cell Environ 17:1101–1112

Somers DE (1999) The physiology and molecular bases of the plant circadian clock. Plant Physiol 121:9–19

Somers DE, Webb AAR, Pearson M, Kay SA (1998) The short-period mutant, *toc1*, alters circadian clock regulation of multiple outputs throughout development in *Arabidopsis thaliana*. Development 125:485–494

Staiger D (2002) Circadian rhythms in *Arabidopsis*: time for nuclear proteins. Planta 214:334–344

Staiger D, Heintzen C (1999) The circadian system of *Arabidopsis thaliana*. Chronobiol Int 16:1–16

Wilkins MB (1992) Circadian rhythms: their origin and control. New Phytol 121:347–375

Wyka TP, Lüttge U (2003) Contribution of C_3 carboxylation to the circadian rhythm of carbon dioxide uptake in a Crassulacean acid metabolism plant *Kalanchoë daigremontiana*. J Exp Bot 54:1471–1479

Wyka TP, Bohn A, Duarte HM, Kaiser F, Lüttge U (2004) Perturbations of malate accumulation and the endogenous rhythms of gas exchange in the Crassulacean acid metabolism plant *Kalanchoë daigremontiana*: testing the tonoplast-as-oscillator model. Planta 219:705–713

Synthesis

Ulrich Lüttge

The critical question has been asked if it makes sense to devote a book-monograph to a single genus of neotropical trees. The diversity of tropical tree species is very large indeed. Why just select one, and why just select *Clusia*? One strong motivation was given by the fact that *Clusia*s are the only trees performing crassulacean acid metabolism (Chap. 1). This in itself would not justify singling out *Clusia* for its own book. However, it turned out that this fact prompted a number of plant scientists for the first time in almost 25 years to study all possible aspects of *Clusia*'s biology in presumably quite a unique way, not only covering a very large number of aspects but also integrating them in a remarkable synthesis. It happened that in the research teams, phytogeographers, phytosociologists and ecologists became familiar with physiological approaches and physiologists took an immediate interest in community structures, aiming to advance from autecology to physiological synecology. The field-laboratory-field-laboratory-field-etc. ping pong worked effectively in all teams involved in making observations and in increasing curiosity in the field, to develop interpretations and hypotheses and to test them in laboratories and phyto-chambers, to come back to the field and check the validity of plant comportment under controlled conditions for performance in the natural environments and to increase the precision of observations. This is an ongoing process in *Clusia* research and, although we try to present here the most recent and up to date knowledge on *Clusia*, I know that while I am writing these lines new discoveries are being made and new interpretations are being formed. In the book we present a reviewing style of chapters often enriched by the presentation of new and unpublished information, which is particularly the case in the fields of nurse plants (Chap. 4), reproductive biology (Chap. 5), phylogeny (Chap. 6), population genetics (Chap. 7), mycorrhiza (Chap. 10) and biological timing (Chap. 11). We are happy to present our material in this way because it underlines the dynamics of the field of *Clusia*-research.

With the integrating synthesis of the knowledge available, *Clusia* is really becoming a general example for almost all facets of plant biology. I prefer

Ecological Studies, Vol. 194
U. Lüttge (Ed.)
Clusia: A Woody Neotropical Genus
of Remarkable Plasticity and Diversity

"example" and hesitate to say "model", because the current model fashion bears many dangers vis à vis biological diversity. Anatomy and morphology bear out structure-function relationships (Chap. 2), e.g. particularly in relation to plant architecture and modes of photosynthesis. The phytogeography of *Clusia* displays complicated relations of plant distribution in tropical America (Chap. 3). On a smaller scale the nurse plant concept can be advanced with the example of *Clusia* revealing new aspects of the built up of community structures (Chap. 4). Reproductive biology of *Clusia* unravels relations of specific reproductive mechanisms to plant success at the community level (Chap. 5). Molecular phylogeny and population genetics show relations of genetic plasticity and diversity (Chaps. 6 and 7) which can be related to ecophysiological flexibility (Chap. 9). Thus, *Clusia* may provide a unique example for the discussion of the mechanisms of diversity where stress is an important factor but not a single dominating environmental parameter and rather a multi-parameter interaction and network is decisive. Photosynthetic activities of *Clusia*s cover all possible ramifications with different modes of photosynthesis and photosynthetic physiotypes, all variants of crassulacean acid metabolism (CAM) and its metabolic diversity, photorespiration, photoprotection and photoinhibition (Chap. 8). Mineral nutrition (Chap. 9) relates to the biotic interactions of mycorrhiza (Chap. 10). Responses to environmental factors show basic features (Chap. 8) as well as specific adaptations (Chap. 9). Anecdotic reports on the performance of *Clusia*s in comparison to other shrubs and trees of similar life form in more than a dozen of different habitats begin to add up to a general picture, where CAM appears to be par-

Fig. Syn.1. The mistletoe *Phthirusa ovata* growing on *Clusia criuva* in a gallery forest in the cerrado near Brasilia, Brazil. In the right hand picture one can see the mistletoe shoot winding around the *Clusia* shoot ticularly beneficial for performance at the community level due to its intrinsic flexibility. However, at the same time new questions emerge as to which traits in addition to modes of photosynthesis may govern behaviour (Chap. 9). The biological clock of *Clusia* has unique dynamics of rapid dampening. This raises questions in relation to endogenous rhythmicity in different modes of photosynthesis. It reiterates the fundamental considerations of the function of endogenous biological timing and the intriguing question if rhythmic timing may not even be a hindrance for flexibility under varying environmental stress situations (Chap. 11).

One aspect that did not come up in this book was the question of the biological interactions of *Clusia* with potential predators. In the field *Clusia* bushes and trees mostly look very healthy and intact. We have very rarely seen traces of predation. Possibly the strong latex production of all *Clusia*s is quite protective. *Clusia*s may, however, be heavily invaded by mistletoes (Fig. Syn.1). Measurements were performed of the mistletoe *Phthirusa ovata* (Pohl) Eichl. on *Clusia criuva* Camb. in the gallery forest along a river in the cerrados near Brasilia, Brazil (see Sect. 9.4.2.4). It was seen that at low photosynthetically active radiation the actual apparent photosynthetic electron transport rates, ETR (Eq. 8.1, Sect. 8.1), were similar for both the *Clusia* host and the mistletoe, but intrinsic photosynthetic capacity as given by maximum ETR obtained from light saturation curves was superior in the host. The mistletoe had more negative $\delta^{13}C$ values (Eq. 8.3, Sect. 8.2) than the host, which implies that it had a lower overall water use efficiency and worked at higher average internal CO_2-concentration (Table Syn.1). These are very typical mistletoe host relationships (Lüttge et al. 1998).

We also found that sometimes *Clusia* may be strongly affected by gall wasps (Fig. Syn.2). This can be quite specific. In the Atlantic rain forest in Brazil (Sect. 9.4.2.10) we found that among four species of *Clusia* only *C. aemygdioi* Silva et Weinberg was regularly seen to be the host of gall wasps (Figs. Syn.2B and C). Using a chlorophyll fluorescence imaging system we have studied photosynthetic capacity of the leaves across the galls which shows an interesting profile (Fig. Syn.3). Relative efficiency of photosystem II, Φ_{PSII}, is particularly high in a ring around the gall and higher there than in the normal leaf tissue, while it sharply drops in the gall tissue itself. Possibly pho-

Table Syn.1. Photosynthesis of the mistletoe *Phytirusa ovata* in relation to its host *Clusia criuva* in a gallery forest in the cerrado near Brasilia, Brazil. Data given are the differences of values of mistletoe minus host. (After data of Lüttge et al. 1998)

Apparent electron transport rate at ca. 100 μmol quanta m^{-2} s^{-1} [μmol m^{-2} s^{-1}]	10
Maximum intrinsic apparent electron transport rate [μmol m^{-2} s^{-1}]	−125
$\delta^{13}C$ [‰]	−5.18
Internal CO_2 partial pressure [Pa/MPa]	80

A
B
C

tosynthesis around the galls is stimulated for providing supplies to the gall larvae. (The unpublished data of this work remain to be processed.)

The final if not the initial question always asked in these days is "what is all this good for"? It is good for our monographic knowledge of *Clusia*. It is good – as I tried to allude to above – for our general knowledge of plant biology. *Clusia* does not appear to be a commercial bestseller. It is used practically in various ways. With its great adaptive flexibility and as a nurse plant it is appre-

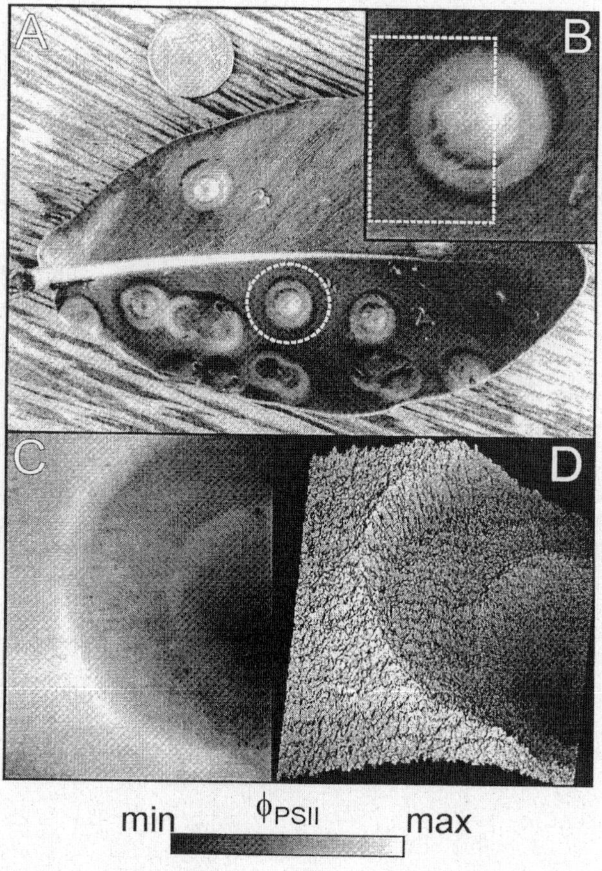

Fig. Syn.3.A Leaf of *C. aemygdioi* with galls. **B** The developed gall encircled in **A. C** Chlorophyll fluorescence image of the relative efficiency of photosystem II (Φ_{PSII}) of the part of the gall marked in **B**; *the bar* indicates minimal and maximal values of Φ_{PSII}. **D** The intensity of Φ_{PSII} is also shown as a three dimensional profile, where the heights represent the relative intensity. (Originals by courtesy of Heitor M. Duarte)

Fig. Syn.2A–C. Gall wasp infected leaves of *Clusia*: **A** *C. arrudae* Planch. et Triana in the Cerro do Cipó, Brazil (see Sect. 9.4.2.3); **B,C** *C. aemygdioi* in the Atlantic rain forest, Brazil (see Sect. 9.4.2.10)

Fig. Syn.4A–D. *Clusia* used: **A** for afforestation of a secondary savanna; **B** as ornamental tree at a gas station in Costa Rica; **C** in the city centre of Rio de Janeiro; **D** near the yacht harbour of Waikiki ciated in afforestation programmes (Fig. Syn.4A), it is used in horticulture (Sternberg et al. 1987) and is often liked as an ornamental plant in the tropics, be that in the garden of a gas station in Costa Rica (Fig. Syn.4B), in the city centre of Rio de Janeiro (Fig. Syn.4C) or on hotel terraces in Waikiki (Fig. Syn.4D).

References

Lüttge U, Haridasan M, Fernandes GW, Mattos EA de, Trimborn P, Franco AC, Caldas LS, Ziegler H (1998) Photosynthesis of mistletoes in relation to their hosts at various sites in tropical Brazil. Trees 12:167–174

Sternberg L da SL, Ting IP, Price D, Hann J (1987) Photosynthesis in epiphytic and rooted *Clusia rosea* Jacq. Oecologia 72:457–460

Taxonomic Index

Additional listings of taxa are found in:
- Appendices 1 and 2, Chapter 3;
- Table 5.1;
- Figure 6.1. and accompanying page 97;
- Figures 6.3 to 6.32, pages 102–105

Subject Index

Q

R

S

Ecological Studies

Volumes published since 2002

Volume 154
Geoecology of Antarctic Ice-Free Coastal Landscapes (2002)
L. Beyer and M. Bölter (Eds.)

Volume 155
Conserving Biological Diversity in East African Forests: A Study of the Eastern Arc Mountains (2002)
W.D. Newmark

Volume 156
Urban Air Pollution and Forests: Resources at Risk in the Mexico City Air Basin (2002)
M.E. Fenn, L. I. de Bauer, and T. Hernández-Tejeda (Eds.)

Volume 157
Mycorrhizal Ecology (2002, 2003)
M.G.A. van der Heijden and I.R. Sanders (Eds.)

Volume 158
Diversity and Interaction in a Temperate Forest Community: Ogawa Forest Reserve of Japan (2002)
T. Nakashizuka and Y. Matsumoto (Eds.)

Volume 159
Big-Leaf Mahogany: Genetic Resources, Ecology and Management (2003)
A. E. Lugo, J. C. Figueroa Colón, and M. Alayón (Eds.)

Volume 160
Fire and Climatic Change in Temperate Ecosystems of the Western Americas (2003)
T. T. Veblen et al. (Eds.)

Volume 161
Competition and Coexistence (2002)
U. Sommer and B. Worm (Eds.)

Volume 162
How Landscapes Change: Human Disturbance and Ecosystem Fragmentation in the Americas (2003)
G.A. Bradshaw and P.A. Marquet (Eds.)

Volume 163
Fluxes of Carbon, Water and Energy of European Forests (2003)
R. Valentini (Ed.)

Volume 164
Herbivory of Leaf-Cutting Ants: A Case Study on *Atta colombica* in the Tropical Rainforest of Panama (2003)
R. Wirth, H. Herz, R.J. Ryel, W. Beyschlag, B. Hölldobler

Volume 165
Population Viability in Plants: Conservation, Management, and Modeling of Rare Plants (2003)
C.A Brigham, M.W. Schwartz (Eds.)

Volume 166
North American Temperate Deciduous Forest Responses to Changing Precipitation Regimes (2003)
P. Hanson and S.D. Wullschleger (Eds.)

Volume 167
Alpine Biodiversity in Europe (2003)
L. Nagy, G. Grabherr, Ch. Körner, D. Thompson (Eds.)

Volume 168
Root Ecology (2003)
H. de Kroon and E.J.W. Visser (Eds.)

Volume 169
Fire in Tropical Savannas: The Kapalga Experiment (2003)
A.N. Andersen, G.D. Cook, and R.J. Williams (Eds.)

Volume 170
Molecular Ecotoxicology of Plants (2004)
H. Sandermann (Ed.)

Volume 171
Coastal Dunes: Ecology and Conservation (2004)
M.L. Martínez and N. Psuty (Eds.)

Volume 172
Biogeochemistry of Forested Catchments in a Changing Environment: A German Case Study (2004)
E. Matzner (Ed.)

Volume 173
Insects and Ecosystem Function (2004)
W. W. Weisser and E. Siemann (Eds.)

Volume 174
Pollination Ecology and the Rain Forest: Sarawak Studies (2005)
D. Roubik, S. Sakai, and A.A. Hamid (Eds.)

Volume 175
Antarctic Ecosystems: Environmental Contamination, Climate Change, and Human Impact (2005)
R. Bargagli

Volume 176
Forest Diversity and Function: Temperate and Boreal Systems (2005)
M. Scherer-Lorenzen, Ch. Körner, and E.-D. Schulze (Eds.)

Volume 177
A History of Atmospheric CO_2 and its Effects on Plants, Animals, and Ecosystems (2005)
J.R. Ehleringer, T.E. Cerling, and M.D. Dearing (Eds.)

Volume 178
Photosynthetic Adaptation: Chloroplast to Landscape (2005)
W.K. Smith, T.C. Vogelmann, and C. Chritchley (Eds.)

Volume 179
Lamto: Structure, Functioning, and Dynamics of a Savanna Ecosystem (2006)
L. Abbadie et al. (Eds.)

Volume 180
Plant Ecology, Herbivory, and Human Impact in Nordic Mountain Birch Forests (2005)
F.E. Wielgolaski (Ed.) and P.S. Karlsson, S. Neuvonen, D. Thannheiser (Ed. Board)

Volume 181
Nutrient Acquisition by Plants: An Ecological Perspective (2005)
H. BassiriRad (Ed.)

Volume 182
Human Ecology: Biocultural Adaptations in Human Cummunities (2006)
H. Schutkowski

Volume 183
Growth Dynamics of Conifer Tree Rings: Images of Past and Future Environments (2006)
E.A. Vaganov, M.K. Hughes, and A.V. Shashkin

Volume 184
Reindeer Management in Northernmost Europe: Linking Practical and Scientific Knowledge in Social-Ecological Systems (2006)
B.C. Forbes, M. Bölter, L. Müller-Wille, J. Hukkinen, F. Müller, N. Gunslay, and Y. Konstantinov (Eds.)

Volume 185
Ecology and Conservation of Neotropical Montane Oak Forests (2006)
M. Kappelle (Ed.)

Volume 186
Biological Invasions in New Zealand (2006)
R.B. Allen and W.G. Lee (Eds.)

Volume 187
Managed Ecosystems and CO_2: Case Studies, Processes, and Perspectives (2006)
J. Nösberger, S.P. Long, R. J. Norby, M. Stitt, G.R. Hendrey, and H. Blum (Eds.)

Volume 188
Boreal Peatland Ecosystem (2006)
R.K. Wieder and D. H. Vitt (Eds.)

Volume 189
Ecology of Harmful Algae (2006)
E. Granéli and J.T. Turner (Eds.)

Volume 190
Wetlands and Natural Resource Management (2006)
J.T.A. Verhoeven, B. Beltman, R. Bobbink, and D.F. Whigham (Eds.)

Volume 191
Wetlands: Functioning, Biodiversity Conservation, and Restoration (2006)
R. Bobbink, B. Beltman, J.T.A. Verhoeven, and D.F. Whigham (Eds.)

Volume 192
Geological Approaches to Coral Reef Ecology (2007)
R.B. Aronson (Ed.)

Volume 193
Biological Invasions (2007)
W. Nentwig (Ed.)

Volume 194
***Clusia*: A Woody Neotropical Genus of Remarkable Plasticity and Diversity** (2007)
U. Lüttge (Ed.)

Plant Ecology
Springer